Student Solutions Manual

for

Introduction to Business Statistics

Fifth Edition

Ronald M. Weiers

Eberly College of Business and Information Technology
Indiana University of Pennsylvania

THOMSON

BROOKS/COLE

Australia • Canada • Mexico • Singapore • Spain • United Kingdom • United States

Printed in the United States of America
1 2 3 4 5 6 7 08 07 06 05 04

Printer: West Group

ISBN: 0-534-49272-X

For more information about our products, contact us at:
Thomson Learning Academic Resource Center
1-800-423-0563

For permission to use material from this text or product, submit a request online at
http://www.thomsonrights.com.
Any additional questions about permissions can be submitted by email to **thomsonrights@thomson.com.**

Thomson Brooks/Cole
10 Davis Drive
Belmont, CA 94002-3098
USA

Asia
Thomson Learning
5 Shenton Way #01-01
UIC Building
Singapore 068808

Australia/New Zealand
Thomson Learning
102 Dodds Street
Southbank, Victoria 3006
Australia

Canada
Nelson
1120 Birchmount Road
Toronto, Ontario M1K 5G4
Canada

Europe/Middle East/South Africa
Thomson Learning
High Holborn House
50/51 Bedford Row
London WC1R 4LR
United Kingdom

Latin America
Thomson Learning
Seneca, 53
Colonia Polanco
11560 Mexico D.F.
Mexico

Spain/Portugal
Paraninfo
Calle/Magallanes, 25
28015 Madrid, Spain

TABLE OF CONTENTS

CHAPTER 1
A PREVIEW OF BUSINESS STATISTICS

SECTION EXERCISES

1.1 d/p/e In ancient times, statistics was mainly employed for counting people or possessions in order to facilitate taxation.

1.3 d/p/m Descriptive statistics are used to summarize and describe a set of data. Inferential statistics are used to make generalizations, estimates, forecasts, or other judgments about the population from which the data (sample) is taken. Inferential statistics are involved when a state senator surveys some of her constituents in order to obtain guidance on how she should vote. She is using statistics to make judgments based on the data.

1.5 d/p/m This information represents inferential statistics since we are using information collected from a sample of 20 adults to make inferences about all adults.

1.7 d/p/e Discrete quantitative variables can only take on certain values along an interval, with the values having gaps between them. Discrete variables are applicable when we want to count the number of times something occurs. Continuous quantitative variables can take on any value along an interval. Continuous variables are applicable when there are no gaps between the exact values which these variables can take on, such as weight, height, volume, or distance.

1.9 d/p/m
a. This information is on the ordinal scale. The industries are viewed in terms of rank instead of the distance between them. However, we do not have a unit of measurement to describe how many more strikes Industry A has than Industry B.
b. This information is on the ratio scale. The industries can be viewed in terms of rank (C has lost fewer days per worker). There is a unit of measurement enabling us to describe how many more days Industry D has lost per worker. There is an absolute zero point and multiples make sense.

1.11 d/p/m The restaurant might be able to make decisions on the quality of food, if there are enough employees, and the cleanliness of the restaurant.

1.13 d/p/m The company may manipulate the data to show what it wants to show; for example, by distorting the scale on a graph to make a large deficit appear very small. Another possibility is to use a sample that is not representative of the population.

Note: In this solutions manual, exercises are categorized according to type, tools required, and level of difficulty:

Type:	Tools:	Difficulty:
d = definitional/conceptual	p = pencil	e = easy
c = computational	a = calculator	m = moderate
p = problem	c = computer	d = difficult

For example, "c/a/m" refers to an exercise that is computational in nature, requires a pocket calculator, and is judged to be moderate in difficulty. The classifications are of necessity subjective. We have attempted to specify the most basic tool that could be practical for the task. In some cases (e.g., simple regression), the pocket calculator can be used even though a computer statistical package (if available) is preferred. In other cases (e.g., multiple regression and analysis of variance), the computer is automatically specified as the required tool.

CHAPTER EXERCISES

1.15 d/p/m This information would represent inferential statistics, since a sample is used to make generalizations about the population.

1.17 d/p/m The amount of money paid out in premiums and the number of policies purchased in the previous year would play a role in how much you pay for your policy. Data such as the number of accidents, speeding tickets, or other moving violations you've had in the past few years could also play a role.

1.19 d/p/m It will help them to be better consumers of statistics. Individuals will be able to protect themselves from others who might be either incompetent or unethical.

1.21 d/p/m
a. The population would be all 40 students who are enrolled in the English class. The sample would be the 5 students in the class that Roger questioned.
b. This is probably not a representative sample, since Roger chose 5 students that always sit in the back. They may all be friends or may have different opinions than those sitting in the front.

CHAPTER 2
VISUAL DESCRIPTION OF DATA

SECTION EXERCISES

2.1 d/p/e A data array is a more orderly presentation, listing data in increasing or decreasing numerical order. With the data array, we can readily: (1) notice the highest and lowest values, (2) identify groups of similar values, and (3) easily see differences between values in the data.

2.3 c/p/e
a. 1043, 1061, 1172, 1344, 1656, 1928, 2483, 2742, 2967, 3066, 3435, 3439, 3608, 5192, 7402
b. Los Angeles is highest, with Chicago also in the high portion of the range of shipment amounts. Pittsburgh is lowest, with Miami and St. Louis among the cities at the low end of the range.

2.5 d/p/e A frequency distribution is a table that divides the data values into classes and shows the number of observed values that fall into each class. By summarizing and reporting data into a frequency distribution, the data can be readily understood and interpreted.

2.7 c/p/m
a. 87.50 thousand   b. lower limit is 35, upper limit is under 45   c. 10 years   d. 40 years

2.9 c/p/m
a. 160.01 thousand   b. lower limit is 45, upper limit is under 55   c. 10 years   d. 50 years

2.11 d/p/e Mutually exclusive means a given data value can fall into only one class. Exhaustive means that the set of classes includes all possible data values.

2.13 c/p/m The number of data values in each category is shown below the graph. The number within each category is shown above the bar. For example, there are 3 data values within the 40 - under 50 category.

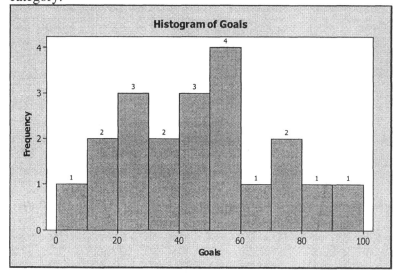

3

2.15 c/a/m

Age (Years)	Percent	Cumulative Frequency
Under 18	0.46	2.06
18 - under 25	26.73	122.44
25 - under 35	45.70	328.26
35 - under 45	19.43	415.76
45 - under 55	5.22	439.28
55 - under 65	1.84	447.55
65 or older	0.62	450.36

2.17 c/c/m The data array and frequency distribution are shown below. None of the portfolio values seems especially large or especially small compared to the others. Although $19,289 is the smallest, it doesn't appear to be a great deal smaller than others at the small end of the values. Likewise, the largest value ($80,331) does not appear to be a great deal larger than others at the large end of the values.

```
19289   20803   23284   23844   27871   31609   32287   32997
33664   34142   36266   36752   37164   38158   38199   40724
41314   41346   41810   42211   42293   42394   42809   43376
43928   45705   46301   46604   46677   46717   48942   48949
49418   49746   50807   50977   51613   51671   53140   53227
53763   53824   54106   55502   55991   55995   56168   56454
57003   57077   57488   58424   59201   59384   60408   60584
61007   61246   61408   61967   62107   62213   62221   66490
67189   67534   67822   67907   68393   68465   68660   69176
70505   70899   73020   73766   73844   74047   76703   80331
```

The number of data values in each category is shown above the bar. For example, there are 10 data values in the $30,000 - under $40,000 category.

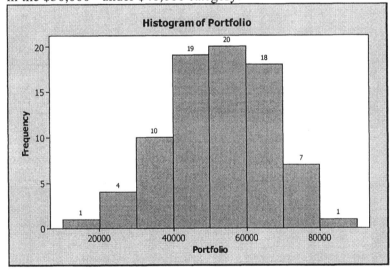

2.19 c/c/m The data array and frequency distribution are shown below. Compared to the other times, the person who received only 5 e-mails seems rather low compared to the others, while the two persons who received 35 e-mails seem to be getting a relatively large quantity of e-mail compared to their colleagues.

```
 5   10   12   14   14   14   15   15   15   15
15   16   16   16   16   16   17   17   17   17
17   17   17   17   17   17   17   18   18   18
18   18   18   18   18   18   18   18   18   18
18   19   19   19   19   19   20   20   20   20
20   20   20   20   20   21   21   21   21   21
21   21   21   21   22   22   22   22   22   22
22   22   22   22   23   23   23   23   23   23
23   23   24   24   24   24   24   25   25   25
26   27   27   27   28   29   29   29   35   35
```

The number of data values in each category is shown above the bar. For example, there are 41 data values in the 20 - under 25 e-mails category.

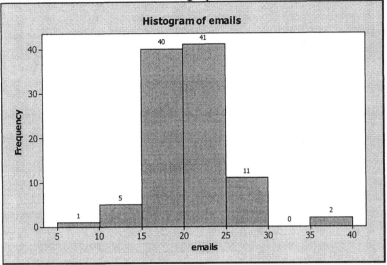

2.21 c/p/m

Stem (Tens)	Leaf (Units)
1	13799
2	24446889
3	0023379
4	1123366689
5	22334677799
6	0113

2.23 c/p/d It is not possible to determine the exact values because there is not enough detail to determine the units digit. One possible data array:

613	695	770	899	955
630	702	864	900	962
633	708	866	903	981
652	721	884	909	
677	755	897	936	

2.25 c/c/m
Stem-and-Leaf Display: Seconds

```
Stem-and-leaf of Seconds   N  = 80
Leaf Unit = 0.10
     1     9 1
     1    10
     1    11
     1    12
     1    13
     3    14 79
     8    15 45557
    12    16 0589
    21    17 145677999
    29    18 04556689
   (12)   19 122344457778
    39    20 0112333456789
    26    21 00357778
    18    22 033359
    12    23 135
     9    24 22249
     4    25 66
     2    26 56
```

5

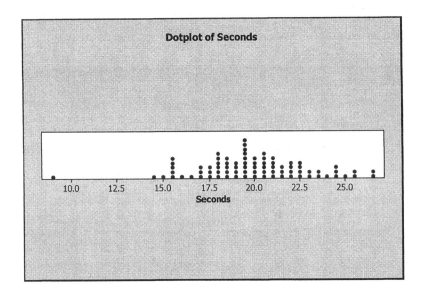

2.27 c/c/m

```
Stem-and-Leaf Display: Appraisal

Stem-and-leaf of Appraisa   N  = 60
Leaf Unit = 10
     1      2  1
     5      2  2233
    12      2  4445555
    20      2  66666777
    25      2  99999
    (8)     3  00000001
    27      3  22233333
    19      3  444455555
    10      3  677
     7      3  888999
     1      4  1
```

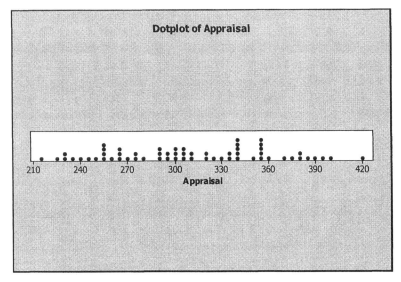

2.29 d/p/m A histogram graphically displays class intervals as well as class frequencies. A bar chart displays the frequencies for a set of categories or classes. Histograms are appropriate for quantitative data, while bar charts are better for qualitative data.

2.31 c/a/m Less-than Ogive for 50 U.S. States:

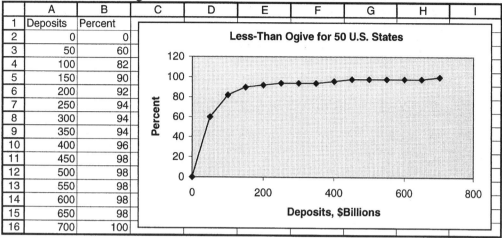

	A	B	C	D	E	F	G	H	I
1	Deposits	Percent							
2	0	0							
3	50	60							
4	100	82							
5	150	90							
6	200	92							
7	250	94							
8	300	94							
9	350	94							
10	400	96							
11	450	98							
12	500	98							
13	550	98							
14	600	98							
15	650	98							
16	700	100							

2.33 c/p/m

	A	B	C	D	E	F	G	H
1	Year	Income						
2	1990	41151						
3	1991	43056						
4	1992	44251						
5	1993	45161						
6	1994	47012						
7	1995	49687						
8	1996	51518						
9	1997	53350						
10	1998	56061						
11	1999	59981						
12	2000	65381						
13								

2.35 c/p/m

McDonald's net income per share and dividends per share:

	A	B	C	D	E	F	G	H	I	J
1	Year	income/share	dividends/share							
2	1992	0.65	0.10							
3	1993	0.73	0.11							
4	1994	0.84	0.12							
5	1995	0.99	0.13							
6	1996	1.11	0.15							
7	1997	1.17	0.16							
8	1998	1.14	0.18							
9	1999	1.44	0.20							
10	2000	1.49	0.22							
11	2001	1.27	0.23							
12	2002	0.70	0.24							
13										
14										
15										
16										

2.37 c/p/d

a. Ownership of a telephone and subscription to cable TV are two different variables, not mutually exclusive categories.

b. Pie Charts.

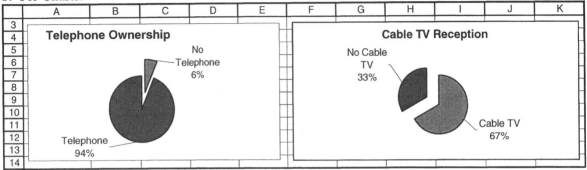

2.39 c/a/m The ship representing 2002 is approximately 2.6 (i.e., 348/134) times as high, 2.6 times as wide, and will have about 6.7 times the area of the ship representing 1980.

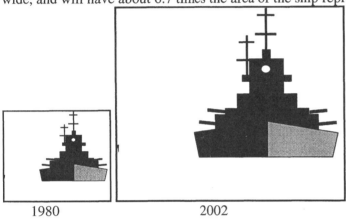

1980   2002

2.41 d/p/e In a positive linear relationship, y tends to increase linearly with increases in x. In a negative linear relationship, y tends to decrease linearly as x increases.

2.43 p/p/m We can use Excel to plot the data for us. As we can see from the graph, there does appear to be a slight direct relationship between the variables -- printers with a higher speed for printing text also tend to have a higher speed for printing graphics. Note: The line in the graph below was actually fitted by Excel. Lines "eyeballed" by humans may vary.

	A	B
1	ppmtext	ppmgraph
2	4.7	0.7
3	7.3	1.3
4	5.9	0.9
5	7.7	7.2
6	5.4	0.9
7	4.8	0.6
8	4.1	1.1
9	4.8	1.0
10	6.4	1.1
11	6.2	1.6

2.45 p/p/m We can use Excel to plot the data for us. As can be seen from the graph, there does appear to be a relationship between the number of businesses in a state and the number of business failures in the state, and the relationship appears to be a direct relationship; that is, as the number of businesses increases, the number of business failures increases. However, there are two rather distinct "clusters", with a group of smaller values quite distant from the two points with larger data values. Thus, it is difficult to determine if the relationship might be linear. Note: The line in the graph below was actually fitted by Excel. Lines "eyeballed" by humans may vary.

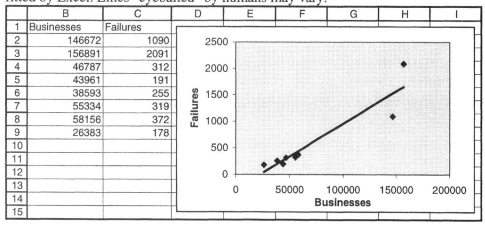

	B	C	D	E	F	G	H	I
1	Businesses	Failures						
2	146672	1090						
3	156891	2091						
4	46787	312						
5	43961	191						
6	38593	255						
7	55334	319						
8	58156	372						
9	26383	178						
10								
11								
12								
13								
14								
15								

2.47 p/c/m

a. Scatter diagram with best-fit linear equation:

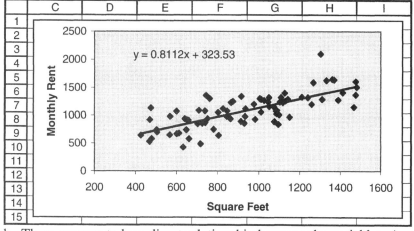

b. There appears to be a direct relationship between the variables. Apartments with more square feet of living area tend to have a higher monthly rental fee.

c. The slope of the equation is +0.8112. Each additional square foot of living area tends to be accompanied by an additional $0.8112 in monthly rental fee.

d. One of the apartments appears to have about 1300 square feet of living space, but the rent is relatively high (over $2000) for an apartment of this size. The location may be very attractive.

2.49 d/p/e Simple tabulation involves just one variable – e.g., we may express a count of how many persons in a sample are males and how many are females. In cross tabulation, we express a count of how many people are in combinations of categories – e.g., how many males are in the under-21 age group, how many females are in the 21-40 age group, and so on.

2.51 c/p/e

a. Simple tabulation. There are 17 vehicles with a Diesel engine, 13 with a gasoline engine.

Tabulated Statistics: Engine

```
Rows: Engine

        Count

1         17
2         13
All       30
```

b. Cross tabulation – e.g., there are 9 vehicles that have a Diesel engine and air conditioning.

Tabulated Statistics: Engine, AC

```
Rows: Engine        Columns: AC

            1         2       All

1           8         9        17
2           5         8        13
All        13        17        30

  Cell Contents --
                  Count
```

c. Average mpg according to type of engine and whether the vehicle has air conditioning – e.g., average fuel economy for the 9 vehicles having a Diesel engine and air conditioning is 32.089 miles per gallon. Diesel engines get a higher mpg than gasoline engines (33.918 versus 22.085) and cars without air conditioning get a higher mpg than cars with air conditioning (31.731 versus 26.541). The most efficient combination is the Diesel engine without air conditioning.

Tabulated Statistics: Engine, AC

```
Rows: Engine        Columns: AC

            1         2       All

1           8         9        17
        35.975    32.089    33.918

2           5         8        13
        24.940    20.300    22.085

All        13        17        30
        31.731    26.541    28.790

  Cell Contents --
                  Count
                mpg:Mean
```

2.53 c/c/m

a. Simple tabulation. Among the 60 schools, 25 offer the bachelors as the highest degree, 24 offer the masters degree as their highest degree, and 11 offer doctorate degrees.

	F	G	H	I	J
1	Count of HighDegr	HighDegr			
2		1	2	3	Grand Total
3	Total	25	24	11	60

b. Cross tabulation – e.g., 13 of the 60 schools are privately supported and offer the bachelors degree as their highest degree.

	F	G	H	I
6	Count	Pub/Priv		
7	HighDegr	1	2	Grand Total
8	1	12	13	25
9	2	12	12	24
10	3	6	5	11
11	Grand Total	30	30	60

c. Average tuition and fees according to highest degree offered and public versus private support – e.g., for publicly supported schools that offer the bachelors as the highest degree, the average for tuition and fees is $5323. Private schools that offer a doctorate degree average $16,092 in tuition and fees. There seems to be a strong relationship between the categorization variables and the amount for tuition and fees.

	F	G	H	I
14	Average of TuitFees	Pub/Priv		
15	HighDegr	1	2	Grand Total
16	1	5323	9516	7503
17	2	8167	11965	10066
18	3	13133	16092	14478
19	Grand Total	8023	11592	9807

2.55 c/c/m

a. Simple tabulation. Among the 35 cities, 3 received a grade of "2" on financial management, 27 received a "3", and 5 received a "4".

	H	I	J	K	L
1	Count	Finance			
2		2	3	4	Grand Total
3	Total	3	27	5	35

b. Cross tabulation – e.g., 13 of the 35 cities received a "3" on financial management and a "2" on information technology.

	H	I	J	K	L
6	Count	Finance			
7	Info Tech	2	3	4	Grand Total
8	1	0	4	0	4
9	2	2	13	0	15
10	3	1	10	3	14
11	4	0	0	2	2
12	Grand Total	3	27	5	35

c. Average population according to grades on financial management and information technology. Cities receiving a "2" on each measure have an average population of 277,257. Cities receiving a "4" on each measure were more than twice as large, with an average population of 676,347. There does seem to be some relationship between the population and the grades received. In general, it seems that larger cities tended to receive higher grades on these measures.

	H	I	J	K	L
15	Average of Population	Finance			
16	Info Tech	2	3	4	Grand Total
17	1		630715		630715
18	2	277257	901745		818480
19	3	496938	1650610	431477	1306962
20	4			676347	676347
21	Grand Total	350484	1138950	529425	984292

CHAPTER EXERCISES

2.57 c/a/m

a. 1216 cities   b. 2614 cities   c. 229 cities; 8.5%   d. 175,000

e. Relative frequency distribution:

population	percent
10,000 - under 25,000	53.54
25,000 - under 50,000	24.03
50,000 - under 100,000	13.54
100,000 - under 250,000	6.42
250,000 - under 500,000	1.38
500,000 - under 1,000,000	0.75
1,000,000 or more	0.34

2.59 c/a/m

a. Data array.

```
482    506    507    537    555    612    626    652    659    666
676    678    701    703    738    744    763    775    775    784
797    823    829    830    838    859    861    863    875    885
889    898    900    902    917    940    960    961    976   1010
1029   1036   1046   1052   1058   1081   1091   1136   1143   1221
```

b. Stem-and-leaf display.

```
Stem (100s) |  Leaf (tens)
        4   |  8
        5   |  0035
        6   |  1255677
        7   |  003467789
        8   |  22335667889
        9   |  0014667
       10   |  12345589
       11   |  34
       12   |  2
```

c. and d. Frequency distribution.

class	frequency	class mark	class width
400 - under 500	1	450	100
500 - under 600	4	550	100
600 - under 700	7	650	100
700 - under 800	9	750	100
800 - under 900	11	850	100
900 - under 1000	7	950	100
1000 - under 1100	8	1050	100
1100 - under 1200	2	1150	100
1200 - under 1300	1	1250	100

e. Minitab histogram and Excel relative frequency polygon.

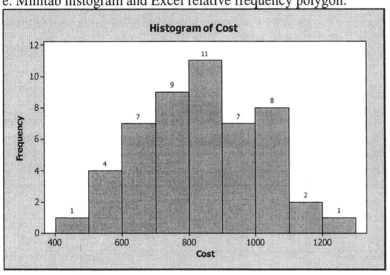

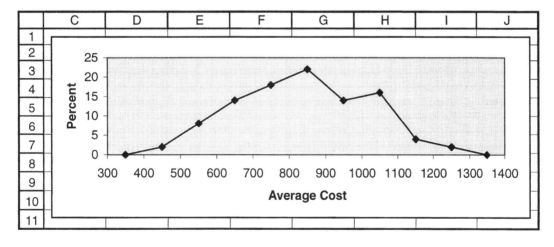

2.61 c/a/m

Homes in the West Region seem to be more expensive.

Price of Home in $ 1000 units	Total U.S. %	West Region %
Under 100	8.27	2.50
100 - under 150	27.34	19.17
150 - under 200	24.37	23.75
200 - under 300	24.37	32.50
300 or over	15.66	22.08

2.63 c/p/d

a. No. 185 192 207 240 246 250 251 256 259 289 296 305
     308 313 314 320 336 342 353 356 381

b. The left-most column shows the number of values in that stem or lower-value stems for values less than 280, or the number of values in that stem or higher-value stems for values greater than 299. The number (2) indicates there are two values in the stem for 280-299 and the median is in that stem.

2.65 d/p/m Guns, human figures wielding knives, and coffins are some of the possible symbols that would tend to convey a negative image.

2.67 c/c/m California ($1221) has the highest average cost and is nearly $100 beyond the next-highest states.

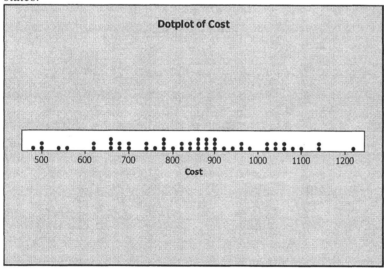

14

2.69 c/c/m
a. Considering only the scatter diagram, the schools given higher scores by the academicians tend to receive higher scores from the lawyers/judges as well.

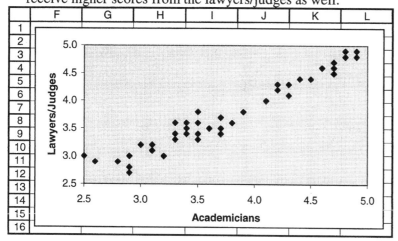

b. The slope of the fitted equation is positive, reflecting the same observation made in part (a).

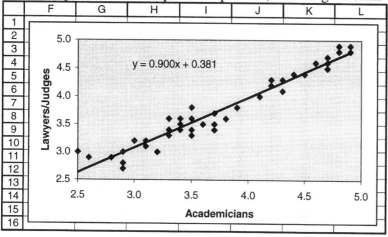

2.71 c/c/m The scatter diagram and fitted equation are shown together.

a. If we consider only the points in the scatter diagram, it is obvious that higher U.S. saving rates tend to be accompanied by higher saving rates in Canada.

b. The fitted linear equation has a positive slope and supports this observation

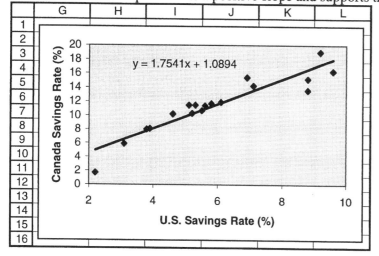

2.73 c/c/m

a. Mechanical supplier 1 supplied the mechanical components for 50 of the air compressors and mechanical supplier 2 provided the mechanical components for the other 50.

	E	F	G	H
1	Count	MechSup		
2		1	2	Grand Total
3	Total	50	50	100

b. Crosstab showing how many compressors had each combination of mechanical supplier and electrical supplier.

	E	F	G	H
5	Count	MechSup		
6	ElectSup	1	2	Grand Total
7	1	17	17	34
8	2	17	17	34
9	3	16	16	32
10	Grand Total	50	50	100

c. Table showing average pressure (psi) for each combination of mechanical supplier and electrical supplier. Compressors associated with mechanical supplier 2 and electrical supplier 1 had the highest average pressure (408.74). Those associated with mechanical supplier 1 and electrical supplier 3 had the lowest average pressure (388.11).

	E	F	G	H
12	Average of PSI	MechSup		
13	ElectSup	1	2	Grand Total
14	1	404.43	408.74	406.58
15	2	400.32	396.60	398.46
16	3	388.11	394.02	391.07
17	Grand Total	397.81	399.90	398.86

d. Overall, the compressors associated with electrical supplier 3 had an average pressure of only 391.07 psi. Regardless of which mechanical supplier was involved, compressors associated with electrical supplier 3 had the lowest average output. For electrical supplier 3, note the 388.11 psi average when associated with mechanical supplier 1 and the 394.02 psi average when associated with mechanical supplier 2.

CHAPTER 3
STATISTICAL DESCRIPTION OF DATA

SECTION EXERCISES

3.1 c/a/m $\overline{x}$ = 61.664/7 = \$8.809 trillion. Median = \$8.782 trillion, the middle value in data array.

3.3 c/a/m $\overline{x}$ = 1141/20 = 57.05 visitors. Median = 57.50, average of 10th and 11th values in data array. The mode is 63. There were three different days with 63 visitors.

3.5 c/a/m $\overline{x}$ = 167.07/20 = 8.35. Median = 8.60, average of 10th and 11th values in data array.

3.7 c/a/m $\overline{x}$ = 1167/30 = 38.9 yrs. Median = 26.5 yrs., average of 15th and 16th values in data array.

3.9 c/a/m $\overline{x}$ = 90(0.35) + 78(0.45) + 83(0.20) = 83.2

3.11 d/p/d
a. Mean will be higher since salaries are usually skewed to the right. Management will emphasize the mean to make the present situation look brighter.
b. The union representative will wish to make the present situation look worse and therefore will emphasize the median.

3.13 c/c/m The Minitab and Excel printouts are shown below.

Descriptive Statistics: PSI

Variable	N	Mean	Median	TrMean	StDev	SE Mean
PSI	100	398.86	396.75	398.64	19.58	1.96

Variable	Minimum	Maximum	Q1	Q3
PSI	351.70	454.50	385.18	414.53

	E	F
1	*PSI*	
2		
3	Mean	398.86
4	Standard Error	1.96
5	Median	396.75
6	Mode	403.90
7	Standard Deviation	19.58
8	Sample Variance	383.50
9	Kurtosis	-0.24
10	Skewness	0.23
11	Range	102.80
12	Minimum	351.70
13	Maximum	454.50
14	Sum	39885.60
15	Count	100

The mean exceeds the median. The distribution is positively skewed.

3.15 c/c/m
```
Descriptive Statistics: age by gender

Variable    gender          N      Mean    Median    TrMean     StDev
age         1              50     40.62     39.00     40.70     11.03
            2              50     41.08     41.50     41.05      9.95

Variable    gender    SE Mean   Minimum   Maximum        Q1        Q3
age         1            1.56     19.00     60.00     32.75     50.25
            2            1.41     19.00     63.00     35.00     46.50
```

The mean age for female employees is less than that for males. The median for the female employees is also lower. For females, the mean age exceeds the median and the distribution is positively skewed.

3.17 c/a/m Range = 75 - 36 = 39 visitors. MAD = 207.00/20 = 19.71 visitors.
$s^2 = 2922.95/19 = 153.84$, and $s = \sqrt{153.84} = 12.40$ visitors.

3.19 c/a/m
a. $\mu = 38.4/7 = \$5.486$ billion. Median = \$2.40 billion. Range = 18.1 - 1.3 = \$16.8 billion.
   Midrange = (18.1 + 1.3)/2 = \$9.7 billion.
b. MAD = 30.657/7 = \$4.38 billion.
c. $\sigma^2 = 222.6086/7 = 31.80$, $\sigma = \$5.64$ billion.

3.21 c/a/m
a. $\bar{x} = 272/10 = 27.2$ mpg. Median = (27 + 29)/2 = 28 mpg. Range = 40 - 10 = 30 mpg.
   Midrange = (10 + 40)/2 = 25 mpg.
b. MAD = 56/10 = 5.6 mpg.
c. $s^2 = 583.6/9 = 64.84$, and $s = \sqrt{64.84} = 8.052$ mpg.

3.23 c/a/m First quartile is in ranked position (11 + 1)/4 = 3; Q_1 = first quartile = 7
Second quartile is in ranked position 2(11 + 1)/4 = 6; Q_2 = second quartile = 18
Third quartile is in ranked position 3(11 + 1)/4 = 9; Q_3 = third quartile = 30
Interquartile range = 30 - 7 = 23; quartile deviation = 23/2 = 11.5

3.25 c/c/m   a. and c. The Excel and Minitab descriptive statistics are shown below.

	C	D
1	*Seconds*	
2		
3	Mean	23.3498
4	Standard Error	0.7764
5	Median	22.8600
6	Mode	22.7400
7	Standard Deviation	5.4897
8	Sample Variance	30.1372
9	Kurtosis	0.6938
10	Skewness	0.6721
11	Range	26.13
12	Minimum	13.40
13	Maximum	39.53
14	Sum	1167.49
15	Count	50

Variable	N	Mean	Median	TrMean	StDev	SE Mean
Seconds	50	23.350	22.860	23.089	5.490	0.776

Variable	Minimum	Maximum	Q1	Q3
Seconds	13.400	39.530	19.095	26.718

The midrange is $(13.40 + 39.53)/2 = 26.465$

b. The mean absolute deviation must be calculated separately. It is $215.809/50 = 4.316$ seconds.

3.27 c/c/m   a. and c. The Excel and Minitab descriptive statistics are shown below.

	C	D
1	*meters*	
2		
3	Mean	90.7713
4	Standard Error	0.9347
5	Median	91.4
6	Mode	85.6
7	Standard Deviation	8.3606
8	Sample Variance	69.9000
9	Kurtosis	-0.1468
10	Skewness	0.0717
11	Range	40.4
12	Minimum	71.8
13	Maximum	112.2
14	Sum	7261.7
15	Count	80

Descriptive Statistics: meters

Variable	N	Mean	Median	TrMean	StDev	SE Mean
meters	80	90.771	91.400	90.742	8.361	0.935

Variable	Minimum	Maximum	Q1	Q3
meters	71.800	112.200	85.025	96.275

The midrange is $(71.8 + 112.2)/2 = 92.0$

b. The mean absolute deviation must be calculated separately. It is $536.3575/80 = 6.70$ meters.

3.29 c/a/e
a. At least $(1-(1/2.5^2))*100 = 84\%$
b. At least $(1-(1/3^2))*100 = 88.89\%$
c. At least $(1-(1/5^2))*100 = 96\%$

3.31 c/a/m Standardized data values: -2.14, -0.77, -0.52, -0.02, -0.02, 0.22, 0.35, 0.60, 0.72, and 1.59; 90% of them are within 2.0 standard deviations of the mean. Chebyshev's Theorem states that at least $(1 – (1/2^2))*100 = 75\%$ should fall within that interval, and these results support the theorem.

3.33 c/a/m Using the empirical rule:
a. 68%. This is the percentage of values that are within ±1 standard deviation of the mean.
b. 2.5%, or 50% - 47.5%; 95% of the values are within ±2 standard deviations of the mean.
c. 84%, or 50% (the area to the left of the mean) plus 34% (the area from the mean to 580).
d. 13.5%, obtained by 47.5% (the area between the mean and 680) minus 34% (the area between the mean and 580).

3.35 c/a/m Coefficient of variation $= s/\bar{x} = (87/315)*100 = 27.62\%$ for Barnsboro. Coefficient of variation $= s/\bar{x} = (1800/8350)*100 = 21.56\%$ for Wellington. Barnsboro has greater relative dispersion.

3.37 c/c/m
a. Box-and-whisker plot and listing of key descriptors. The distribution is positively skewed.

	A	B	C	D	E	F	G	H
1	**Box Plot**							
2								
3	*absent*							
4	Smallest = 1							
5	Q1 = 8							
6	Median = 9							
7	Q3 = 12							
8	Largest = 17							
9	IQR = 4							
10	Outliers: 1, 1,							
11								
12				BoxPlot				
13								
14								
15								
16								
17								
18								
19								
20								
21	0	5		10		15		20
22								

b. A portion of the data and standardized data, and descriptive statistics for the 100 standardized values.

	D	E	F	G	H
1	absent	StdAbsent		*StdAbsent*	
2	8	-0.3980			
3	10	0.2239		Mean	0.0000
4	13	1.1566		Standard Error	0.1000
5	8	-0.3980		Median	-0.0871
6	13	1.1566		Mode	-0.3980
7	10	0.2239		Standard Deviation	1.0000
8	11	0.5348		Sample Variance	1.0000
9	7	-0.7089		Kurtosis	-0.0461
10	1	-2.5743		Skewness	-0.2065
11	11	0.5348		Range	4.9745
12	4	-1.6416		Minimum	-2.5743
13	8	-0.3980		Maximum	2.4002
14	13	1.1566		Sum	0.0000
15	8	-0.3980		Count	100
16	11	0.5348			

3.39 c/a/d

a. Frequency distribution with classes having widths of 1:

class	m_i	f_i	f_im_i	$f_im_i^2$
6 - under 7	6.5	1	6.5	42.25
7 - under 8	7.5	6	45.0	337.50
8 - under 9	8.5	7	59.5	505.75
9 - under 10	9.5	6	57.0	541.50
			sum = 168.0	sum = 1427.0

The estimates are $\bar{x} = 168.0 / 20 = 8.40$ and $s^2 = \dfrac{1427.0 - (20)(8.40)^2}{19} = 0.8316$, $s = 0.912$

b. The mean and standard deviation for the actual data were 8.353 and 0.868, respectively.

c. Frequency distribution with classes having widths of 0.5:

class	m_i	f_i	f_im_i	$f_im_i^2$
6.0 - under 6.5	6.25	0	0	0
6.5 - under 7.0	6.75	1	6.75	45.56
7.0 - under 7.5	7.25	4	29.00	210.25
7.5 - under 8.0	7.75	2	15.50	120.13
8.0 - under 8.5	8.25	2	16.50	136.13
8.5 - under 9.0	8.75	5	43.75	382.81
9.0 - under 9.5	9.25	5	46.25	427.81
9.5 - under 10.0	9.75	1	9.75	95.06
		sum = 20	sum = 167.50	sum = 1417.75

The estimates are now $\bar{x} = 167.50 / 20 = 8.375$ and $s^2 = \dfrac{1417.75 - (20)(8.375)^2}{19} = 0.7862$, $s = 0.887$

The approximations have improved.

d. If each data value were the midpoint of its own class, the approximate values would be identical to the exact values.

3.41 c/a/d

m_i	f_i	f_im_i	$f_im_i^2$
5	25	125	625
15	17	255	3,825
25	15	375	9,375
35	9	315	11,025
45	10	450	20,250
55	4	220	12,100
	sum = 80	sum = 1740	sum = 57,200

Approximate values: $\bar{x} = 1740 / 80 = 21.75$ and $s^2 = \dfrac{57,200 - (80)(21.75)^2}{79} = 245.00$, $s = 15.65$

3.43 c/a/e Because the variables are inversely related, r will be negative. Thus, r will be the negative square root of 0.64, or r = -0.8.

3.45 c/c/m

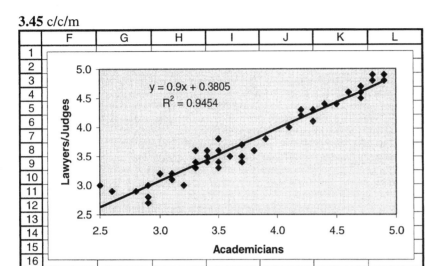

Ratings from the academicians explain 94.54% of the variation in the ratings of the lawyers/judges. The coefficient of correlation is the positive (since the slope is positive) square root of 0.9454, or r = 0.972

3.47 c/c/m

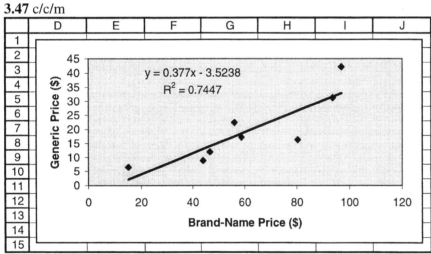

The equation explains 74.47% of the variation in the generic prices. The coefficient of correlation is the positive (since the slope is positive) square root of 0.7447, or r = 0.863.

CHAPTER EXERCISES

3.49 c/a/d $\bar{x}$ = (5(50) + 2(30) + 4(60) + 10(20))/(50 + 30 + 60 + 20) = \$4.69

3.51 c/a/m $\bar{x} = \sum x / n = 924/8 = 115.5$ Median = (116 + 121)/2 = 118.5; There is no mode.

3.53 d/p/m The distribution is not symmetrical. It is positively skewed.

3.55 d/p/m

a. Since all values should be increased by 0.1, the sample mean will increase by 0.1 to 3.1 lbs. Since the relative variation is unchanged, the sample standard deviation will still be 0.5 lbs.

b. Using the empirical rule, this would be 4.1 lbs., obtained by $3.1 + 2(0.5)$. Approximately 95% of the data values will lie within 2 standard deviations of the mean.

3.57 c/a/m

a. $\bar{x} = \sum x / n = 33.59 / 16 = 2.10$ tons, Median $= (2.08 + 2.15)/2 = 2.115$ tons.

   Range $= 2.31 - 1.85 = 0.46$ tons   Midrange $= (1.85 + 2.31)/2 = 2.08$ tons.

b. $MAD = \sum |x_i - \bar{x}| / n = 2.19 / 16 = 0.137$ tons

c. $s^2 = \dfrac{\sum (x_i - \bar{x})^2}{n-1} = 0.3669 / 15 = 0.02446, \quad s = \sqrt{s^2} = 0.156$ tons

3.59 c/a/m The median is approximately 120 watts. The first quartile is approximately 116 watts. The third quartile is approximately 124 watts. The range is approximately $130 - 110 = 20$ watts. The distribution appears to be symmetrical.

3.61 c/a/m   Exercise 3.57: coefficient of variation $= (s/\bar{x})*100 = (0.156/2.10)*100 = 7.43\ \%$

   Exercise 3.60: coefficient of variation $= (s/\bar{x})*100 = (0.0684/0.0736)*100\% = 92.9\%$

There is greater variation for the data in exercise 3.60.

3.63 c/a/m Median $= (24 + 25)/2 = 24.5$ pages.   First Quartile $= 22(0.75) + 22(0.25) = 22$ pages. Third Quartile $= 29(0.25) + 35(0.75) = 33.5$ pages.

Variable	N	Mean	Median	TrMean	StDev	SE Mean
pages	20	25.65	24.50	25.72	8.01	1.79

Variable	Minimum	Maximum	Q1	Q3
pages	11.00	39.00	22.00	33.50

3.65 c/c/m

a. Descriptive statistics.

	C	D
1	*Utility*	
2		
3	Mean	1644.000
4	Standard Error	13.953
5	Median	1651.000
6	Mode	1765.000
7	Standard Deviation	220.624
8	Sample Variance	48674.916
9	Kurtosis	1.495
10	Skewness	0.113
11	Range	1635
12	Minimum	1016
13	Maximum	2651
14	Sum	411000
15	Count	250

b. Boxplot with interpretation statistics.

	A	B	C	D	E	F	G	H
1	**Box Plot**							
2								
3	*Utility*							
4	Smallest = 1016							
5	Q1 = 1495.75							
6	Median = 1651							
7	Q3 = 1782.5							
8	Largest = 2651							
9	IQR = 286.75							
10	Outliers: 2651, 1057, 1016,							

BoxPlot

c. As shown in part (b), there are two outlier households ($1057 and $1016) at the low end and one ($2651) at the high end of utility expenditures. Energy-conservation officials may wish to examine these households for habits or characteristics that should either be emulated or avoided.

3.67 c/c/m

a. Descriptive statistics.

	C	D
1	*SAT*	
2		
3	Mean	517.96
4	Standard Error	5.51
5	Median	519.50
6	Mode	437.00
7	Standard Deviation	110.26
8	Sample Variance	12158.00
9	Kurtosis	0.43
10	Skewness	-0.12
11	Range	673
12	Minimum	159
13	Maximum	832
14	Sum	207182
15	Count	400

Descriptive Statistics: SAT

Variable	N	Mean	Median	TrMean	StDev	SE Mean
SAT	400	517.96	519.50	518.57	110.26	5.51

Variable	Minimum	Maximum	Q1	Q3
SAT	159.00	832.00	448.25	589.00

b. Boxplot with interpretation statistics

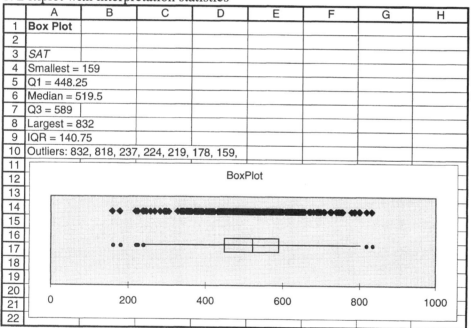

c. A test-taker would have to score 589 on the math portion to be higher than 75% of the sample members. He or she would have to score 449 (448.25, rounded up) to be higher than 25% of the sample members. These correspond to the third and first quartiles, respectively.

3.69 c/c/m

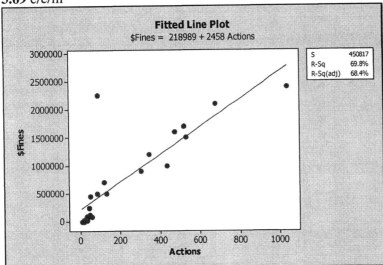

Through the linear estimation equation, the number of actions explains 69.8% of the variation in fine amounts. Because the slope is positive, the coefficient of correlation is the positive square root of 0.698, or r = 0.84.

CHAPTER 4
DATA COLLECTION AND SAMPLING METHODS

SECTION EXERCISES

4.1 d/p/e Primary data are generated by the researcher for the purpose of studying the problem at hand; secondary data already exist and are not generated for the sole purpose of studying the problem at hand. Internal secondary data come from sources within the company or industry being studied; sources such as accounting and financial statements and annual reports are examples of internal secondary data. External secondary data are compiled from sources outside the company or industry being studied.

4.3 d/p/m
a. These data would be secondary to the firm itself because the information was not generated to examine a specific problem.
b. These data would also be secondary to a competitor. Again, the data were not generated by the competitor to examine a specific problem.

4.5 d/p/m There are three major approaches to carrying out survey research: direct mail, personal interview, and telephone interview.

4.7 d/p/m This question is biased and will most likely lead to many false negative answers.

4.9 d/p/m A compiled mailing list consists of persons who are alike in some way, such as gender, occupation, or age. A response mailing list, on the other hand, includes persons who share some activity such as subscribing to the same magazine or making donations to the same political party.

4.11 d/p/m This phenomonon represents a type of response error.

4.13 d/p/m They would not want to employ the telephone survey method, as this would prevent respondents from actually being able to see and touch the alternative personal digital assistant models.

4.15 d/p/m
a. The dependent variable is the rate of response; the independent variable is the monetary incentive.
b. The control group in this experiment is the group that does not receive the monetary incentive for responding to the survey. (They are not exposed to the treatment.)

4.17 d/p/m Among others, the general cleanliness of the exterior, whether it is part of a national chain of reputable restaurants, and many other possibilities.

4.19 d/p/e Secondary data are collected by someone other than the researcher, for purposes other than the problem or decision at hand. Internal secondary data were gathered by someone within the firm or organization. External secondary data were gathered by someone outside the firm or organization.

4.21 d/p/m Corporate news, product offerings, prices, and specifications are often available from a company's Web site. The browser home pages that offer stock quotes also typically offer corporate information, assessments, and recent news articles pertaining to a company's future prospects. The U.S. Bureau of the Census Web site (http://www.census.gov) is one of the most powerful starting points to external secondary information of all types. Search engines like google.com allow us to find details about practically anything within just a few seconds. We can also tap others' expertise through monitoring and participating in internet discussion groups.

4.23 d/p/m The *Encyclopedia of Associations* identifies industry-related trade associations that collect and disseminate data that are related to their industry. It is also an excellent source for identifying private organizations having special interests that may be related to the researcher's informational needs.

4.25 p/c/m Results will vary, of course, depending on the county in which a college or university is located. As an example, for the University of Pittsburgh, a portion of the Bureau of the Census page dealing with Allegeheny County, Pennsylvania is shown below:

Allegheny County QuickFacts from the US Census Bureau - Netscape

File Edit View Go Communicator Help

Back Forward Reload Home Search Netscape Print Security Shop Stop

Bookmarks Location: http://quickfacts.census.gov/qfd/states/42/42003.html What's Related

People QuickFacts	Allegheny County	Pennsylvania
Population, 2000	1,281,666	12,281,054
Population, percent change, 1990 to 2000	-4.1%	3.4%
Persons under 5 years old, percent, 2000	5.5%	5.9%
Persons under 18 years old, percent, 2000	21.9%	23.8%
Persons 65 years old and over, percent, 2000	17.8%	15.6%
White persons, percent, 2000 (a)	84.3%	85.4%
Black or African American persons, percent, 2000 (a)	12.4%	10.0%
American Indian and Alaska Native persons, percent, 2000 (a)	0.1%	0.1%
Asian persons, percent, 2000 (a)	1.7%	1.8%
Native Hawaiian and Other Pacific Islander, percent, 2000 (a)	Z	Z
Persons reporting some other race, percent, 2000 (a)	0.3%	1.5%
Persons reporting two or more races, percent, 2000	1.1%	1.2%
Female population, percent, 2000	52.6%	51.7%
Persons of Hispanic or Latino origin, percent, 2000 (b)	0.9%	3.2%
White persons, not of Hispanic/Latino origin, percent, 2000	83.8%	84.1%
High school graduates, persons 25 years and over, 1990	732,566	5,878,654
College graduates, persons 25 years and over, 1990	209,645	1,412,746
Housing units, 2000	583,646	5,249,750

Document: Done

a. The median household income (not shown in the screen capture above) is reported as $38,893.

b. The Bureau of the Census reports that 17.8% of the 1,281,666 county residents are over 65 years of age, but this page does not indicate the exact number of social security recipients. Another page within the site lists the number of social security recipients as 183,614.

c. Per-capita retail sales (not shown here) is reported as $10,099.

4.27 p/c/m Results will vary, depending on when you view the company's online annual report. Shown below is the Coca-Cola information based on the company's 2000 Annual Report.

See Coca-Cola.com, McDonalds.com, Dell.com, and Compaq.com for the latest information.

4.29 p/c/m Depending on your last name, you might find either a very large number or just a few listings. Even Idaho, with its relatively small population, has more than 250 "Smith" listings.

4.31 p/c/m This can be a real eye-opener. Suffice to say that there are a great many devices out there that most of us have never dreamed of. One of the sites listed by Google is http://www.spyzone.com and motion detectors, night vision equipment, covert cameras, tracking devices, and listening probes are only a few of the many devices available for personal or industrial use.

4.33 d/p/m A sampling error is an error that occurs because a sample has been taken instead of a census. It can be reduced by using a larger sample size. A nonsampling error is a bias because it is a directional error. Nonsampling errors cannot be reduced by simply increasing the size of the sample. It is necessary to take some action to eliminate the cause of the error.

4.35 d/p/e A parameter is a characteristic of a population while a statistic is a characteristic of a sample. We typically use a statistic to estimate a parameter.

4.37 d/p/e Nonsampling errors cannot be reduced by simply increasing the size of a sample. To reduce these kinds of errors, it is necessary to take some action that will eliminate the underlying source of the error, such as adjusting a machine or replacing parts that have become worn through repeated use.
In business research, nonsampling errors can arise from improper research methods or procedures, such as an interviewer whose personality causes respondents to provide answers that are biased in some way.

4.39 d/p/m This is a systematic sample since we are selecting every k = 13th invoice.

4.41 d/p/d The quota sample and the stratified sample are similar in that the population is divided up into layers or "strata". However, in a stratified sample a simple random sample is taken of members from each stratum, while the members of the various strata are not chosen using a probability sampling technique in a quota sample. The quota sample is far inferior to the stratified sample in terms of representation.

4.43 d/p/m
a. The agent should use a sample so that relatively few of the watches will have to be broken.
b. The company should use a sample because it would be too difficult and expensive to find every individual who consumes Cheerios.
c. The producers should use a sample because a census would take an inordinate amount of time and money.
d. They should use a census because there are only a few companies who manufacture submarines.

4.45 d/p/m Periodicity is when the order in which the population appears includes a cyclical variation in which the length of the cycle is the same as the value of "k" that is used in selecting the sample. This could present a problem since we might not end up with a sample that is representative of the population -- e.g., if we wanted to sample 10 months out of the past 10 years, we would not want to take a systematic sample with k = 12. Otherwise, we would end up choosing the same month in all 10 years.

4.47 c/c/m Using Excel, the 30 values in the simple random sample are as shown below.

	C	D	E
1	1774	1612	1581
2	1604	1747	1533
3	1845	1611	1943
4	1656	1495	1835
5	1949	1834	1609
6	1476	1832	1476
7	1521	1612	1943
8	1454	1721	1456
9	1830	1660	1215
10	1666	1929	1566

4.49 c/c/m Using Excel, the descriptive statistics for all 100 persons, along with descriptive statistics for each of the two simple random samples with n = 20.

	E	F	G	H	I	J
1	*tickets*		*sample 1*		*sample 2*	
2						
3	Mean	4.60	Mean	5.35	Mean	4.70
4	Standard Error	0.21	Standard Error	0.52	Standard Error	0.45
5	Median	5.00	Median	5.00	Median	5.00
6	Mode	5.00	Mode	5.00	Mode	5.00
7	Standard Deviation	2.11	Standard Deviation	2.32	Standard Deviation	2.00
8	Sample Variance	4.44	Sample Variance	5.40	Sample Variance	4.01
9	Kurtosis	0.89	Kurtosis	2.18	Kurtosis	1.97
10	Skewness	0.47	Skewness	1.21	Skewness	0.37
11	Range	12	Range	10	Range	9
12	Minimum	0	Minimum	2	Minimum	1
13	Maximum	12	Maximum	12	Maximum	10
14	Sum	460	Sum	107	Sum	94
15	Count	100	Count	20	Count	20

a. The mean of the first sample is 5.35 tickets.
b. The mean of the second sample is 4.70 tickets.
c. The values of the means are not identical. Because of unavoidable sampling error, it is not very likely that the mean of two such samples would be identical. The mean for sample 2 is closer to 4.60 tickets, the mean for the 100 persons from which the two samples were taken.

4.51 d/p/e The union official is most concerned with representing the union the way the majority of members want it to be represented. Thus, an appropriate question for the researcher to answer is, "Which of the three stances would you like to see the union take in the upcoming negotiations?"

4.53 d/p/m Sampling error can be traced to the decision to survey a sample of the companies rather than all 25. Response error occurs when companies exaggerate their ability to supply desired components on time and in the quantity desired. Finally, nonresponse error can be traced to the differences between the 8 companies who responded to the survey and the 7 who did not respond.

4.55 d/p/m A multiple choice question has several possible answers; a dichotomous question has two possible answers. An open-ended question will have no "set" responses. Some examples:
a. Multiple choice: On average, how often do you read The Wall Street Journal?
   [ ] daily   [ ] 3 to 4 times per week   [ ] 1 to 2 times per week   [ ] never
b. Dichotomous: Do you have your own subscription to The Wall Street Journal? [ ] Yes   [ ] No
c. Open-ended: Which features do you consider to be most important in The Wall Street Journal?

4.57 d/p/m External validity refers to the extent to which the results of an experiment can be generalized to other settings. This experiment would probably have external validity for salespersons working in snowy regions; however, it is not clear that a snowmobile will be much of an incentive to someone living in Florida, Arizona, or other warm climates.

4.59 d/p/m
a. The Professional Bowlers Association is one such organization. Two others are the National Bowling Council and the Youth Bowling Council.
b. The Child Abuse Prevention Network and the National Clearinghouse on Child Abuse and Neglect Information are two sources of information on child abuse.
c. The National Safety Council and the Industrial Accident Prevention Association are two sources of information on industrial accidents.
d. The Antique Auto Club and the Horseless Carriage Club of America are two organizations with information about antique and classic cars. There are a lot of more specialized clubs, such as the Mercury Restorers Club of America and the Vintage Volkswagen Club of America.

4.61 d/p/m There are many considerations in evaluating suitability of data in any study. Do the data cover a long enough time frame? Are they presented in the proper series (daily, weekly, monthly, etc.), or can they be modified to the proper timeliness without losing too much information? Were the data generated by someone who was hoping to "prove" a specific result?
If the data are not free, are the costs justified by the value of the data, and will they allow the project to stay within set budgetary constraints? Special skepticism should be reserved for data generated by unknown sources who have something to gain or lose based upon the findings of the study. In acquiring data, it is best to remember the Latin phrase "caveat emptor." (Let the buyer beware.)

4.63 p/c/m A portion of the Bureau of the Census page dealing with Kalamazoo County, Michigan.

People QuickFacts	Kalamazoo County	Michigan
Population, 2000	238,603	9,938,444
Population, percent change, 1990 to 2000	6.8%	6.9%
Persons under 5 years old, percent, 2000	6.5%	6.8%
Persons under 18 years old, percent, 2000	24.1%	26.1%
Persons 65 years old and over, percent, 2000	11.4%	12.3%
White persons, percent, 2000 (a)	84.6%	80.2%
Black or African American persons, percent, 2000 (a)	9.7%	14.2%
American Indian and Alaska Native persons, percent, 2000 (a)	0.4%	0.6%
Asian persons, percent, 2000 (a)	1.8%	1.8%
Native Hawaiian and Other Pacific Islander, percent, 2000 (a)	Z	Z
Persons reporting some other race, percent, 2000 (a)	1.3%	1.3%
Persons reporting two or more races, percent, 2000	2.2%	1.9%
Female population, percent, 2000	51.6%	51.0%
Persons of Hispanic or Latino origin, percent, 2000 (b)	2.6%	3.3%
White persons, not of Hispanic/Latino origin, percent, 2000	83.5%	78.6%
High school graduates, persons 25 years and over, 1990	112,359	4,485,883
College graduates, persons 25 years and over, 1990	36,535	1,014,047
Housing units, 2000	99,250	4,234,279

a. From another portion of this site, we can find that, for those who are 25 years or over, 83.4% are high school graduates.
b. As shown above, 11.4% of the 238,603 county residents are 65 or older.
c. In a lower portion of the page shown above, the median household income is reported as $41,517.

4.65 p/c/m Results will vary, of course. For listings in the telephone directory white pages, it is possible to obtain a great deal of information, including such things as exact location maps and the names and descriptions of nearby businesses.

4.67 d/p/m Because many people who participate in these activities are not available between 6 P.M. and 9 P.M., the sample will be biased toward nonparticipants. For this reason, the sample will be neither random nor representative.

4.69 c/a/m
a. Since 200(0.26) = 52.0, we should question 52 college graduates.
b. We should question 200 - 52 = 148 persons who are not college graduates.

4.71 d/p/m Quota sampling is being used. The population is divided up into layers, or "strata", then the members of the various strata are chosen using a nonprobability sampling technique.

4.73 p/c/m As shown below, the mean of the simple random sample of 30 employees was 38.02 hours, somewhat greater than the national average of 34.2 hours. It seems possible that the 300 Acme Eyebolt workers might be putting in a longer workweek than their production counterparts nationwide.

	E	F
1	*Sample*	
2		
3	Mean	38.02
4	Standard Error	0.40
5	Median	37.95
6	Mode	37.30
7	Standard Deviation	2.21
8	Sample Variance	4.87
9	Kurtosis	1.54
10	Skewness	0.42
11	Range	11.1
12	Minimum	32.9
13	Maximum	44
14	Sum	1140.7
15	Count	30

CHAPTER 5
PROBABILITY: REVIEW OF BASIC CONCEPTS

SECTION EXERCISES

5.1 d/p/m This is a subjective probability, since it represents the degree to which the president believes that an event will occur.

5.3 d/p/m The sample space consists of two possible outcomes: (1) IBM's net profit increases next year, and (2) it does not. The possible events that could occur are:
B = IBM's net profit increases next year
B' = IBM's net profit does not increase next year
P(B) is approximately 0.9, therefore P(B') = 1 - 0.9 = 0.1 since P(B) + P(B') = 1.

5.5 d/p/m You could check a large number of electric meters and find the proportion of times that the right-hand digit is a seven.

5.7 d/p/m The information in text Table 5.1 would suggest that insurance premiums would decrease for men and increase for women. This would occur since women were probably paying lower rates than men before, because mortality rates are lower for females than for males.

5.9 c/p/m   If the odds in favor of an event happening are A:B, the probability is A/(A + B). Since A = 4 and B = 7, the probability is 4/(4+7) = 4/11 = 0.36.

5.11 d/p/m

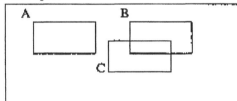

5.13 c/a/m
a. (A and A') contains zero victims. You can't have a blood alcohol level of 0% and have a blood alcohol level > 0% at the same time.
b. (C or F) contains 6 + 41 + 77 + 35 + 29 + 8 = 196 victims.
c. (A' and G') contains 7 + 8 + 8 + 6 + 41 + 77 = 147 victims.
d. (B or G') contains 454 - (47 + 35) = 372 victims since the 47 and the 35 are the only values not contained in (B or G').

5.15 d/p/m parts a, b, and c:

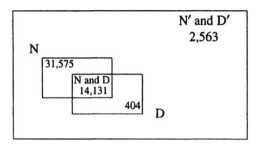

d. Region (N or D). There are 31,575 + 14,131 + 404 = 46,110 gas well completions in this region.
e. Region (N' or D'). There are 31,575 + 404 + 2,563 = 34,542 gas well completions in this region.

5.17 c/p/m Let F = Opening Franchise and R = Raising Speed limit. The two events F and R are not mutually exclusive.

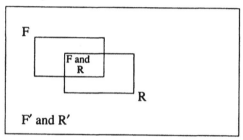

5.19 c/a/m P(South or Midwest or High) = (11.2 + 9.2 + + 4.2 + 6.4)/35.6 = 31.0/35.6 = 0.871.
P(West and Low) = 2.6/35.6 = 0.073.

5.21 c/a/m

	A	B	C	D
Number Who Sign Up	x = 0	x = 1	x = 2	x ≥ 3
Relative Frequency	0.35	0.30	0.25	0.10

The probability that at least two people sign up is P(C or D) = P(C) + P(D) = 0.25 + 0.10 = 0.35.
The probability that no more than one person signs up is P(A or B) = P(A) + P(B) = 0.35 + 0.30 = 0.65.

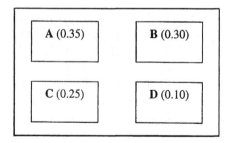

5.23 c/a/m Let F = franchisees, C = company, and A = affiliates. Since A, C, and F are mutually exclusive events, P(F or A) = 17,864/31,108 + 4244/31,108 = 0.711.

5.25 c/a/m
a. Events A, B, C, D, and E are mutually exclusive, so their respective rectangles will not overlap.
b. P(A) = 0.13.
c. P(A or B) = 0.13 + 0.22 = 0.35.
d. P(A or D') = P(D') = 1 – 0.17 = 0.83, since A is contained in D'.

5.27 c/a/m

	Purchasing	Financial	Total
Male	67,000	245,000	312,000
Female	33,000	150,000	183,000
Total	100,000	395,000	495,000
	Purchasing	Financial	Total
Male	0.135	0.495	0.630
Female	0.067	0.303	0.370
Total	0.202	0.798	1.000

P(Purchasing Manager or Male) = P(PM) + P(M) - P(PM and M)
= 0.202 + 0.63 - 0.135 = 0.697

5.29 c/a/m

a. $P(D\ or\ N) = P(D) + P(N) - P(D\ and\ N) = \dfrac{14{,}535}{48{,}673} + \dfrac{45{,}706}{48{,}673} - \dfrac{14{,}131}{48{,}673} = \dfrac{46{,}110}{48{,}673} = 0.947$

b. $P(D'\ or\ N') = P(D') + P(N') - P(D'\ and\ N') = \dfrac{34{,}138}{48{,}673} + \dfrac{2{,}967}{48{,}673} - \dfrac{2{,}563}{48{,}673} = \dfrac{34{,}542}{48{,}673} = 0.710$

c. $P(D\ or\ N') = P(D) + P(N') - P(D\ and\ N') = \dfrac{14{,}535}{48{,}673} + \dfrac{2{,}967}{48{,}673} - \dfrac{404}{48{,}673} = \dfrac{17{,}098}{48{,}673} = 0.351$

d. $P(D'\ or\ N) = P(D') + P(N) - P(D'\ and\ N) = \dfrac{34{,}138}{48{,}673} + \dfrac{45{,}706}{48{,}673} - \dfrac{31{,}575}{48{,}673} = \dfrac{48{,}269}{48{,}673} = 0.992$

5.31 c/p/d It is possible for P(A) = 0.7, P(B) = 0.6, and P(A and B) = 0.35. See the Venn diagram below.

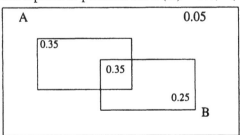

Events A and B cannot be mutually exclusive because if they were, then P(A or B) = P(A) + P(B) = 0.7 + 0.6 = 1.3. However, this is impossible since probabilities cannot be greater than one.
Events A and B cannot be independent because if they were, then P(A and B) = P(A)*P(B) = 0.7*0.6 = 0.42. However, the problem says that P(A and B) = 0.35, not 0.42.

5.33 c/a/m Define the following events:

A = Household with income under $10,000 owns a dishwasher

B = Household with income over $40,000 owns a dishwasher

From the statement of the problem, we know $P(A') = 0.843$ and $P(B') = 0.218$.

Therefore, $P(A) = 1 - 0.843 = 0.157$ and $P(B) = 1 - 0.218 = 0.782$. Events A and B are independent.

a. $P(A' \text{ and } B') = P(A')*P(B') = (0.843)(0.218) = 0.184$

b. $P(A \text{ and } B) = P(A)*P(B) = (0.157)(0.782) = 0.123$

c. $P(A \text{ and } B') = P(A)*P(B') = (0.157)(0.218) = 0.034$

d. $P(B \text{ and } A') = P(B)*P(A') = (0.782)(0.843) = 0.659$

5.35 c/a/m Define the following event: A_i = part i lasts through the warranty period.

Since the parts operate independently, P(appliance works satisfactorily throughout warranty)

$= P(A_1 \text{ and } A_2 \text{ and } A_3 \text{ and ... and } A_{16}) = P(A_1)*P(A_2)*P(A_3)* \ldots *P(A_{16}) = (0.99)^{16} = 0.851$.

5.37 c/a/d Define the following events:

A_1 = first individual's tax return prepared by H & R Block, A_2 = second individual's tax return prepared by H & R Block. From the statement of the problem, we know $P(A_1) = 0.137$,

$P(A_2) = 0.137$, and A_1 and A_2 are independent events.

a. $P(A_1 \text{ and } A_2) = P(A_1)*P(A_2) = 0.137*0.137 = 0.019$

b. $P(A_1' \text{ and } A_2') = P(A_1')*P(A_2') = 0.863*0.863 = 0.745$

c. $P(A_1 \text{ and } A_2') + P(A_1' \text{ and } A_2)$

$= P(A_1)*P(A_2') + P(A_1')*P(A_2) = 0.137*0.863 + 0.863*0.137 = 0.236$

5.39 c/a/m Define the following events: A = takes cab A, B = takes cab B, S = stalls at the light.

From the statement of the problem, we know $P(A) = 0.5$, $P(B) = 0.5$, $P(S \mid A) = 0.25$,

and $P(S \mid B) = 0.1$. See the tree diagram below.

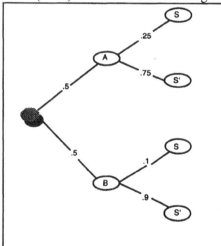

$P(S) = P(A \text{ and } S) + P(B \text{ and } S) = P(A)*P(S \mid A) + P(B)*P(S \mid B) = 0.5*0.25 + 0.5*0.1 = 0.175$

5.41 c/a/m Given the fact that they are entering the restaurant, it is likely that they enjoy Wendy's food; i.e., they are probably not among the 40% who dislike Wendy's hamburgers. Also, because they are a couple, their tastes may be similar. The probability may be greater than 0.36.

5.43 c/a/m

	A	B	C	D
Number Who Sign Up	x = 0	x = 1	x = 2	x ≥ 3
Relative Frequency	0.35	0.30	0.25	0.10

If the advisor has at least one person sign up for the advanced class, the probability that at least three people have signed up is $0.10/(0.30 + 0.25 + 0.10) = 0.10/0.65 = 0.154$

5.45 c/a/m

a. P(boy) = 0.51
b. With medical intervention, P(have a boy | wanted a boy) = 0.85.
c. With medical intervention, P(have a girl | wanted a girl) = 0.77.

5.47 c/a/d

a. Events: F = fair coin, F' = unbalanced coin. Since we randomly select a coin, the prior probability of selecting the fair coin is 0.5, P(F) = 0.5.
b. Events: H = coin lands heads, H' = coin lands tails. The posterior probability we want to find is P(F | H). The tree diagram at the left, below, is the order in which things actually occurred. The tree diagram at the right is the "reversed" tree diagram.

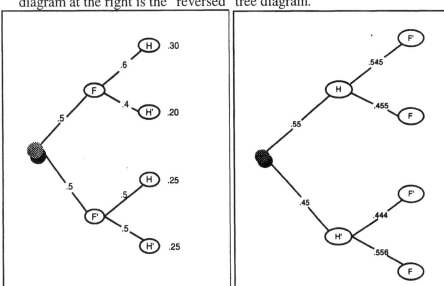

Using the "chronological" tree diagram at the left, we can determine the probabilities shown in the "reversed" tree diagram at the right.

$$P(H) = P(F' \text{ and } H) + P(F \text{ and } H) = 0.30 + 0.25 = 0.55 \qquad P(F|H) = \frac{P(F \text{ and } H)}{P(H)} = \frac{0.25}{0.55} = 0.455$$

Using Bayes' Theorem:

$$P(F|H) = \frac{P(F \text{ and } H)}{P(H)} = \frac{P(F)P(H|F)}{[P(F)*P(H|F)] + [P(F')*P(H|F')]} = \frac{0.5*0.5}{(0.5*0.5) + (0.5*0.6)} = 0.455$$

5.49 c/a/d

a. Events: A = item is from machine A, B = item is from machine B, and C = item is from machine C. The prior probability the item came from machine C is 0.1, P(C) = 0.1.

b. Events: D = the item is defective and D' = the item is not defective. The posterior probability we want to find is P(C | D). The tree diagram in the order in which things actually occurred is given below:

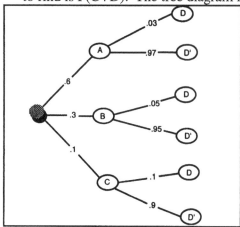

$$P(D) = P(A \text{ and } D) + P(B \text{ and } D) + P(C \text{ and } D) = 0.018 + 0.015 + 0.010 = 0.043$$

$$P(C|D) = \frac{P(C \text{ and } D)}{P(D)} = \frac{0.01}{0.043} = 0.233$$

Using Bayes' Theorem:

$$P(C|D) = \frac{P(C \text{ and } D)}{P(D)} = \frac{P(C)*P(D|C)}{[P(A)*P(D|A)] + [P(B)*P(D|B)] + [P(C)*P(D|C)]}$$

$$= \frac{0.1*0.1}{(0.6*0.03) + (0.3*0.05) + (0.1*0.1)} = 0.233$$

5.51 c/a/m There are k = 8 independent events, each of which can occur in n = 2 different ways. Using the principle of multiplication, the total number of possibilities is $n^k = 2^8 = 256$.

5.53 c/a/m There are 4 different specifications in the ad. The first specification has $n_1 = 2$ choices, the second has $n_2 = 3$ choices, the third has $n_3 = 2$ choices, and the fourth has $n_4 = 3$ choices. Using the principle of multiplication, there are $n_1*n_2*n_3*n_4 = 2*3*2*3 = 36$ different versions of the ad.

5.55 d/a/m Permutations take order into account while combinations do not. Thus, the number of combinations will be less than the number of permutations, as there are r! permutations for every possible combination. The number of combinations is 1/r! as large as the number of permutations.

5.57 c/a/m The number of possible combinations of n = 6 workers that can fill r = 3 openings is:

$$\binom{n}{r} = \frac{n!}{r!(n-r)!} = \binom{6}{3} = \frac{6!}{3!(6-3)!} = \frac{6*5*4*3!}{3!(3*2*1)} = 20$$

5.59 c/a/m Since order is important, we need to find the number of possible permutations of n = 35 customers taken r = 20 at a time.

$$\frac{n!}{(n-r)!} = \frac{35!}{(35-20)!} = \frac{35!}{15!} = 7.9019*10^{27}$$

CHAPTER EXERCISES

5.61 c/a/m Define the following: A = win $1000 per week for life, and B = win a free hamburger.

a. If the odds in favor are A to B, then the probability is A/(A + B).

$$P(A) = \frac{1}{1 + 200,000,000} = 0.000000005$$

b. P(B) = 1/(1+15) = 0.0625

c. P(A or B) = P(A) + P(B) = 0.000000005 + 0.0625 = 0.062500005 (A and B are mutually exclusive events.)

d. P(A and B) = 0 since these are mutually exclusive events. Sheila will receive only one coupon.

5.63 c/a/m

a. There are 1000 different numbers that all have an equal chance of occurring. Since Sam is buying only one ticket, the probability he wins is 1/1000, or 0.001.

b. The probability found in part (a) is a classical probability since it represents the proportion of times that an event can be theoretically expected to occur.

c. The number drawn each day is independent of the number drawn on any other day. Therefore, the number drawn one day will have no effect on the probability of a number occurring the next day.

5.65 c/a/d

a. The number of possible combinations of n = 3 with r = 1 person convicted is

$$\binom{n}{r} = \frac{n!}{r!(n-r)!} = \binom{3}{1} = \frac{3!}{1!2!} = 3.$$

b. The number of possible combinations of n = 3 with r = 2 people convicted is

$$\binom{n}{r} = \frac{n!}{r!(n-r)!} = \binom{3}{2} = \frac{3!}{2!1!} = 3.$$

c. The number of possible combinations of n = 3 with r = 3 people convicted is

$$\binom{n}{r} = \frac{n!}{r!(n-r)!} = \binom{3}{3} = \frac{3!}{3!0!} = 1.$$

5.67 c/a/d Define the following events: A = system A is "down" and B = system B is "down". From the statement of the problem, we know P(A) = 0.1 and P(B) = 0.05. A and B are independent events.

a. P(A and B) = P(A)*P(B) = (0.1)(0.05) = 0.005

b. P(A' and B') = P(A')*P(B') = (1 - 0.1)(1 - 0.05) = 0.855

c. P(A and B') + P(A' and B) = P(A)*P(B') + P(A')*P(B) = (0.1)*(1 - 0.05) + (1 - 0.1)(0.05) = 0.14

d. P(A' or B') = P(A') + P(B') - P(A' and B') = 0.90 + 0.95 - 0.855 = 0.995

5.69 c/a/m Define these events: A = less than \$20,000   B = \$20,000 - under \$40,000
C = \$40,000 - under \$60,000   D = \$60,000 - under \$80,000   E = \$80,000 or more

a. P(A or B or C or D) = P(A) + P(B) + P(C) + P(D) = 0.1 + 0.2 + 0.3 + 0.25 = 0.85

b. $P(D|A') = \dfrac{P(D \text{ and } A')}{P(A')} = \dfrac{0.25}{1-0.1} = 0.278$

c. For each household:

$$P[E|(C \text{ or } D \text{ or } E)] = \dfrac{P[E \text{ and } (C \text{ or } D \text{ or } E)]}{P(C \text{ or } D \text{ or } E)} = \dfrac{0.15}{0.30 + 0.25 + 0.15} = \dfrac{0.15}{0.70} = 0.214$$

Since the 2 households are independent, the probability that this occurs for both households is $(0.214)^2 = 0.046$.

5.71 c/a/m
a. P(drilled in U.S.) = 7/10, or 0.7
b. P(drilled in U.S. or dry) = P(drilled in U.S.) + P(dry) - P(drilled in U.S. and dry)
$$= 7/10 + 7/10 - 5/10 = 9/10 = 0.9$$
c. P(dry | drilled in Nigeria) = P(dry and drilled Nigeria)/P(drilled Nigeria) = (2/10)/(3/10) = 0.667

5.73 c/a/m The probability that any individual vehicle will be stolen is 1/189, so the probability that any individual will not be stolen is 188/189. If we randomly select 5 vehicles, whether one is stolen is independent of the other vehicles being stolen.
a. P(none of the vehicles is stolen) = $(188/189)^5 = 0.974$
b. P(all the vehicles are stolen) = $(1/189)^5 = 4.147*10^{-12}$

c. Number of combinations of n = 5 vehicles with r = 2 stolen: $\displaystyle \binom{5}{2} = \dfrac{n!}{r!(n-r)!} = \dfrac{5!}{2!3!} = 5(4)/2! = 10$

5.75 c/a/d P(adult has had CPR training) = 0.10; P(adult has not had CPR training) = 1 - 0.10 = 0.90.
For n independent adults, P(all of them have not had CPR) = 0.9^n. Therefore, P(at least one has had CPR)
= $1 - 0.9^n$. Find n so that $1 - 0.9^n \geq 0.90$ => $0.9^n < 0.10$.
When n = 22, then $0.9^n = 0.0985$, which is < 0.10.

5.77 c/a/m The probability that all three individuals would have a Sears appliance is $(2/3)^3 = 0.296$.
Since the probability that none of the individuals would have a Sears appliance is $(1/3)^3 = 0.037$, the
probability that at least one of them will have a Sears appliance is 1 - 0.037 = 0.963.

5.79 c/a/d Define the following events: A = machine is in adjustment, D = unit is defective.

a. P(A) = 0.9

b. The posterior probability we want to find is P(A | D). The chronological (left) and reversed tree diagrams are shown below.

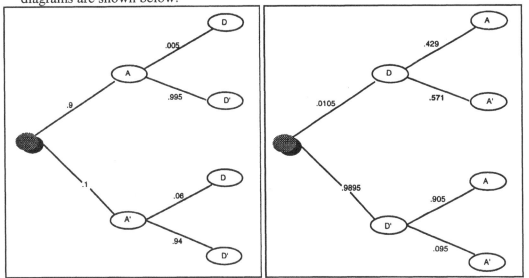

$$P(D) = P(A \text{ and } D) + P(A' \text{ and } D) = 0.0045 + 0.0060 = 0.0105$$

$$P(A|D) \frac{P(A \text{ and } D)}{P(D)} = \frac{0.0045}{0.0105} = 0.429$$

Using Bayes' Theorem:

$$P(A|D) = \frac{P(A)*P(D|A)}{[P(A)P(D|A)] + [P(A')P(D|A')]} = \frac{0.9*0.005}{(0.9*0.005) + (0.1*0.06)} = \frac{0.0045}{0.0105} = 0.429$$

5.81 c/a/m The number of possible combinations of n = 10 officers taken r = 3 at a time, disregarding their order, is:

$$\binom{n}{r} = \binom{10}{3} = \frac{10!}{3!7!} = \frac{10*9*8}{3*2*1} = 120$$

5.83 c/a/m Since order is important, the number of possible permutations of n = 12 faculty members to teach r = 3 courses is:

$$\frac{n!}{(n-r)!} = \frac{12!}{9!} = \frac{12*11*10*9}{9!} = 12*11*10 = 1320$$

5.85 c/a/m There are n = 2 integers to choose from (0 or 1) for the k = 51 digit sequence. Using the principle of multiplication, the number of different possibilities for the network's password is $n^k = 2^{51} \approx 2.25*10^{15}$.

CHAPTER 6
DISCRETE PROBABILITY DISTRIBUTIONS

SECTION EXERCISES

6.1 d/p/e A random variable is described as "random" because we don't know ahead of time exactly what value it will have following the experiment.

6.3 d/p/m
a. Discrete. There can't be a fraction of a customer.
b. Continuous. Although the total cannot include a fraction of a penny, there are so many possibilities that, for practical purposes, this random variable can be considered continuous.
c. Discrete. Although the passage of time is continuous, a digital watch reports it as discrete; e.g., there is a one-second gap between the 10:25:12 display and the 10:25:13 display.
d. Continuous. This is an exact measurement that could take on any value within an interval.

6.5 c/a/m Mean: $\mu = E(x) = \sum x_i P(x_i) = 3(0.2) + 8(0.7) + 10(0.1) = 7.2$

Variance: $\sigma^2 = E[(x - \mu)^2] = \sum (x_i - \mu)^2 P(x_i)$

$= (3 - 7.2)^2(0.2) + (8 - 7.2)^2(0.7) + (10 - 7.2)^2(0.1) = 4.76$

$\sigma = \sqrt{\sigma^2} = \sqrt{4.76} = 2.18$

x_i	$P(x_i)$	$x_i P(x_i)$	$(x_i - \mu)^2 P(x_i)$
3	0.20	0.60	3.53
8	0.70	5.60	0.45
10	0.10	1.00	0.78
	1.0	7.20	4.76

6.7 c/a/m Mean: $\mu = E(x) = \sum x_i P(x_i) = 0(0.1) + 1(0.3) + 2(0.3) + 3(0.2) + 4(0.1) = 1.9$

Variance: $\sigma^2 = E[(x - \mu)^2] = \sum (x_i - \mu)^2 P(x_i)$

$= (0 - 1.9)^2(0.1) + (1 - 1.9)^2(0.3) + (2 - 1.9)^2(0.3) + (3 - 1.9)^2(0.2) + (4 - 1.9)^2(0.1)$

$= 1.29$ and $\sigma = \sqrt{\sigma^2} = \sqrt{1.29} = 1.14$

x_i	$P(x_i)$	$x_i P(x_i)$	$(x_i - \mu)^2 P(x_i)$
0	0.10	0.00	0.361
1	0.30	0.30	0.243
2	0.30	0.60	0.003
3	0.20	0.60	0.242
4	0.10	0.40	0.441
		1.90	1.290

6.9 p/a/m For x = Marlin's monetary outcome for staging this scam, the probability distribution of x is:

x_i	$P(x_i)$	$x_i P(x_i)$
-$500	0.1	-50
$0	0.3	0
$100	0.6	60
	1.0	$10

Marlin's expected monetary outcome is $E(x) = \sum x_i P(x_i) = -500(0.1) + 0(0.3) + 100(0.6) = \$10$.

6.11 p/a/m The discrete random variable is x = the number of additional employees that must be hired. The probability distribution for x is:

x_i	$P(x_i)$	$x_iP(x_i)$
1	0.1	0.1
2	0.5	1.0
5	0.4	2.0
	1.0	3.1

Mean: $\mu = E(x) = \sum x_i P(x_i) = 1(0.1) + 2(0.5) + 5(0.4) = 3.1$ new employees

6.13 p/a/m Let x = profit per disc; the profit will be $9 minus $1 times the number of dots that show up on the rolled die.

No. of dots	Sale Price	x_i = Net Profit	$P(x_i)$	$x_iP(x_i)$
1	$8	$3.00	0.166667	0.50
2	$7	$2.00	0.166667	0.33
3	$6	$1.00	0.166667	0.17
4	$5	$0.00	0.166667	0.00
5	$4	- $1.00	0.166667	-0.17
6	$3	-$2.00	0.166667	-0.33
			1.0	$0.50

$\mu = E(x) = \sum x_i P(x_i) = \$0.50$ profit per sale

6.15 p/a/m

If Transco does not renovate the plant, the expected profit will be

   $\mu = E(x) = \sum x_i P(x_i) = 35(.3) + 38(.5) + 42(.2) = \$37.9$ million

If Transco carries out minor plant renovations, the expected profit will be

   $\mu = E(x) = \sum x_i P(x_i) = 28(.3) + 45(.5) + 55(.2) = \$41.9$ million

If Transco carries out major plant renovations, the expected profit will be

   $\mu = E(x) = \sum x_i P(x_i) = 21(.3) + 40(.5) + 70(.2) = \$40.3$ million

Transco's expected profit will be highest if it carries out minor plant renovations.

6.17 d/p/m As n becomes larger, the shape of the distribution tends to become more bell-shaped. Such a shape is characteristic of the normal distribution, discussed in Chapter 7.

6.19 c/a/m

a. $\mu = E(x) = n\pi = 12(0.3) = 3.6$

b. $\sigma^2 = E[(x - \mu)^2] = n\pi(1 - \pi) = 12(0.3)(1 - 0.3) = 2.52$ and $\sigma = \sqrt{\sigma^2} = \sqrt{2.52} = 1.59$

c. Using Minitab:

```
Binomial with n = 12 and p = 0.300000
        x      P( X = x)
        0         0.0138
        1         0.0712
        2         0.1678
        3         0.2397
        (partial printout)
```

Using a pocket calculator and the binomial formula in the text:

$$P(x) = \frac{n!}{x!(n-x)!}\pi^x(1-\pi)^{n-x} \quad \text{and} \quad P(3) = \frac{12!}{3!(12-3)!}(0.3)^3(0.7)^{12-3} = 0.2397$$

d. Using either the cumulative binomial tables or the Minitab printout shown below:

Using the cumulative binomial tables in the back of the text.
$P(2 \le x \le 8) = P(x \le 8) - P(x \le 1) = 0.9983 - 0.0850 = 0.9133$

Using Minitab, refer to the printout and find $P(x \le 8) - P(x \le 1)$

```
Binomial with n = 12 and p = 0.300000
        x      P( X <= x)
        0         0.0138
        1         0.0850
        2         0.2528
        3         0.4925
        4         0.7237
        5         0.8822
        6         0.9614
        7         0.9905
        8         0.9983
        9         0.9998
       10         1.0000
```

e. $P(x > 3) = 1 - P(x \le 3) = 1 - 0.4925 = 0.5075$

6.21 p/a/m From exercise 6.10, n = 5 and the probability of receiving a coupon is 0.60. Use the table of individual binomial probabilities or the Minitab printout below.

```
Binomial with n = 5 and p = 0.600000
        x      P( X = x)
        0         0.0102
        1         0.0768
        2         0.2304
        3         0.3456
        4         0.2592
        5         0.0778
```

a. $P(x = 0) = 0.0102$ b. $P(x = 2) = 0.2304$ c. $P(x = 3) = 0.3456$ d. $P(x = 5) = 0.0778$

6.23 p/a/m Binomial experiment with n = 3, π = 0.01. Let x = number of Andersons chosen. Shown below are the Minitab and binomial formula approaches.

```
Binomial with n = 3 and p = 0.0100000
        x       P( X = x)
        0         0.9703
        1         0.0294
        2         0.0003
        3         0.0000
```

a. P(x = 0)

Using the binomial formula:

$$P(x) = \frac{n!}{x!(n-x)!} \pi^x (1-\pi)^{n-x} \quad \text{and} \quad P(0) = \frac{3!}{0!(3-0)!}(0.01)^0(0.99)^{3-0} = 0.9703$$

Using Minitab, P(x = 0) = 0.9703 from the first line of the output.

b. Using the individual probabilities in the printout, P(x = 1) = 0.0294
c. Using the individual probabilities in the printout, P(x = 2) = 0.0003
d. Using the individual probabilities in the printout, P(x = 3) = 0.0000

6.25 p/a/m With n = 4, π = 0.52, the probabilities are:

```
Binomial with n = 4 and p = 0.520000
        x       P( X = x)
        0         0.0531
        1         0.2300
        2         0.3738
        3         0.2700
        4         0.0731
```

a. P(x ≤ 3) = 1 - P(x = 4) = 1 - 0.0731 = 0.9269. We are assuming that the population large, so that the chance of selecting a coffee drinker 10 or over is the same on each trial.
b. The probability for 11 p.m. to 7 a.m. would be much larger than 52%. We can still use the binomial probabilities if this probability does not change from trial to trial.
c. The probability would most likely be lower during these daytime hours. Binomial can be used if the population is large and this probability of a coffee drinker does not change from trial to trial.

6.27 p/a/m With n = 4, π = 0.60, the individual probabilities can be found in the binomial tables. Minitab is used below.

```
Binomial with n = 4 and p = 0.600000
        x       P( X = x )
     0.00         0.0256
     1.00         0.1536
     2.00         0.3456
     3.00         0.3456
     4.00         0.1296
```

P(x = 0) = 0.0256;   P(x = 1) = 0.1536;   P(x ≥ 2) = 0.3456 + 0.3456 + 0.1296 = 0.8208
P(1 ≤ x ≤ 3) = 0.1536 + 0.3456 + 0.3456 = 0.8448

6.29 p/a/m With n = 3, π = 0.50, the individual probabilities can be found in the binomial tables or determined with computer assistance, as shown in the Minitab printout below.

```
Binomial with n = 3 and p = 0.500000
        x       P( X = x )
     0.00         0.1250
     1.00         0.3750
     2.00         0.3750
     3.00         0.1250
```

For Alicia, P(x = 0) is 0.1250; P(x = 3) is 0.1250; and P(x = 1) is 0.3750.

6.31 p/a/m With n = 4, π = 0.10, the individual probabilities can be found in the binomial tables or determined with computer assistance, as shown in the Minitab printout below.

```
Binomial with n = 4 and p = 0.100000
        x       P( X = x )
     0.00         0.6561
     1.00         0.2916
     2.00         0.0486
     3.00         0.0036
     4.00         0.0001
```

The probability that x = 0 of the four wheel bearings will fail is 0.6561. The probability that one or more of the bearings will fail and result in the truck breaking down is 1 - 0.6561, or 0.3439.

6.33 p/a/d
a. The probability that Emily will be bumped is 0.00187.
b. For any individual, the probability of not being bumped is 1 - 0.00187 = 0.99813. The probability that x = 0 of the 10 travelers will make the trip without being bumped is 0.99813^{10} = 0.9815. Thus, the probability that at least one person will be bumped is 1 - 0.9815, or 0.0185. The same result can be obtained by using the Minitab probabilities below.

```
Binomial with n = 10 and p = 0.00187000
        x       P( X = x )
     0.00         0.9815
     1.00         0.0184
     2.00         0.0002
     3.00         0.0000
     4.00         0.0000
```

c. With n = 220 and π = 0.00187, computer assistance will be much more convenient than using a pocket calculator, and the Minitab cumulative probabilities are shown below.

```
Binomial with n = 220 and p = 0.00187000
        x       P( X <= x )
     0.00         0.6625
     1.00         0.9355
     2.00         0.9915
     3.00         0.9992
     4.00         0.9999
     5.00         1.0000
     6.00         1.0000
```

If last year's 220-flight total were to be repeated over an extremely large number of years, the probability of x ≥ 4 bumps in any given year would be just 1 - 0.9992, or 0.0008. Yes, the company's experience last year was very unusual.

6.35 c/a/m Using the Poisson formula, with $\lambda = 2.0$, the Poisson probabilities can be computed with the formula or obtained with computer assistance. Using Minitab, the individual Poisson probabilities are shown below.

```
Poisson with mu = 2.00000
        x        P( X = x)
        0         0.1353
        1         0.2707
        2         0.2707
        3         0.1804
        4         0.0902
        5         0.0361
        6         0.0120
        7         0.0034
        8         0.0009
        9         0.0002
       10         0.0000
```

a. Using the formula, $P(x) = \dfrac{\lambda^x e^{-\lambda}}{x!}$, $\quad P(x = 0) = \dfrac{2.0^0 e^{-2}}{0!} = \dfrac{(1)(0.13534)}{1} = 0.1353$

b. Using the formula, $P(x) = \dfrac{\lambda^x e^{-\lambda}}{x!}$, $\quad P(x = 1) = \dfrac{2.0^1 e^{-2}}{1!} = \dfrac{(2)(0.13534)}{1} = 0.2707$

c. Using the formula, $P(x) = \dfrac{\lambda^x e^{-\lambda}}{x!}$, $\quad P(x = 2) = \dfrac{2.0^2 e^{-2}}{2!} = \dfrac{(4)(0.13534)}{2} = 0.2707$

Using the formula, $P(x) = \dfrac{\lambda^x e^{-\lambda}}{x!}$, $\quad P(x = 3) = \dfrac{2.0^3 e^{-2}}{3!} = \dfrac{(8)(0.13534)}{6} = 0.1804$

Thus, $P(x \le 3) = P(x = 0) + P(x = 1) + P(x = 2) + P(x = 3)$
$= 0.1353 + 0.2707 + 0.2707 + 0.1804 = 0.8571$

d. $P(x \ge 2) = 1 - P(x \le 1) = 1 - [P(x = 0) + P(x = 1)] = 1 - [0.1353 + 0.2707]$
$= 1 - 0.4060 = 0.5940$

6.37 p/a/m
a. Let $\pi = 530/100{,}000 = 0.0053$. The expected value of x is $E(x) = \lambda = n\pi$
$= 1000(0.0053) = 5.3$.

b. $P(x = 3) = 0.1239$ has been determined by referring to the individual probabilities in the Minitab printout, below left. The cumulative probabilities are shown in the printout on the right.

```
Poisson with mu = 5.30000        Poisson with mu = 5.30000
        x      P( X = x )                x      P( X <= x )
     0.00       0.0050                 0.00       0.0050
     1.00       0.0265                 1.00       0.0314
     2.00       0.0701                 2.00       0.1016
     3.00       0.1239                 3.00       0.2254
     4.00       0.1641                 4.00       0.3895
     5.00       0.1740                 5.00       0.5635
     6.00       0.1537                 6.00       0.7171
     7.00       0.1163                 7.00       0.8335
     8.00       0.0771                 8.00       0.9106
     9.00       0.0454                 9.00       0.9559
    10.00       0.0241                10.00       0.9800
    11.00       0.0116                11.00       0.9916
    12.00       0.0051                12.00       0.9967
    13.00       0.0021                13.00       0.9988
    14.00       0.0008                14.00       0.9996
    15.00       0.0003                15.00       0.9999
    16.00       0.0001                16.00       1.0000
    17.00       0.0000                17.00       1.0000
```

c. Using the individual probabilities in part b, P(x = 5) = 0.1740.
d. Using the cumulative probabilities in part b, P(x ≤ 8) = 0.9106.
e. Using the cumulative probabilities in part b, P(3 ≤ x ≤ 10) = P(x ≤ 10) - P(x ≤ 2)
   = 0.9800 - 0.1016 = 0.8784.

6.39 p/a/m
a. Let λ = expected value of x, and E(x) = 4.2(500/1000) = 2.1 divorces.
b. P(x = 1) = 0.2572. Using λ = 2.1, the Poisson probabilities can be found by using the computer or by referring to the individual and cumulative Poisson tables in the text.
c. Using the "2.1" column of individual probabilities, P(x = 4) = 0.0992.
d. Using the "2.1" column of cumulative probabilities, P(x ≤ 6) = 0.9941.
e. Using the "2.1" column of cumulative probabilities,
   P(2 ≤ x ≤ 5) = P(x ≤ 5) - P(x ≤ 1) = 0.9796 - 0.3796 = 0.6000.

6.41 p/a/m The expected number of emergency calls tomorrow is E(x) = λ = 3.1 calls. We can refer to the cumulative Poisson probabilities in text Appendix A or use the computer. The Minitab cumulative probabilities are shown below. Because Howard can handle up to 8 calls, the probability that he will receive more calls than he can handle is P(x ≥ 9) = 1 - P(x ≤ 8) = 1 - 0.9953 = 0.0047. Considering this extremely low probability, Howard might want to consider a slight reduction in the level of personnel and equipment he devotes to handling emergency calls.

```
Poisson with mu = 3.10000
        x       P( X <= x )
     0.00        0.0450
     1.00        0.1847
     2.00        0.4012
     3.00        0.6248
     4.00        0.7982
     5.00        0.9057
     6.00        0.9612
     7.00        0.9858
     8.00        0.9953
     9.00        0.9986
    10.00        0.9996
    11.00        0.9999
    12.00        1.0000
```

6.43 p/a/m
a. Let λ = expected value of x. This is nπ = 1000(0.01) = 10.0 foreclosures.
b. P(x = 7) = 0.0901 can be found by using either the Poisson probability tables or the computer.
c. Using the "10.0" column of individual probabilities, P(x = 10) = 0.1251.
d. Using the "10.0" column of cumulative probabilities, P(x ≤ 12) = 0.7916.
e. Using the "10.0" column of cumulative probabilities,
   P(8 ≤ x ≤ 15)= P(x ≤ 15) - P(x ≤ 7) = 0.9513 - 0.2202 = 0.7311.

6.45 p/a/m For a Poisson distribution with λ = 10.1, the cumulative probabilities for x = number of days of work lost during 20,000 worker-hours of production can be obtained using either the Poisson cumulative probability tables or the computer. Shown below is the Minitab printout for x values between 7 and 15.

```
Poisson with mu = 10.1000
       x       P( X <= x )
    7.00          0.2113
    8.00          0.3217
    9.00          0.4455
   10.00          0.5705
   11.00          0.6853
   12.00          0.7820
   13.00          0.8571
   14.00          0.9112
   15.00          0.9477
```

The probability that no more than 8 days are lost is P (x ≤ 8) = 0.3217.
Considering Maxwell, P(x ≥ 14) is 1 - P(x ≤ 13) = 1 - 0.8571, and P(x ≥ 14) = 0.1429. Maxwell Steel's experience is higher than the mean of 10.1, but would not be considered as *very* unusual.

6.47 p/a/m On the basis of 3000 drivers, the mean is 30.0(3000/100,000), or 0.9. For a Poisson distribution with λ = 0.9, the cumulative probabilities for x = number of drivers murdered on the job can be obtained using either the Poisson cumulative probability tables or the computer. Shown below is the Minitab printout for x values between 0 and 4.

```
Poisson with mu = 0.900000
       x       P( X <= x )
    0.00          0.4066
    1.00          0.7725
    2.00          0.9371
    3.00          0.9865
    4.00          0.9977
```

The probability that at least 2 of the city's drivers will be murdered on the job next year is P(x ≥ 2) = 1 - P(x ≤ 1) = 1 - 0.7725, or 0.2275. The expected amount the city will be providing to families of murdered drivers next year is E(x)*50,000 = 0.9(50,000) = $45,000.

6.49 p/a/m Referring to the cumulative probabilities shown in the solution to Exercise 6.48, the probability that the line will break down at least 9 times in a given day is P(x ≥ 9) = 1 - P(x ≤ 8) = 1 - 0.9991, or 0.0009. This many breakdowns in a single day is too rare to be considered a coincidence. Management should be suspicious.

6.51 c/c/m

a. and b. Simulation of 50 observations from a binomial distribution with $\pi = 0.20$ and n = 5. The distribution of simulated values is shown below. The distribution tails off toward the higher values, (positive skewness).

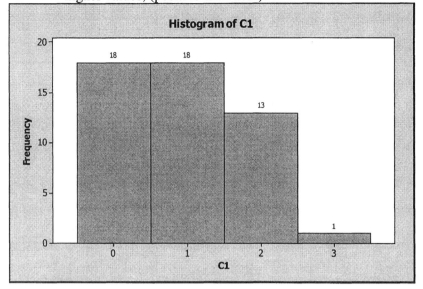

c. Repeating parts a and b, but with 500 simulated observations, the distribution of simulated values is shown below. As with the previous distribution, this one tails off toward the higher values. The amount of skewness does not appear to be appreciably different from the distributions in parts a and b.

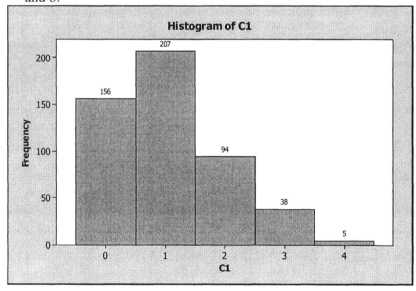

d. Even for very large numbers of observations, the distribution would not become symmetrical; but the observed relative frequencies would approach the theoretical probabilities associated with the underlying binomial distribution with n = 5 and $\pi = 0.2$. This is due to the law of large numbers.

CHAPTER EXERCISES

6.53 p/a/m We can use either the computer or the table of individual binomial probabilities for $\pi = 0.9$ and n = 5. As shown in the individual Minitab probabilities below, the probability that exactly 3 persons will leave a tip is 0.0729.

```
Binomial with n = 5 and p = 0.900000
        x       P( X = x )
     0.00         0.0000
     1.00         0.0005
     2.00         0.0081
     3.00         0.0729
     4.00         0.3281
     5.00         0.5905
```

6.55 p/a/m For a span of 50,000 flying hours, the expected value is 2.1/2, or 1.05 crashes per 50,000 hours. The probability is 0.7174 that no more than one crash will occur in the next 50,000 flying hours.

```
Poisson with mu = 1.05000
        x       P( X <= x )
     0.00         0.3499
     1.00         0.7174
     2.00         0.9103
     3.00         0.9778
     4.00         0.9955
     5.00         0.9992
     6.00         0.9999
     7.00         1.0000
```

6.57 p/a/m Using either the computer or the table of cumulative binomial probabilities for n = 8 and $\pi = 0.4$, $P(x \leq 2) = 0.3154$. The Minitab printout of cumulative probabilities is shown below:

```
Binomial with n = 8 and p = 0.400000
        x       P( X <= x )
     0.00         0.0168
     1.00         0.1064
     2.00         0.3154
     3.00         0.5941
     4.00         0.8263
     5.00         0.9502
     6.00         0.9915
     7.00         0.9993
     8.00         1.0000
```

6.59 d/p/m No, it would not be appropriate. The probability of a success would not be the same from one trial (customer) to the next. For example, if this car were sold to the first of the three customers, the probability of selling the car to either of the other two customers would become zero.

6.61 p/a/m The probability distribution would be binomial, with n = 10 and $\pi = 0.4$. We can either use the computer or refer to the table of individual binomial probabilities for n = 10 and $\pi = 0.4$.

```
Binomial with n = 10 and p = 0.400000
        x       P( X = x )
     0.00         0.0060
     1.00         0.0403
     2.00         0.1209
     3.00         0.2150
     4.00         0.2508
     5.00         0.2007
     6.00         0.1115
     7.00         0.0425
     8.00         0.0106
     9.00         0.0016
    10.00         0.0001
```

6.63 p/a/m The mine will operate all day on those days when $x \leq 2$. Using either the computer or the "1.2" column of the table of Poisson cumulative probabilities, $P(x \leq 2) = 0.8795$. On 87.95% of the days, the mine will be operational throughout the day.

6.65 p/a/m There are sixteen customers on the panel, 80% of 16 yields 12.8 customers. Assuming that $\pi = 0.90$, we want to find the probability that there are at least 13 panel members who come from homes in which gas is used for heating. Using the table of binomial cumulative probabilities with n = 16 and $\pi = 0.90$, $P(x \geq 13) = 1 - P(x \leq 12) = 1 - 0.0684 = 0.9316$, or 93.16%.

6.67 p/a/m If 5 flaws occur per 2000 feet of tubing, the expected number of flaws in 200 feet will be 5(200/2000), or 0.5. The average for the associated Poisson distribution is $\lambda = 0.5$ flaws per 200 feet. Let x = observed number of flaws per 200 feet.

$$P(x) = \frac{e^{-\lambda}\lambda^x}{x!}, \quad P(x = 0) = \frac{e^{-0.5}0.5^0}{0!} = \frac{(0.6065)(1)}{1} = 0.6065$$

This result can also be obtained from the table of Poisson individual probabilities or by using the computer.

6.69 p/a/m Using the binomial distribution, with n = 5 and $\pi = 0.30$, the probability that exactly three of these families will be owners of savings bonds is P(x = 3):

$$P(x) = \frac{n!}{x!(n-x)!}\pi^x(1-\pi)^{n-x}, \quad P(3) = \frac{5!}{3!(5-3)!}(0.3)^3(0.7)^2 = 0.1323$$

We can also use the table of binomial individual probabilities or the computer. Using Minitab:

```
Binomial with n = 5 and p = 0.300000
        x       P( X = x )
     0.00         0.1681
     1.00         0.3602
     2.00         0.3087
     3.00         0.1323
     4.00         0.0284
     5.00         0.0024
     6.00         0.0000
```

6.71 p/a/m
a. The expected value of x = the number of injuries during the next 100,000 student days is $\lambda = 2.4$ injuries.
b. With $\lambda = 2.4$, the probability distribution for the possible values of x will be as shown below.

```
Poisson with mu = 2.40000
       x        P( X = x)
       0          0.0907
       1          0.2177
       2          0.2613
       3          0.2090
       4          0.1254
       5          0.0602
       6          0.0241
       7          0.0083
       8          0.0025
       9          0.0007
      10          0.0002
      11          0.0000
```

6.73 p/a/d The distribution is Poisson with $\lambda = 3.0$ calls per minute. The system can handle as many as 5 calls per minute. Given this information, the probability that the system will be overloaded in the next minute is $P(x \geq 6) = 1 - P(x \leq 5) = 1 - 0.9161 = 0.0839$. This can be found from the table of Poisson cumulative probabilities or by using the computer.

6.75 p/a/m For a binomial distribution with n = 10 and π = 0.90, the cumulative probabilities for x = number of companies actively involved in data warehousing can be obtained using either the binomial cumulative probability table or the computer. Shown below is the Minitab printout for x values between 5 and 10.

```
Binomial with n = 10 and p = 0.900000
      x       P( X <= x )
    5.00         0.0016
    6.00         0.0128
    7.00         0.0702
    8.00         0.2639
    9.00         0.6513
   10.00         1.0000
```

The probability that at least 7 of the companies are actively involved in data warehousing is $P(x \geq 7) = 1 - P(x \leq 6) = 1 - 0.0128$, or 0.9872.

6.77 p/a/m For a binomial distribution with n = 9 and π = 0.10, the cumulative probabilities for x = number of computers that are inoperable can be obtained using either the binomial cumulative probability table or the computer. Shown below is the Minitab printout for x values between 0 and 6.

```
Binomial with n = 9 and p = 0.100000
      x       P( X <= x )
    0.00         0.3874
    1.00         0.7748
    2.00         0.9470
    3.00         0.9917
    4.00         0.9991
    5.00         0.9999
    6.00         1.0000
```

In order to have at least 8 computers that work, the number of inoperable computers must be no more than 1. As shown in the printout above, $P(x \leq 1) = 0.7748$.

6.79 p/a/m For a binomial distribution with n = 40, π = 0.02, and x = the number of tires with surface blemishes, we can use the computer to determine $P(x \geq 5)$. Given the printout below, $P(x \geq 5) = 1 - P(x \leq 4) = 1 - 0.9988$, or 0.0012.

```
Binomial with n = 40 and p = 0.0200000
      x       P( X <= x )
    0.00         0.4457
    1.00         0.8095
    2.00         0.9543
    3.00         0.9918
    4.00         0.9988
    5.00         0.9999
    6.00         1.0000
```

If the true population proportion were really 0.02, it would be extremely unlikely to have 5 or more blemished tires in a sample of 40. The advertising claim is not very believable. Alternatively, we can use the Poisson approximation to the binomial distribution. In this case, the mean of the Poisson distribution would be $E(x) = \lambda = n\pi = 40(0.02) = 0.8$ tires with blemishes.

6.81 p/a/m Allen and his Acura: For a Poisson distribution with $\lambda = 0.91$, and x = number of defects per vehicle, the Poisson individual probabilities are shown below. (Approximate values can be obtained using the Poisson individual probabilities table with $\lambda = 0.9$.) The probability that Allen's Acura will have exactly 2 problems is shown below as 0.1667.

```
Poisson with mu = 0.910000
      x      P( X = x )
   0.00       0.4025
   1.00       0.3663
   2.00       0.1667
   3.00       0.0506
   4.00       0.0115
   5.00       0.0021
   6.00       0.0003
   7.00       0.0000
```

Kevin and his Kia: For a Poisson distribution with $\lambda = 2.51$, and x = number of defects per vehicle, the Poisson cumulative probabilities are shown below. (Approximate values can be obtained using the Poisson cumulative probabilities table with $\lambda = 2.5$.) The probability that Kevin's Kia will have no more than 1 problem is shown below as 0.2853.

```
Poisson with mu = 2.51000
      x      P( X <= x )
   0.00       0.0813
   1.00       0.2853
   2.00       0.5413
   3.00       0.7554
   4.00       0.8898
   5.00       0.9573
   6.00       0.9855
   7.00       0.9957
   8.00       0.9988
   9.00       0.9997
  10.00       0.9999
  11.00       1.0000
```

The probability that neither Allen nor Kevin will have a problem is 0.4025*0.0813 = 0.0327.

CHAPTER 7
CONTINUOUS PROBABILITY DISTRIBUTIONS

SECTION EXERCISES

7.1 d/p/e Discrete probability distributions can be expressed as histograms where the height of the vertical bars is the probability for the various values that the random variable can take on. However, continuous probability distributions are smooth curves. Since the random variable can take on any value along a range, we find the probability that the random variable is within a certain interval by finding the area under the curve within this interval.

7.3 d/p/m The area beneath the probability density function represents the probability of the random variable, x, being between minus infinity and plus infinity. Since x must be between minus infinity and plus infinity (this is a certain event), the area must be 1.

7.5 d/p/m There is an infinite number of normal curves possible. There are two descriptors that decide which specific curve you are talking about: the mean and the standard deviation.

7.7 d/p/m Just superimpose part B of Figure 7.3 over part A so that the means are the same.

7.9 c/a/m x is normally distributed with $\mu = 25$ and $\sigma = 5$. Using the approximate areas beneath the normal curve, as discussed in Section 7.2 of the chapter and shown in Figure 7.4:
a. $P(x \geq 25) = P(x \geq \mu) = 0.5$
b. $P(20 \leq x \leq 30) = P(\mu - \sigma \leq x \leq \mu + \sigma) = 0.683$
c. $P(x \leq 30) = P(x \leq \mu + \sigma) = 0.5 + (0.683/2) = 0.5 + 0.3415 = 0.8415$
d. $P(x = 26.2) = 0$
e. $P(15 \leq x \leq 25) = P(\mu - 2\sigma \leq x \leq \mu) = 0.955/2 = 0.4775$
f. $P(x \geq 15) = P(x \geq \mu - 2\sigma) = 0.5 + (0.955/2) = 0.5 + 0.4775 = 0.9775$

7.11 p/a/m Given x = amount of first mortgage, normally distributed: $\mu = \$245,000$, $\sigma = \$30,000$. Using the approximate areas beneath the normal curve, as discussed in Section 7.2 of the chapter and shown in Figure 7.4:
a. $P(x > 245,000) = P(x > \mu) = 0.5$
b. $P(185,000 < x < 305,000) = P(\mu - 2\sigma < x < \mu + 2\sigma) = 0.955$
c. $P(215,000 < x < 275,000) = P(\mu - \sigma < x < \mu + \sigma) = 0.683$
d. $P(x > 155,000) = P(x > \mu - 3\sigma) = 0.5 + (0.997/2) = 0.5 + 0.4985 = 0.9985$

7.13 p/a/m Given x = bag delivery times, normally distributed with $\mu = 12.1$ mins., $\sigma = 2.0$ mins. Using the approximate areas beneath the normal curve, as discussed in Section 7.2 of the chapter and shown in Figure 7.4:
a. $P(x > 14.1) = P(x > \mu + \sigma) = 0.5 - (0.683/2) = 0.5 - 0.3415 = 0.1585$
b. $P(10.1 < x < 14.1) = P(\mu - \sigma < x < \mu + \sigma) = 0.683$
c. $P(x < 8.1) = P(x < \mu - 2\sigma) = 0.5 - (0.955/2) = 0.5 - 0.4775 = 0.0225$
d. $P(10.1 < x < 16.1) = P(\mu - \sigma < x < \mu + 2\sigma) = (0.683/2) + (0.955/2) = 0.3415 + 0.4775 = 0.819$

7.15 p/a/m Given x = commuting times, normally distributed with μ = 23.0 mins., σ = 5.0 mins. Using the approximate areas beneath the normal curve, as discussed in Section 7.2 of the chapter and shown in Figure 7.4:

a. $P(x > 38.0) = P(x > \mu + 3\sigma) = 0.5 - (0.997/2) = 0.5 - 0.4985 = 0.0015$

b. $P(x < 18.0) = P(x < \mu - \sigma) = 0.5 - (0.683/2) = 0.5 - 0.3415 = 0.1585$ It would be a good idea for him to update his resume.

7.17 d/p/m

a. 25% of the time, z is less than the first quartile. $P(z < Q) = 0.25$. Look up the area 0.5000 - 0.2500 = 0.2500 in the body of the standard normal table; taking the closest value, Q = -0.67. (Q is negative since Q is on the left side of 0.)
   25% of the time, z is greater than the third quartile. $P(z > Q) = 0.25$. Since the normal distribution is symmetric, Q = 0.67.

b. 10% of the time, z is less than the first decile. $P(z < D) = 0.10$. Look up the area 0.5000 - 0.1000 = 0.4000 in the body of the standard normal table; taking the closest value, D = -1.28. (D is negative since D is on the left side of 0.)
   10% of the time, z is greater than the ninth decile. $P(z > D) = 0.10$. Since the normal distribution is symmetric, D = 1.28.

c. 23% of the time, z is less than the 23rd percentile. $P(z < P) = 0.23$. Look up the area 0.5000 - 0.2300 = 0.2700 in the body of the standard normal table; taking the closest value, P = -0.74. (P is negative since P is on the left side of 0.)
   23% of the time, z is greater than the 77th percentile. $P(z > P) = 0.23$. Since the normal distribution is symmetric, P = 0.74.

7.19 c/p/e Given x is normally distributed with μ = 200 and σ = 25.

a. x = 150 $z = (x - \mu)/\sigma = (150 - 200)/25 = -2.00$

b. x = 180 $z = (180 - 200)/25 = -0.80$ c. x = 200 $z = (200 - 200)/25 = 0.00$

d. x = 285 $z = (285 - 200)/25 = 3.40$ e. x = 315 $z = (315 - 200)/25 = 4.60$

7.21 c/a/e

a. $P(0.00 \le z \le 1.10) = 0.3643$ ("1.1" row and the ".00" column of the standard normal table)

b. $P(z \ge 1.10) = 0.1357$. Since the total area to the right of the mean is 0.5000, the area in the right tail of the distribution is 0.5000 - 0.3643 = 0.1357.

c. $P(z \le 1.35) = 0.9115$. Referring to the "1.3" row and the ".05" column of the standard normal table, we find an area of 0.4115. Since the total area to the left of the mean is 0.5000, the area to the left of z = 1.35 is 0.5000 + 0.4115 = 0.9115.

7.23 c/a/e

a. $P(-1.96 \le z \le 1.27) = 0.4750$ (the area from z = 0.00 to z = -1.96) + 0.3980 (the area from z = 0.00 to z = 1.27) = 0.8730

b. $P(0.29 \le z \le 1.00) = 0.3413$ (the area from z = 0.00 to z = 1.00) - 0.1141 (the area from z = 0.00 to z = 0.29) = 0.2272

c. $P(-2.87 \le z \le -1.22) = 0.4979$ (the area from z = 0.00 to z = -2.87) - 0.3888 (the area from z = 0.00 to z = -1.22) = 0.1091

7.25 c/p/m

a. $P(0.00 \le z \le z_0) = 0.20$. Look up the area 0.2000 in the body of the standard normal table; taking the closest value, z_0 will be 0.52.

b. $P(0.00 \le z \le z_0) = 0.48$. Look up the area 0.4800 in the body of the standard normal table; taking the closest value, z_0 will be 2.05.

c. $P(z \le z_0) = 0.54$. Look up the area 0.5400 - 0.5000 = 0.0400 in the body of the standard normal table; taking the closest value, z_0 will be 0.10.

d. $P(z \ge z_0) = 0.30$. Look up the area 0.5000 - 0.3000 = 0.1000 in the body of the standard normal table; taking the closest value, z_0 will be 0.52.

7.27 p/a/m From exercise 7.11, x is normally distributed with $\mu = \$245,000$ and $\sigma = \$30,000$.

a. $P(190,000 < x < 210,000) = P(-1.83 < z < -1.17) = 0.4664 - 0.3790 = 0.0874$

b. $P(x > 210,000) = P(z > -1.17) = 0.5000 + 0.3790 = 0.8790$

c. $P(x < 275,000) = P(z < 1.00) = 0.5000 + 0.3413 = 0.8413$

7.29 p/a/m From exercise 7.15, x is normally distributed with $\mu = 23.0$ minutes and $\sigma = 5$ minutes.

a. $P(x < 22) = P(z < -0.20) = 0.5000 - 0.0793 = 0.4207$

b. $P(22 < x < 26)) = P(-0.20 < z < 0.60) = 0.0793 + 0.2257 = 0.3050$

c. $P(x > 25) = P(z > 0.40) = 0.5000 - 0.1554 = 0.3446$

7.31 p/a/d From exercise 7.11, x is normally distributed with $\mu = \$245,000$ and $\sigma = \$30,000$. The quantity to be determined is the first-mortgage amount (A) such that only 5% of the mortgage customers exceed A dollars.

$$P(x > A) = 0.05 \text{ or } P\left(z > \frac{A - 245,000}{30,000}\right) = 0.05$$

To solve for A, find the value of z which corresponds to a right-tail area of 0.05. Referring to the standard normal table, look up 0.5000 - 0.0500 = 0.4500 in the body of the table. The area from z = 0.00 to z = 1.64 includes an area of 0.4495 and the area from z = 0.00 to z = 1.65 includes an area of 0.4505. The 0.4500 area corresponds to a z that is halfway between z = 1.64 and z = 1.65, so we will interpolate and use z = 1.645.

$P(z > 1.645) = 0.05$. Equating $\dfrac{A - 245,000}{30,000}$ to 1.645, solve for A, and A = \$294,350.

7.33 p/a/m Given x = viewing time is normally distributed with $\mu = 4.28$ hrs., $\sigma = 1.30$ hrs.

a. $P(x < 2.00) = P(z < -1.75) = 0.5000 - 0.4599 = 0.0401$

b. We are looking for the z value corresponding to an area of 0.49 in the standard normal table. The closest value is z = 2.33. Solving (T – 4.28)/1.30 = 2.33, we obtain T = 7.31 hours.

7.35 p/a/d Let x = drying time. The distribution is normal with $\mu = 2.5$ mins. and $\sigma = 0.25$ mins.

The problem is to find A so $P(x < A) = 0.9980$, or $P\left(z < \dfrac{A-2.5}{0.25}\right) = 0.9980$

We need to find the z value that corresponds to an area of 0.9980 – 0.5000 = 0.4980 in our standard normal table. The closest z value is z = 2.88. Thus, $P(z \le 2.88) = 0.9980$ and $P\left(z < \dfrac{A - 2.5}{0.25}\right) = 0.9980$

So $\dfrac{A - 2.5}{0.25} = 2.88$, or $A = 2.5 + 0.25(2.88) = 3.22$ minutes. The timer is set to dry for 3.22 minutes.

7.37 p/a/d The pump lifetimes are currently normally distributed with $\mu = 63{,}000$ miles and $\sigma = 10{,}000$ miles. The mean will now have to be a value such that the area to the left of x = 50,000 is just 0.0200. Since the standard normal distribution table gives areas from the midpoint, we must look for the z value corresponding to an area of 0.4800. The nearest area listed is 0.4798, associated with z = -2.05. We can now set up an expression that can be solved for the value of the necessary mean, M:

This is $-2.05 = \dfrac{50{,}000 - M}{10{,}000}$ and, solving for M, we find that M = 70,500 miles.

7.39 d/p/m The correction expands each possible value of a discrete variable by 0.5 in each direction. This is needed because the binomial distribution is discrete (having gaps between the possible values) while the normal distribution is continuous and can take on any value within an interval.

7.41 c/a/m

a. $\mu = n\pi = 40(0.25) = 10.0$ and $\sigma = \sqrt{n\pi(1-\pi)} = \sqrt{40(0.25)(1-0.25)} = 2.739$

b. $P(x = 8) = P(7.5 \leq x \leq 8.5)$ after continuity correction.

$$P(\frac{7.5-10}{2.739} < z < \frac{8.5-10}{2.739}) = (-0.91 < z < -0.55) = 0.3186 - 0.2088 = 0.1098$$

$P(12 \leq x \leq 16) = P(11.5 \leq x \leq 16.5)$ after continuity correction.

$$P(\frac{11.5-10}{2.739} < z < \frac{16.5-10}{2.739}) = (0.55 < z < 2.37) = 0.4911 - 0.2088 = 0.2823$$

$P(10 \leq x \leq 12) = P(9.5 \leq x \leq 12.5)$ after continuity correction.

$$P(\frac{9.5-10}{2.739} < z < \frac{12.5-10}{2.739}) = (-0.18 < z < 0.91) = 0.0714 + 0.3186 = 0.3900$$

$P(x \geq 14) = P(x \geq 13.5)$ after continuity correction.

$$P(z > \frac{13.5-10}{2.739}) = P(z > 1.28) = 0.5000 - 0.3997 = 0.1003$$

7.43 p/a/d Given x is binomial with $\pi = 0.60$ and n = 20.

a. $\mu = n\pi = 20(0.60) = 12.0$ $\quad \sigma = \sqrt{n\pi(1-\pi)} = \sqrt{20(0.60)(1-0.60)} = 2.19$

b. Referring to the individual binomial tables in the appendix, P(x = 12) = 0.1797.

c. Using the normal approximation to the binomial distribution, $P(x = 12) = P(11.5 \leq x \leq 12.5)$ after continuity correction, $= P(-0.23 \leq z \leq 0.23) = 2(0.0910) = 0.1820$.

d. $P(x \geq 10) = P(x \geq 9.5)$ after continuity correction, $= P(z \geq -1.14) = 0.5000 + 0.3729 = 0.8729$.

7.45 p/a/m Let x = the number of tax returns in this group prepared by H & R Block, x is binomial with $\pi = 0.137$ and n = 1000.

$\mu = n\pi = 1000(0.137) = 137$ $\quad \sigma = \sqrt{n\pi(1-\pi)} = \sqrt{1000(0.137)(1-0.137)} = 10.873$

$P(110 \leq x \leq 140) = P(109.5 \leq x \leq 140.5)$ after continuity correction, $= P(-2.53 \leq z \leq 0.32) = 0.4943 + 0.1255 = 0.6198$.

7.47 p/a/m From exercise 7.45, x is binomial with $\pi = 0.137$ and n = 1000, $\mu = 137$, $\sigma = 10.873$. Without continuity correction, $P(110 \leq x \leq 140) = P(-2.48 \leq z \leq 0.28) = 0.4934 + 0.1103 = 0.6037$. With the continuity correction, $P(110 \leq x \leq 140)$ was equal to 0.6198. The probabilities differ by just 0.0161. The larger n is or the closer π is to 0.5, the less important the correction becomes.

7.49 d/p/e The Poisson distribution describes a discrete random variable which is the number of "rare events" occurring during a given interval of time, space, or distance. For a Poisson process, the exponential distribution describes a continuous random variable, x = the amount of time, space, or distance between occurrences of these rare events.

7.51 d/p/m A Poisson random variable would be x = the number of calls from customers in a minute. The exponential-distribution counterpart would be y = minutes between calls from customers.

7.53 c/a/m With x exponentially distributed with mean = $1/\lambda = 1/0.02 = 50$, $P(x \geq k) = e^{-\lambda k} = e^{-0.02k}$

a. $P(x \geq 30) = e^{-0.02(30)} = e^{-0.6} = 0.5488$ 　　b. $P(x \geq 40) = e^{-(0.02)40} = e^{-0.8} = 0.4493$

c. $P(x \geq 50) = e^{-(0.02)50} = e^{-1.0} = 0.3679$ 　　d. $P(x \geq 60) = e^{-(0.02)60} = e^{-1.2} = 0.3012$

7.55 p/a/m The mean of the corresponding Poisson distribution would be $\lambda = 1/8 = 0.125$, so $P(x \geq 10) = e^{-(0.125)10} = e^{-1.25} = 0.2865$

7.57 p/a/d The variable x = thousands of flying hours between fatal crashes is exponentially distributed with a mean of $1/\lambda = 1/0.0135 = 74.074$ thousand flying hours, and Arnold's Flying Service flies 40 thousand hours each year.

The probability that Arnold's will not experience a fatal crash until at least a year from today, or until at least x = 40 thousand flying hours, is $P(x \geq 40) = e^{-0.0135(40)} = e^{-0.54} = 0.583$.

The probability that Arnold's will not experience a fatal crash until at least two years from today, or until at least x = 80 thousand flying hours, is $P(x \geq 80) = e^{-0.0135(80)} = e^{-1.08} = 0.340$.

This is most easily done with the computer. We want the inverse of the cumulative probability distribution, and the specific cumulative probability of interest is 1 - 0.90, or 0.10. As shown in the Minitab printout below, there is a 0.10 probability that the next fatal crash will occur within the next 7.8045 thousand flying hours.

```
Inverse Cumulative Distribution Function

Exponential with mean = 74.0740
P( X <= x )          x
  0.1000        7.8045
```

Accordingly, there is a 0.90 probability that no fatal crash will occur until *at least* 7.8045 thousand flying hours from now. This can also be expressed as

$P(x \geq 7.8045) = e^{-0.0135(7.8045)} = 0.90$.

7.59 c/p/m

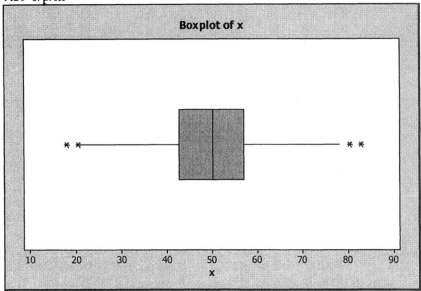

7.61 c/a/m Given x is normally distributed with μ = 150 and σ = 25.

P(x ≤ 140) = P(x ≤ μ - 1.2σ) = P(z ≤ -1.20) = 0.5000 - 0.3849 = 0.1151

We would expect 230.2 of the 2000 observations (11.51%) to have a value of 140 or less. The actual number would not be equal to the expected number. However, the more observations we select the closer we will tend to come to what we expect.

CHAPTER EXERCISES

7.63 p/a/d Let x = the number of jurors that are charge-account holders; x is binomial with n = 9 and π = 0.35. Since nπ = 9(0.35) = 3.15 is less than 5, we cannot use the normal approximation to the binomial. Using the binomial distribution and Minitab cumulative probabilities:

```
Binomial with n = 9 and p = 0.350000
        x       P( X <= x)
        0          0.0207
        1          0.1211
        2          0.3373
        3          0.6089
```

P(x ≥ 4) = 1 – P(x ≤ 3) = 1 – 0.6089 = 0.3911

7.65 p/p/d According to exercise 7.64, 9.01% of the time at least 110 out of 200 persons would say they like soft drink A better than B even if the soft drinks are not really different. Soft drink A might be superior to B since 110 out of 200 persons did say they like soft drink A better than B. However, this could be just a chance occurrence.

7.67 p/a/m Let x = the amount of life insurance coverage for insured households with heads 35-44 years old; x is normally distributed with μ = $186,100 and σ = $40,000

$$P(x < 130,000) = P(z < \frac{130,000 - 186,100}{40,000}) = P(z < -1.40) = 0.5000 - 0.4192 = 0.0808$$

7.69 p/a/m Let x = the daily volume of packages for FedEx; x is normally distributed with $\mu = 3,077,000$ and $\sigma = 400,000$.

$$P(3,000,000 < x < 3,200,000) = P(\frac{3,000,000 - 3,077,000}{400,000} < z < \frac{3,200,000 - 3,077,000}{400,000})$$

$$= P(-0.19 < z < 0.31) = 0.0753 + 0.1217 = 0.1970$$

7.71 p/a/m Let x = first-visit office fee paid by a new patient for self employed physicians; x is normally distributed with $\mu = \$48.90$ and $\sigma = \$12$.

a. $P(x \geq 45) = P(z > \frac{45 - 48.9}{12}) \Rightarrow P(z > -0.33) = 0.5000 + 0.1293 = 0.6293$

b. Assume that the chance of paying at least \$45 per visit is 0.6293 from part a. If n = 4 patients are selected at random, and assuming the chance of paying \$45 per visit is 0.6293 for each patient, the distribution is binomial. $P(x = 4)$ can be found using the computer or by calculating via the binomial formula (below):

$$P(x = 4) = \binom{n}{x}\pi^x(1 - \pi)^{n-x} = \binom{4}{4}0.6293^4(1 - 0.6293)^0 = (1)(0.1568)(1) = 0.1568$$

7.73 p/a/m Let x = the number of persons of voting age in the sample that voted in the 1996 presidential election; x is binomial with n = 30 and $\pi = 0.64$. Since $n\pi = 30(0.64) = 19.2$ and $n(1 - \pi) = 30(1 - 0.64) = 10.8$ are both larger than 5, we can use the normal approximation.

$$\mu = n\pi = 30(0.64) = 19.2, \quad \sigma = \sqrt{n\pi(1 - \pi)} = \sqrt{30(0.64)(1 - 0.64)} = 2.629$$

Applying correction for continuity:

$$P(x \geq 20) = P(x \geq 19.5) = P(z \geq \frac{19.5 - 19.2}{2.629}) = P(z \geq 0.11) = 0.5000 - 0.0438 = 0.4562$$

7.75 p/a/d According to exercise 7.74, if the public relations agency's claim is correct, there is practically no chance whatsoever of only 320 or fewer people out of 500 feeling the company is an industry leader. Therefore, if the survey taken was a good random sample of the residents in a 50-mile radius, the public relations agency's claim does not appear to be correct.

7.77 p/a/m $P(x \geq k) = e^{-\lambda k}$ The mean and standard deviation of the exponential distribution ($1/\lambda$) is:

$$1/\lambda = \frac{1}{0.024} = 41.667 \quad \text{Therefore, } P(x \geq k) = e^{-0.024k} \text{ and}$$

$$P(x \leq 45) = 1 - e^{-0.024(45)} = 1 - e^{-1.08} = 1 - 0.3396 = 0.6604$$

7.79 p/a/m Let x = the number of years until the next highway motorcycle fatality. For the club members, we would expect 0.43 deaths per 1 million miles or 0.86 deaths per 2 million miles (one year of travel). The mean is $1/\lambda = 1/0.43 = 2.326$.

$$P(x \geq k) = e^{-0.86k} \text{ and } P(x \geq 1) = e^{-0.86(1)} = e^{-0.86} = 0.4232$$

$$P(x \geq 2) = e^{-0.86(2)} = e^{-1.72} = 0.1791$$

7.81 p/a/d Let x = the score on the exam; assume x is normally distributed with $\mu = 81$, $\sigma = 8.5$.

$$P(78 \leq x \leq 88) = P(\frac{78 - 81}{8.5} \leq z \leq \frac{88 - 81}{8.5}) = P(-0.35 \leq z \leq 0.82) = 0.1368 + 0.2939 = 0.4307$$

7.83 p/a/d Let x = the actual precooked hamburger weight, and x is normally distributed with $\mu = 5.5$ ounces and $\sigma = 0.15$ ounces. The probability that the journalist will receive a hamburger with a precooked weight less than 5.3 ounces is

$$P(x < 5.3) = P(z < \frac{5.3 - 5.5}{0.15}) = P(z < -1.33) = 0.5000 - 0.4082 = 0.0918$$

The probability that at least 2 of the four customers will receive a hamburger with precooked weight greater than 5.7 ounces:

First, for a single hamburger, $P(x > 5.7) = P(z > \frac{5.7 - 5.5}{0.15}) = P(z > 1.33) = 0.5000 - 0.4082 = 0.0918$.

For a binomial process with n = 4 trials and $\pi = 0.0918$, we can find $P(x \geq 2)$ either with the pocket calculator and the methods of chapter 6, or with the computer. The Minitab printout and cumulative probabilities are shown below.

```
Binomial with n = 4 and p = 0.0918000
       x        P( X <= x )
     0.00          0.6803
     1.00          0.9554
     2.00          0.9971
     3.00          0.9999
     4.00          1.0000
```

and $P(x \geq 2) = 1 - P(x \leq 1) = 1 - 0.9554 = 0.0446$.

7.85 p/a/d The quantity to be determined is the mean weight (M) that would result in just 2% of the packages containing less than 20 ounces. First, find the value of z corresponding to a left-tail area of 0.0200. Referring to the standard normal table, we find this to be z = -2.05.
Now substitute $\mu = M$, $\sigma = 0.3$, and z = -2.05 into the z-score formula:

$$z = \frac{x - \mu}{\sigma} \Rightarrow -2.05 = \frac{20.0 - M}{0.3} \quad \text{Solving for M: M = 20.615 ounces, the necessary mean.}$$

7.87 p/a/d

a. The mean of the Poisson distribution is $\lambda = 1.25/10,000$, or 0.000125 punctures per mile. Its inverse is the mean of the corresponding exponential distribution, $1/\lambda = 8000$ miles between punctures. The probability they will not have to change any tires during their vacation is:

$$P(x \geq k) = e^{-\lambda k} = e^{-0.000125k} \quad \text{and} \quad P(x \geq 6164) = e^{-0.000125(6164)} = e^{-0.7705} = 0.4628$$

b. The probability that they will not experience a puncture before getting to Denver (2016 miles):

$$P(x \geq 2016) = e^{-0.000125(2016)} = e^{-0.2520} = 0.7772$$

c. The probability is 0.80 that they will experience a puncture before M miles.
This is most easily done with the computer. We want the inverse of the cumulative probability distribution, and the cumulative probability of interest is 0.80. As shown in the Minitab printout below, there is a 0.80 probability that the next puncture will occur within the next 12,900 miles.

```
Inverse Cumulative Distribution Function

Exponential with mean = 8000.00
P( X <= x )        x
    0.8000      1.29E+04
```

Accordingly, there is a 0.20 probability that no puncture until *at least* 12,900 miles from now.
This can also be expressed as: $P(x \geq 12,900) = e^{-0.000125(12,900)} = 0.20$

7.89 p/a/m For x = minutes between patrol car visits, the mean is 20 minutes and x is exponentially distributed. The probability of the alarm shutting off before the next patrol car arrives is P(x > 15). Using Minitab to find this probability, the printout and cumulative probabilities are shown below.

Cumulative Distribution Function

```
Exponential with mean = 20.0000
       x       P( X <= x )
  10.0000         0.3935
  15.0000         0.5276
  20.0000         0.6321
```

and P(x > 15) = 1 - P(x ≤ 15) = 1.000 - 0.5276 = 0.4724

CHAPTER 8
SAMPLING DISTRIBUTIONS

SECTION EXERCISES

8.1 d/p/m The sampling distribution of a statistic is a probability distribution of the sample statistic for all possible samples of that particular size. Therefore, a sampling distribution is a probability distribution. It is for a statistic (which is a random variable) with a sample size specified.

8.3 c/a/e The population proportion, $\pi = 180/1000 = 0.18$; sample proportion, $p = 150/1000 = 0.15$; sample size, $n = 1000$.

8.5 c/a/d

a. $\mu = 1(0.5) + 2(0.5) = 1.5$

   $\sigma^2 = E(x - \mu)^2 = (1 - 1.5)^2(0.5) + (2 - 1.5)^2(0.5) = 0.25$

   $\sigma = \sqrt{\sigma^2} = \sqrt{0.125} = 0.5$

b. and c.

sample	$\overline{x}$ = mean of this sample	probability of selecting this sample
1,1	(1 + 1)/2 = 1.0	1/4
1,2	(1 + 2)/2 = 1.5	1/4
2,1	(2 + 1)/2 = 1.5	1/4
2,2	(2 + 2)/2 = 2.0	1/4

d. The sampling distribution of the mean (n = 2) is:

$\overline{x}$:	1.0	1.5	2.0
$P(\overline{x})$:	0.25	0.50	0.25

   $\mu_{\overline{x}} = 1(0.25) + 1.5(0.50) + 2(0.25) = 1.5$

   $\sigma_{\overline{x}}^2 = E(\overline{x} - \mu)^2 = (1 - 1.5)^2(0.25) + (1.5 - 1.5)^2(0.50) + (2 - 1.5)^2(0.25) = 0.125$

   $\sigma_{\overline{x}} = \sqrt{0.125} = 0.354$

e. Repeating parts (b) through (d) using a sample of size n = 3.

sample	$\overline{x}$ = mean of this sample	probability of selecting this sample
1,1,1	(1 + 1 + 1)/3 = 1	1/8
1,1,2	(1 + 1 + 2)/3 = 4/3	1/8
1,2,1	(1 + 2 + 1)/3 = 4/3	1/8
1,2,2	(1 + 2 + 2)/3 = 5/3	1/8
2,1,1	(2 + 1 + 1)/3 = 4/3	1/8
2,1,2	(2 + 1 + 2)/3 = 5/3	1/8
2,2,1	(2 + 2 + 1/3 = 5/3	1/8
2,2,2	(2 + 2 + 2)/3 = 2	1/8

The sampling distribution of the mean (n = 3) is:

$\overline{x}$:	1	4/3	5/3	2
$P(\overline{x})$:	1/8	3/8	3/8	1/8

$\mu_{\overline{x}} = 1(1/8) + 4/3(3/8) + 5/3(3/8) + 2(1/8) = 1.5$

$$\sigma_{\bar{x}}^2 = E(\bar{x} - \mu)^2 = (1 - 1.5)^2(1/8) + (4/3 - 1.5)^2(3/8) + (5/3 - 1.5)^2(3/8) + (2 - 1.5)^2(1/8) = 0.0833$$

$$\sigma_{\bar{x}} = \sqrt{0.0833} = 0.289$$

The larger sample size does not affect the mean of the sampling distribution. The standard error of the sampling distribution gets smaller when the sample size is increased.

8.7 c/a/m

a. $\sigma_{\bar{x}} = \dfrac{\sigma}{\sqrt{n}} = \dfrac{100}{\sqrt{16}} = 25$

b. $\sigma_{\bar{x}} = \dfrac{\sigma}{\sqrt{n}} = \dfrac{100}{\sqrt{100}} = 10$

c. $\sigma_{\bar{x}} = \dfrac{\sigma}{\sqrt{n}} = \dfrac{100}{\sqrt{400}} = 5$

d. $\sigma_{\bar{x}} = \dfrac{\sigma}{\sqrt{n}} = \dfrac{100}{\sqrt{1000}} = 3.162$

8.9 c/a/m Since x is normally distributed with $\mu = 80$ and $\sigma = 10$, $\bar{x}$ is normally distributed with

$\mu_{\bar{x}} = 80$ and $\sigma_{\bar{x}} = \dfrac{\sigma}{\sqrt{n}} = \dfrac{10}{\sqrt{25}} = 2.$

a. $P(\bar{x} > 78) = P(z > \dfrac{\bar{x} - \mu}{\sigma/\sqrt{n}}) = P(z > \dfrac{78 - 80}{2}) = P(z > -1.00) = 0.5000 + 0.3413 = 0.8413$

b. $P(79 < \bar{x} < 85) = P(\dfrac{79 - 80}{2} < z < \dfrac{85 - 80}{2}) = P(-0.50 < z < 2.50) = 0.1915 + 0.4938 = 0.6853$

c. $P(\bar{x} < 85) = P(z < \dfrac{85 - 80}{2}) = P(z < 2.50) = 0.5000 + 0.4938 = 0.9938$

8.11 c/a/m Given $\mu = 62$, $\sigma = 15$, $n = 36$, and $\sigma_{\bar{x}} = \dfrac{\sigma}{\sqrt{n}} = \dfrac{15}{\sqrt{36}} = 2.5$

$P(\bar{x} < 65) = P(z < 1.20) = 0.5000 + 0.3849 = 0.8849$

$P(55 \le \bar{x} \le 65) = P(-2.80 \le z \le 1.20) = 0.4974 + 0.3849 = 0.8823$

8.13 c/a/m Given $\mu = 100$ horsepower with $\sigma = 10$, $n = 4$, and $\sigma_{\bar{x}} = \dfrac{\sigma}{\sqrt{n}} = \dfrac{10}{\sqrt{4}} = 5$

$P(\text{Crane not performing properly}) = P(\bar{x} < 380/4) = P(\bar{x} < 95) = P(z < (95-100)/5) = P(z < -1.00) = 0.5000 - 0.3413 = 0.1587$. The engineer's concern appears to be justified in that there is a good chance that the crane will not perform properly.

8.15 p/a/m Let x = consumption of wattage of drill with $\mu = 295$ and $\sigma = 12$ watts. Samples of size 9 are selected, where the distribution of $\bar{x}$, from the central limit theorem, has $\mu = 295$ and

$\sigma_{\bar{x}} = \dfrac{\sigma}{\sqrt{n}} = \dfrac{12}{\sqrt{9}} = 4.$

$P(\text{truckload of 5000 drills rejected}) = P(\bar{x} > 300) = P(z > \dfrac{300 - 295}{4})$

$= P(z > 1.25) = 0.5000 - 0.3944 = 0.1056$

Thus, there is a 10.56% chance the truckload is rejected.

8.17 p/a/m Given x is exponentially distributed with $\mu = \sigma = 3.5$. Since $n = 36$, $\bar{x}$ is approximately normally distributed by the central limit theorem with $\mu_{\bar{x}} = 3.5$ and

$$\sigma_{\bar{x}} = \frac{\sigma}{\sqrt{n}} = \frac{3.5}{\sqrt{36}} = 0.5833$$

$$P(\bar{x} \geq 4) = P(z \geq \frac{4 - 3.5}{0.5833}) = P(z \geq 0.86) = 0.5000 - 0.3051 = 0.1949$$

8.19 p/a/m Since $n = 50$, $\bar{x}$ is approximately normally distributed by the central limit theorem with

$$\mu_{\bar{x}} = 4.9 \text{ and } \sigma_{\bar{x}} = \frac{\sigma}{\sqrt{n}} = \frac{3.5}{\sqrt{50}} = 0.495$$

$$P(\bar{x} \leq 5.7) = P(z \leq \frac{5.7 - 4.9}{0.495}) = P(z \leq 1.62) = 0.5000 + 0.4474 = 0.9474$$

If the sample size had been only $n = 8$ patients, the central limit theorem would not apply. The solution would have required the assumption that the underlying population is normally distributed.

8.21 p/a/d Since $n = 30$, $\bar{x}$ is approximately normally distributed by the central limit theorem with

$$\mu_{\bar{x}} = 75 \text{ and } \sigma_{\bar{x}} = \frac{\sigma}{\sqrt{n}} = \frac{10}{\sqrt{30}} = 1.826$$

$$P(\bar{x} > 70) = P(z > \frac{70 - 75}{1.826}) = P(z > -2.74) = 0.5000 + 0.4969 = 0.9969$$

$$P(\bar{x} > 80) = P(z > \frac{80 - 75}{1.826}) = P(z > 2.74) = 0.5000 - 0.4969 = 0.0031$$

8.23 c/a/m With $\pi = 0.265$ and $n > 30$ (i.e., $n \geq 31$), $n\pi \geq 8.215$ and $n(1 - \pi) \geq 22.785$. Since both $n\pi$ and $n(1 - \pi)$ are ≥ 5, the sampling distribution of p will be approximately normally distributed with a mean of $\pi = 0.265$ and a standard error of no more than

$$\sigma_p = \sqrt{\frac{\pi(1 - \pi)}{n}} = \sqrt{\frac{0.265(1 - 0.265)}{31}} = 0.079$$

8.25 c/a/m With $\pi = 0.40$ and $n = 300$, both $n\pi$ and $n(1 - \pi)$ are ≥ 5. The sampling distribution of p will be approximately normally distributed.

$$\sigma_p = \sqrt{\frac{\pi(1 - \pi)}{n}} = \sqrt{\frac{0.40(1 - 0.40)}{300}} = 0.0283$$

a. $P(p \geq 0.35) = P(z \geq \frac{p - \pi}{\sigma_p}) = P(z \geq \frac{0.35 - 0.40}{0.0283}) = P(z \geq -1.77)$

$= 0.5000 + 0.4616 = 0.9616$

b. $P(0.38 \leq p \leq 0.42) = P(\frac{0.38 - 0.40}{0.0283} \leq z \leq \frac{0.42 - 0.40}{0.0283}) = P(-0.71 \leq z \leq 0.71)$

$= 0.2611 + 0.2611 = 0.5222$

c. $P(p \leq 0.45) = P(z \leq \frac{0.45 - 0.40}{0.0283}) = P(z \leq 1.77) = 0.5000 + 0.4616 = 0.9616$

8.27 p/a/m

a. $\pi = 0.426 =$ population proportion who are females.

b. p = 70/200 = 0.35

c. $\sigma_p = \sqrt{\dfrac{\pi(1-\pi)}{n}} = \sqrt{\dfrac{0.426(1-0.426)}{200}} = 0.035$

d. $P(p \geq 0.35) = P(z \geq \dfrac{0.350 - 0.426}{0.035}) = P(z \geq -2.17) = 0.5000 + 0.4850 = 0.9850$

8.29 p/a/m Let $\pi = 0.20$ of orders returned. Given n = 100, the distribution of p will be normal with

$E(p) = \pi = 0.20$ and $\sigma_p = \sqrt{\dfrac{\pi(1-\pi)}{n}} = \sqrt{\dfrac{0.2(1-0.2)}{100}} = 0.04$

Find the $P(10$ or more of the 100 orders are returned$) = P(p \geq \dfrac{10}{100})$

$= P(p \geq 0.10) = P(z \geq \dfrac{0.10 - 0.20}{0.04}) = P(z \geq -2.5) = 0.5000 + 0.4938 = 0.9938$

8.31 p/a/d With $E(p) = \pi = 0.05$ and n = 400, $\sigma_p = \sqrt{\dfrac{\pi(1-\pi)}{n}} = \sqrt{\dfrac{0.05(1-0.05)}{400}} = 0.0109$

and the probability that at least 60 of the drivers would happen to be sick on a given day is

$P(p \geq \dfrac{60}{400})$, or $P(p \geq 0.15) = P(z \geq \dfrac{0.15 - 0.05}{0.0109}) = P(z \geq 9.18) = 0.5000 - 0.5000$, or 0.0000.

The probability that such a large proportion of drivers would happen to call in sick on a given day is (to four decimal places) 0.0000. The school district's claim is much more credible.

8.33 d/p/m A finite population is one which is a set size; it is not infinitely large. In this case, we need to reduce the standard error for the sampling distribution. The larger the percentage of the population that is sampled, the smaller the sampling error will be.

8.35 d/p/d The finite population correction factor will reduce the size of the standard error. Whenever n < 0.05N, the correction will have little effect.

8.37 p/a/m Let $\bar{x}$ = average income of the fund-raising group; $\bar{x}$ is approximately normally distributed by the central limit theorem (since n = 30) with $\mu = 58,000$ and

$\sigma_{\bar{x}} = \dfrac{\sigma}{\sqrt{n}} * \sqrt{\dfrac{N-n}{N-1}} = \dfrac{10,000}{\sqrt{30}} * \sqrt{\dfrac{200-30}{200-1}} = 1687.4748$

$P(\bar{x} \geq 60,000) = P(z \geq \dfrac{60,000 - 58,000}{1687.4748}) = P(z \geq 1.19) = 0.5000 - 0.3830 = 0.1170$

8.39 p/a/m Let p = proportion of defectives in the sample, n = 40, and $\pi = 0.20$; p will be approximately normally distributed since $n\pi$ and $n(1 - \pi)$ are both ≥ 5.

$\sigma_p = \sqrt{\dfrac{\pi(1-\pi)}{n}} * \sqrt{\dfrac{N-n}{N-1}} = \sqrt{\dfrac{0.2(1-0.2)}{40}} * \sqrt{\dfrac{300-40}{300-1}} = 0.0590$

$P(p \geq 0.15) = P(z \geq \dfrac{0.15 - 0.20}{0.059}) = P(z \geq -0.85) = 0.5000 + 0.3023 = 0.8023$

8.41 c/c/m The population is exponentially distributed with $\mu = 6$ and $\sigma = 6$. The expected mean of the sample means is $\mu_{\bar{x}} = 6$. Although the population is not normally distributed, n = 30, and we can invoke the central limit theorem and the standard error of the sample means would be estimated as

$$\sigma_{\bar{x}} = \frac{\sigma}{\sqrt{n}} = \frac{6}{\sqrt{30}} = 1.095.$$ For 150 Minitab-generated samples, each with n = 30, the observed mean of the

sample means is 6.0053. The standard error of the sample means is 1.1687. Both values are very close to their expected values.

```
Descriptive Statistics: xbar

Variable            N        Mean      Median      TrMean       StDev     SE Mean
xbar              150      6.0053      5.9447      5.9859      1.1687      0.0954

Variable      Minimum     Maximum          Q1          Q3
xbar           2.9801      9.9940      5.3037      6.7290
```

With the larger sample size, the histogram of sample means is much more symmetrical, and approximately normal.

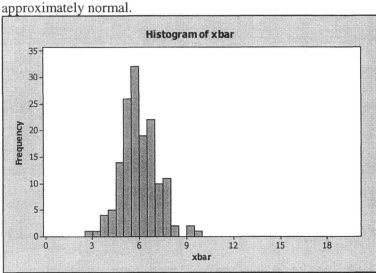

8.43 c/c/m The population is normally distributed with $\mu = 80$ and $\sigma = 20$. The expected mean of the sample means is $\mu_{\bar{x}} = 80$. We can estimate the standard error of the sample means as

$\sigma_{\bar{x}} = \dfrac{\sigma}{\sqrt{n}} = \dfrac{20}{\sqrt{30}} = 3.65$. For 150 Minitab-generated samples, each with n = 30, the observed mean of the sample means is 79.921. The standard error of the sample means is 3.616. Both values are very close to their expected values.

```
Descriptive Statistics: xbar
```

Variable	N	Mean	Median	TrMean	StDev	SE Mean
xbar	150	79.921	79.800	79.881	3.616	0.295

Variable	Minimum	Maximum	Q1	Q3
xbar	71.651	91.158	77.247	82.390

The histogram of sample means is approximately normal.

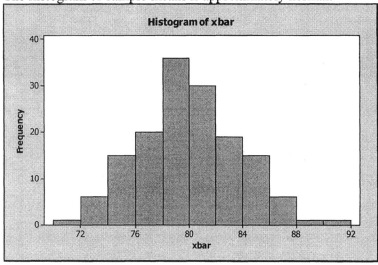

CHAPTER EXERCISES

8.45 p/a/d From exercise 8.44, $\pi = 0.55$ and $\sigma_p = 0.0287$ (assuming the manager's claim is true.)

a. $P(p \le 0.49) = P(z \le \dfrac{0.49 - 0.55}{0.0287}) = P(z \le -2.09) = 0.5000 - 0.4817 = 0.0183$

b. Since only 49% of the voters in the sample favored the candidate and the probability of no more than 49% favoring the candidate is just 0.0183 if the manager's claim is correct, the manager's claim is probably not correct.

8.47 p/a/m

a. Since n = 5 is less than 30, we would need to assume that the population is normally distributed in order to use the standard normal table to find $P(\bar{x} < 50)$.

b. Let $\bar{x}$ = the average fee for the 36 customers. Since $n = 36$, $\bar{x}$ is approximately normally distributed by the central limit theorem with $\mu = 55.93$ and $\sigma_{\bar{x}} = \dfrac{\sigma}{\sqrt{n}} = \dfrac{20}{\sqrt{36}} = 3.333$.

$P(\bar{x} < 50) = P(z < \dfrac{50 - 55.93}{3.333}) = P(z < -1.78) = 0.5000 - 0.4625 = 0.0375$

8.49 p/a/d Let $\bar{x}$ = the average time to produce a unit; $\bar{x}$ is approximately normally distributed by the central limit theorem (since n = 36) with $\mu = 25$ and $\sigma_{\bar{x}} = \dfrac{\sigma}{\sqrt{n}} = \dfrac{3}{\sqrt{36}} = 0.5$.

a. When $\bar{x} = 26.2$, $z = \dfrac{\bar{x} - \mu}{\sigma_{\bar{x}}} = \dfrac{26.2 - 25}{0.5} = 2.40$

b. $P(\bar{x} \geq 26.2) = P(z \geq \dfrac{26.2 - 25}{0.5}) = P(z \geq 2.40) = 0.5000 - 0.4918 = 0.0082$

c. When the machine is properly calibrated, the probability that the mean for a random sample of n = 36 will be at least 26.2 seconds is only 0.0082. Since this did occur, the machine is probably not properly calibrated.

8.51 d/p/m In order to determine the $P(\bar{x} \geq 220)$ when n = 5, we would need to assume that the population is approximately normally distributed. Since n is less than 30, the central limit theorem does not apply.

8.53 p/a/d p will be approximately normally distributed with $\pi = 0.20$, and

$$\sigma_p = \sqrt{\dfrac{\pi(1-\pi)}{n}} = \sqrt{\dfrac{0.20(1-0.20)}{100}} = 0.04$$

a. p = 10/100 = 0.10

b. $P(p \leq 0.10) = P(z \leq \dfrac{0.10 - 0.20}{0.04}) = P(z \leq -2.50) = 0.5000 - 0.4938 = 0.0062$

c. Since, if π is really 0.20, the probability of only making 10 sales out of 100 contacts is very small (0.0062), we would not accept Charlie's explanation that he "just had a bad day."

8.55 p/a/m Let p = proportion of vehicles with a defect in the sample; p will be approximately normally distributed with $\pi = 200/500 = 0.40$ and

$$\sigma_p = \sqrt{\dfrac{\pi(1-\pi)}{n}} * \sqrt{\dfrac{N-n}{N-1}} = \sqrt{\dfrac{0.40(1-0.40)}{500}} * \sqrt{\dfrac{500-50}{500-1}} = 0.0658$$

$$P(p \leq 15/50) = P(p \leq 0.30) = P(z \leq \dfrac{0.30 - 0.40}{0.0658}) = P(z \leq -1.52) = 0.5000 - 0.4357 = 0.0643$$

8.57 p/a/m Assuming that the distribution of completion times has $\mu = 38$ minutes and $\sigma = 5$ minutes, the distribution of $\bar{x}$ is normal with $\sigma_{\bar{x}} = \dfrac{\sigma}{\sqrt{n}} = \dfrac{5}{\sqrt{50}} = 0.7071$.

$$P(\bar{x} > 45) = P(z > \dfrac{45 - 38}{0.7071}) = P(z > 9.90)$$

This is well beyond the limits of our printed standard normal distribution table. Using the computer, we could verify that this probability (to four decimal places) is 0.0000.

8.59 p/a/m Let p = proportion of males in a sample of travelers; p will be approximately normally distributed with $\pi = 0.76$ and

$$\sigma_p = \sqrt{\frac{\pi(1-\pi)}{n}} = \sqrt{\frac{0.76(1-0.76)}{200}} = 0.0302$$

$$P(p \geq 0.80) = P(z \geq \frac{0.80 - 0.76}{0.0302}) = P(z \geq 1.32) = 0.5000 - 0.4066 = 0.0934$$

The proportion of males in the most recent group of 200 guests is a little higher than 0.76, but does not seem to be extraordinarily high. There is a 0.0934 probability that a proportion at least this high could occur simply by chance.

8.61 p/a/m Assuming that the distribution of bulb lifetimes is normal with $\mu = 2000$ hours and $\sigma = 100$ hours, the distribution of $\overline{x}$ is normal with $\sigma_{\overline{x}} = \frac{\sigma}{\sqrt{n}} = \frac{100}{\sqrt{16}} = 25$ hours.

$$P(\overline{x} \geq 2050) = P(z \geq \frac{2050 - 2000}{25}) = P(z \geq 2.00) = 0.5000 - 0.4772 = 0.0228$$

The probability is 0.0228 that the vendor will get the contract.

CHAPTER 9
ESTIMATION FROM SAMPLE DATA

SECTION EXERCISES

9.1 d/p/e A point estimate is a single number that estimates the value of the population parameter, while an interval estimate includes a range of possible values which are likely to include the population parameter.

9.3 d/p/m When the interval estimate is associated with a degree of confidence that it actually includes the population parameter, it is referred to as a confidence interval.

9.5 d/p/m In order for s^2 to be an unbiased estimator of σ^2, we must use $(n - 1)$ as the divisor when we calculate the variance of the sample. However, s will not be an unbiased estimator of σ.

9.7 c/a/m

a. $\bar{x} = \dfrac{\sum x}{n} = \dfrac{21}{8} = 2.625$ b. $s^2 = \dfrac{n\sum x^2 - (\sum x)^2}{n(n-1)} = \dfrac{8(87) - 21^2}{8(8-1)} = 4.554$

9.9 d/p/e The accuracy of a point estimate is the difference between the observed sample statistic and the actual value of the population parameter being estimated.

9.11 c/a/m

a. point estimate of π: $p = \dfrac{450}{1000} = 0.45$ b. confidence interval for π: 0.419 to 0.481

c. confidence level: 95%; confidence coefficient: 0.95

d. accuracy: for 95% of such intervals, the sample proportion would not differ from the actual population proportion by more than $(0.481 - 0.419)/2 = 0.031$.

9.13 d/p/m In this case, we need to assume that the population is normally distributed and the population standard deviation is known.

9.15 c/a/m

a. For a confidence level of 90%, z = 1.645 (look up 0.90/2 = 0.4500 in the body of the standard normal table). The 90% confidence interval for μ is:

$\bar{x} \pm z\dfrac{\sigma}{\sqrt{n}} = 240 \pm 1.645\dfrac{10}{\sqrt{30}} = 240 \pm 3.003$, or between 236.997 and 243.003

b. For a confidence level of 95%, z = 1.96 (look up 0.95/2 = 0.4750 in the body of the standard normal table). The 95% confidence interval for μ is:

$\bar{x} \pm z\dfrac{\sigma}{\sqrt{n}} = 240 \pm 1.96\dfrac{10}{\sqrt{30}} = 240 \pm 3.578$, or between 236.422 and 243.578

We could also obtain these confidence intervals by using the Excel worksheet template tmzint.xls, on the disk that came with the text. Just enter the values for $\bar{x}$ (240), n (30), σ (0.10), and the confidence level desired (0.90 for 90%, 0.95 for 95%). Excel provides the lower and upper confidence limits:

	A	B	C	D
1	Confidence interval for the population mean,			
2	using the z distribution and known			
3	(or assumed) pop. std. deviation, sigma:			
4				
5	Sample size, n:			30
6	Sample mean, xbar:			240.000
7	Known or assumed pop. sigma:			10.0000
8	Standard error of xbar:			1.82574
9				
10	Confidence level desired:			0.90
11	alpha = (1 - conf. level desired):			0.10
12	z value for desired conf. int.:			1.6449
13	z times standard error of xbar:			3.003
14				
15	Lower confidence limit:			236.997
16	Upper confidence limit:			243.003

	A	B	C	D
1	Confidence interval for the population mean,			
2	using the z distribution and known			
3	(or assumed) pop. std. deviation, sigma:			
4				
5	Sample size, n:			30
6	Sample mean, xbar:			240.000
7	Known or assumed pop. sigma:			10.0000
8	Standard error of xbar:			1.82574
9				
10	Confidence level desired:			0.95
11	alpha = (1 - conf. level desired):			0.05
12	z value for desired conf. int.:			1.9600
13	z times standard error of xbar:			3.578
14				
15	Lower confidence limit:			236.422
16	Upper confidence limit:			243.578

9.17 p/a/m Using Minitab:

```
One-Sample Z: score

The assumed sigma = 4
Variable          N      Mean    StDev   SE Mean          90.0% CI
score            10     85.00    14.79      1.26  (    82.92,    87.08)


One-Sample Z: score
The assumed sigma = 4

Variable          N      Mean    StDev   SE Mean          95.0% CI
score            10     85.00    14.79      1.26  (    82.52,    87.48)
```

9.19 p/a/m For a confidence level of 95%, z = 1.96 (in the standard normal distribution, 95% of the area is between z = -1.96 and z = 1.96). The 95% confidence interval for μ is:

$$\overline{x} \pm z \frac{\sigma}{\sqrt{n}} = 150 \pm 1.96 \frac{3}{\sqrt{35}} = 150 \pm 0.994 \text{, or between 149.006 and 150.994}$$

The confidence level is 95% that the population average torque being applied during the assembly process is between 149.006 and 150.994 lbs.-ft.

9.21 d/p/m If the sample size had been n = 5, the central limit theorem would not apply. Therefore, in order for the sampling distribution of the sample mean to be approximately normally distributed, we would have to assume that the population is normally distributed.

9.23 p/c/m Using Data Analysis Plus, the data file for this exercise, and the specified value for σ (0.25 fluid ounces), the 90% confidence interval for the population mean is shown as from 99.897 to 100.027. The 100.0 fluid ounces value is within this interval, so the mean content could be 100.0 fluid ounces. Also shown is the corresponding Minitab printout.

	A	B	C
1	z-Estimate: Mean		
2			Fl_Oz
3	Mean		99.962
4	Standard Deviation		0.233
5	Observations		40
6	SIGMA		0.25
7	LCL		99.897
8	UCL		100.027

```
One-Sample Z: Fl_Oz

The assumed sigma = 0.25
Variable          N      Mean    StDev   SE Mean          90.0% CI
Fl_Oz            40   99.9618   0.2325    0.0395  ( 99.8967,100.0268)
```

9.25 d/p/e As the sample size increases, the t distribution converges on the standard normal distribution. The two distributions are identical as the sample size approaches infinity.

9.27 c/p/e Referring to the 0.10 column and the d.f. = 28 row of the t table, the value of t corresponding to an upper tail area of 0.10 is t = 1.313.

9.29 c/a/m For d.f. = 85:
a. $P(t \geq A) = 0.10$. Referring to the 0.10 column and the d.f. = 85 row of the t table, A = 1.292.
b. $P(t \leq A) = 0.025$. Referring to the 0.025 column and the d.f. = 85 row of the t table, the value of t corresponding to a right-tail area of 0.025 is t = 1.988. Since the curve is symmetrical, the value of t for a left-tail area of 0.025 is A = -1.988.
c. $P(-A \leq t \leq A) = 0.98$. In this case, each tail will have an area of (1 - 0.98)/2 = 0.01. Referring to the 0.01 column and the d.f. = 85 row of the t table, A = 2.371.

9.31 c/a/m First, compute the sample mean and standard deviation: $\bar{x} = \dfrac{\sum x}{n} = \dfrac{1007}{20} = 50.35$

$$s^2 = \frac{n\sum x^2 - \left(\sum x\right)^2}{n(n-1)} = \frac{20(52,099) - 1007^2}{20(20-1)} = 73.5026 \text{ and } s = \sqrt{73.5026} = 8.5734$$

a. For a confidence level of 95%, the right-tail area of interest is (1 - 0.95)/2 = 0.025 with d.f. = n - 1 = 20 - 1 = 19. From the 0.025 column and the d.f. = 19 row of the t table, t = 2.093.

The 95% confidence interval for μ is: $\bar{x} \pm t\dfrac{s}{\sqrt{n}} = 50.35 \pm 2.093\dfrac{8.5734}{\sqrt{20}} = 50.35 \pm 4.012$,

or between 46.338 and 54.362.

b. For a confidence level of 99%, the right-tail area of interest is (1 - 0.99)/2 = 0.005 with d.f. = 19. Referring to the 0.005 column and the d.f. = 19 row of the t table, t = 2.861. The 99% confidence

interval for μ is: $\bar{x} \pm t\dfrac{s}{\sqrt{n}} = 50.35 \pm 2.861\dfrac{8.5734}{\sqrt{20}} = 50.35 \pm 5.485$, or from 44.865 to 55.835.

Shown below are Minitab printouts with the 95% and 99% confidence intervals for the population mean.

```
One-Sample T: x
Variable          N      Mean     StDev    SE Mean       95.0% CI
x                20     50.35      8.57       1.92   (  46.34,   54.36)

One-Sample T: x
Variable          N      Mean     StDev    SE Mean       99.0% CI
x                20     50.35      8.57       1.92   (  44.87,   55.83)
```

9.33 p/a/m Given n = 50, $\bar{x}$ = 25, and s = 10, d.f. = n - 1 = 49.

a. The 95% confidence level for μ is: $\bar{x} \pm t\dfrac{s}{\sqrt{n}} = 25 \pm 2.010\dfrac{10}{\sqrt{50}} = 25 \pm 2.84$, or 22.16 to 27.84.

Using Excel worksheet template tmtint.xls, provided with the text, we obtain a comparable result:

	A	B	C	D
1	Confidence interval for the population mean,			
2	using the t distribution:			
3				
4	Sample size, n:			50
5	Sample mean, xbar:			25.000
6	Sample standard deviation, s:			10.0000
7	Standard error of xbar:			1.414
8				
9	Confidence level desired:			0.95
10	alpha = (1 - conf. level desired):			0.05
11	degrees of freedom (n - 1):			49
12	t value for desired conf. int:			2.0096
13	t times standard error of xbar:			2.842
14				
15	Lower confidence limit:			22.158
16	Upper confidence limit:			27.842

b. The interval constructed in part (a) would still be appropriate, as the distribution of the sample means approximates a t-distribution regardless of the distribution of the original population.

9.35 p/a/m Using Minitab:

```
One-Sample T: amps

Variable       N      Mean    StDev   SE Mean        95.0% CI
amps          16    29.131    1.080     0.270  ( 28.556,  29.707)
```

We are 95% confident that the population mean amperage is within the interval shown above.

9.37 p/a/m Given n = 20, $\bar{x}$ = 1535 and s = 30. Degrees of freedom, d.f. = n - 1 = 19.

The 95% confidence interval for μ is: $\bar{x} \pm t\dfrac{s}{\sqrt{n}} = 1535 \pm 2.093\dfrac{30}{\sqrt{20}} = 1535 \pm 14.04$,

or between 1520.96 and 1549.04. Using Excel worksheet template tmtint.xls, the computer-assisted 95% confidence interval is shown below.

	A	B	C	D
1	Confidence interval for the population mean,			
2	using the t distribution:			
3				
4	Sample size, n:			20
5	Sample mean, xbar:			1535.0
6	Sample standard deviation, s:			30.0
7	Standard error of xbar:			6.708
8				
9	Confidence level desired:			0.95
10	alpha = (1 - conf. level desired):			0.05
11	degrees of freedom (n - 1):			19
12	t value for desired conf. int:			2.093
13	t times standard error of xbar:			14.040
14				
15	Lower confidence limit:			1520.96
16	Upper confidence limit:			1549.04

9.39 p/a/m Although the exercise can be done with the pocket calculator and formulas, we will use the computer and the Estimators.xls workbook that accompanies Data Analysis Plus on the CD that came with the text. As shown below, the 98% confidence interval for the population mean is from 10.15 to 13.55 minutes. We are 98% confident that the population mean is within this interval and, since 13.0 minutes is within the interval, we would conclude that the population mean might be 13.0. To that extent, a sample of the same size as this one would not seem unusual if it had a mean of 13.0.

	A	B	C	D	E
1	t-Estimate of a Mean				
2					
3	Sample mean	11.85	Confidence Interval Estimate		
4	Sample standard deviation	3	11.85	plus/minus	1.70
5	Sample size	20	Lower confidence limit		10.15
6	Confidence level	0.98	Upper confidence limit		13.55

9.41 p/c/m As shown in the Minitab printout below, the 95% confidence interval for the population mean is from 21.924 to 22.676 hours. We are 95% confident that the population mean is within this interval and, since 22.9 hours is not within the interval, we would conclude that the population mean could not be 22.9. To that extent, a sample of the same size as this one might seem unusual if it had a mean of 22.9.

```
One-Sample T: hours
Variable          N      Mean     StDev    SE Mean        95.0% CI
hours            40    22.300     1.175      0.186   ( 21.924,  22.676)
```

9.43 p/a/m For a confidence level of 95%, z = 1.96. The 95% confidence interval for π is:

$$p \pm z\sqrt{\frac{p(1-p)}{n}} = 0.46 \pm 1.96\sqrt{\frac{0.46(1-0.46)}{1000}} = 0.46 \pm 0.031, \text{ or from } 0.429 \text{ to } 0.491$$

Using the computer and the Estimators.xls workbook that accompanies Data Analysis Plus:

	A	B	C	D	E
1	z-Estimate of a Proportion				
2					
3	Sample proportion	0.46	Confidence Interval Estimate		
4	Sample size	1000	0.46	plus/minus	0.031
5	Confidence level	0.95	Lower confidence limit		0.429
6			Upper confidence limit		0.491

9.45 p/a/m Using the Estimators.xls workbook that accompanies Data Analysis Plus:

	A	B	C	D	E
1	z-Estimate of a Proportion				
2					
3	Sample proportion	0.20	Confidence Interval Estimate		
4	Sample size	200	0.200	plus/minus	0.047
5	Confidence level	0.90	Lower confidence limit		0.153
6			Upper confidence limit		0.247

Based on this confidence interval, the 0.50 value falls far above the upper limit, and it would not seem credible that "over 50% of the students would like a new mascot."

9.47 p/a/m The 95% confidence interval for π is:

$$p \pm z\sqrt{\frac{p(1-p)}{n}} = 0.40 \pm 1.96\sqrt{\frac{0.40(1-0.40)}{1000}} = 0.40 \pm 0.03 \text{, or from } 0.37 \text{ to } 0.43$$

Using the computer and the Estimators.xls workbook that accompanies Data Analysis Plus:

	A	B	C	D	E
1	z-Estimate of a Proportion				
2					
3	Sample proportion	0.40	Confidence Interval Estimate		
4	Sample size	1000	0.40	plus/minus	0.03
5	Confidence level	0.95	Lower confidence limit		0.37
6			Upper confidence limit		0.43

9.49 p/a/m The 95% confidence interval for π is:

$$p \pm z\sqrt{\frac{p(1-p)}{n}} = 0.602 \pm 1.96\sqrt{\frac{0.602(1-0.602)}{1800}} = 0.602 \pm 0.0226 \text{, or from } 0.5794 \text{ to } 0.6246$$

Using the computer and the Estimators.xls workbook that accompanies Data Analysis Plus:

	A	B	C	D	E
1	z-Estimate of a Proportion				
2					
3	Sample proportion	0.602	Confidence Interval Estimate		
4	Sample size	1800	0.6020	plus/minus	0.0226
5	Confidence level	0.95	Lower confidence limit		0.5794
6			Upper confidence limit		0.6246

9.51 p/a/d

a. The 99% confidence interval for π is:

$$p \pm z\sqrt{\frac{p(1-p)}{n}} = 0.57 \pm 2.58\sqrt{\frac{0.57(1-0.57)}{900}} = 0.57 \pm 0.043 \text{, or from } 0.527 \text{ to } 0.613$$

Using the computer and the Estimators.xls workbook that accompanies Data Analysis Plus:

	A	B	C	D	E
1	z-Estimate of a Proportion				
2					
3	Sample proportion	0.57	Confidence Interval Estimate		
4	Sample size	900	0.570	plus/minus	0.043
5	Confidence level	0.99	Lower confidence limit		0.527
6			Upper confidence limit		0.613

b. It does appear to be a "sure thing" that the contract will be approved by the union since the 99% confidence interval in part (a) only contains values over 0.50. Therefore, more than half of the employees should vote for the contract.

9.53 p/a/m The exercise can be done with a pocket calculator and formulas, but we will use the computer and the Estimators.xls xls workbook that accompanies Data Analysis Plus.
The 90% confidence interval for π is from 0.925 to 0.975.

	A	B	C	D	E
1	z-Estimate of a Proportion				
2					
3	Sample proportion	0.95	Confidence Interval Estimate		
4	Sample size	200	0.950	plus/minus	0.025
5	Confidence level	0.90	Lower confidence limit		0.925
6			Upper confidence limit		0.975

9.55 p/a/m The exercise can be done with a pocket calculator and formulas, but we will use the computer and the Estimators.xls workbook that accompanies Data Analysis Plus.
The 95% confidence interval for π is from 0.566 to 0.634. Ms. McCarthy must get at least 65% of the union vote, but 0.65 exceeds the range of values in the confidence interval. This suggests that she will not obtain the necessary level of union support she needs.

	A	B	C	D	E
1	z-Estimate of a Proportion				
2					
3	Sample proportion	0.60	Confidence Interval Estimate		
4	Sample size	800	0.600	plus/minus	0.034
5	Confidence level	0.95	Lower confidence limit		0.566
6			Upper confidence limit		0.634

9.57 p/a/m Using Data Analysis Plus, we find that 40.0% of the 200 potential investors in this sample consider themselves to be "someone who enjoys taking risks." The 99% confidence interval for π is from 0.311 to 0.489.

	A	B
1	z-Estimate: Proportion	
2		RespCode
3	Sample Proportion	0.400
4	Observations	200
5	LCL	0.311
6	UCL	0.489

9.59 d/p/e One way of estimating the population standard deviation is to use a relatively small-scale pilot study from which the sample standard deviation is used as a point estimate. A second approach is to use the results of a similar study done in the past. We can also estimate σ as 1/6 the approximate range of data values.

9.61 p/a/m For the 99% level of confidence, z = 2.58. The maximum likely error is e = 1.0 and the estimated population standard deviation is σ = 3.7. The sample size needed is:

$$n = \frac{z^2\sigma^2}{e^2} = \frac{2.58^2(3.7)^2}{1.0^2} = 91.13 \text{, rounded up to 92}$$

9.63 p/a/m For the 95% level of confidence, z = 1.96. The maximum likely error is e = 0.03, or 3 percentage points. Assuming the candidate has no idea regarding the actual value of the population proportion, we will use p = 0.5 to calculate the necessary sample size.

$$n = \frac{z^2 p(1-p)}{e^2} = \frac{1.96^2(0.5)(1-0.5)}{0.03^2} = 1067.11 \text{, rounded up to 1068}$$

9.65 p/a/m For the 90% level of confidence, z = 1.645. The maximum likely error is e = 0.02 and we will estimate the population proportion with p = 0.15. The number of owners who must be included in the sample is:

$$n = \frac{z^2 p(1-p)}{e^2} = \frac{1.645^2 (0.15)(1-0.15)}{0.02^2} = 862.55 \text{, rounded up to 863}$$

9.67 p/a/d For the 95% level of confidence, z = 1.96. The maximum acceptable error is e = 0.04, or 4 percentage points. Using p = 0.5 (the most conservative value to use when determining sample size), the sample size used was:

$$n = \frac{z^2 p(1-p)}{e^2} = \frac{1.96^2 (0.5)(1-0.5)}{0.04^2} = 600.25 \text{, rounded up to 601}$$

9.69 d/p/e

a. The finite population correction will lead to a narrower confidence interval than if an infinite population had been assumed, since the standard error is reduced.

b. The finite population correction will lead to a smaller required sample size than if an infinite population had been assumed, since the standard error is reduced.

9.71 c/a/m The population is finite with N = 1200, n = 600, p = 0.55.
The 95% confidence level for the population proportion, π, is given below.

$$p \pm z\left(\sqrt{\frac{p(1-p)}{n}} * \sqrt{\frac{N-n}{N-1}}\right) = 0.55 \pm 1.96\sqrt{\frac{0.55(1-0.55)}{600}} * \sqrt{\frac{1200-600}{1200-1}} = 0.55 \pm 0.028,$$

or from 0.522 to 0.578.

The 99% confidence level for the population proportion, π, is given below.

$$p \pm z\left(\sqrt{\frac{p(1-p)}{n}} * \sqrt{\frac{N-n}{N-1}}\right) = 0.55 \pm 2.58\sqrt{\frac{0.55(1-0.55)}{600}} * \sqrt{\frac{1200-600}{1200-1}} = 0.55 \pm 0.037,$$

or from 0.513 to 0.587.
Each interval includes only values that are greater than 0.500, so it seems likely that the fee increase will be passed.

9.73 c/a/m Given a population size N = 800, and e = 0.03. Since no estimate has been made regarding the actual population proportion, we will be conservative and use p = 0.5. The sample size necessary to have a 95% confidence level is given below with z = 1.96:

$$n = \frac{p(1-p)}{\dfrac{e^2}{z^2} + \dfrac{p(1-p)}{N}} = \frac{0.5(1-0.5)}{\dfrac{0.03^2}{1.96^2} + \dfrac{0.5(0.5)}{800}} = 457.22 \text{, rounded up to 458}$$

9.75 c/a/m For a 95% confidence interval for the population proportion, z = 1.96. Since no estimate has been made regarding the actual population proportion, we will be conservative and use p = 0.5 with e = 0.03 and N = 100.

$$n = \frac{p(1-p)}{\dfrac{e^2}{z^2} + \dfrac{p(1-p)}{N}} = \frac{0.5(1-0.5)}{\dfrac{0.03^2}{1.96^2} + \dfrac{0.5(0.5)}{100}} = 91.43 \text{, rounded up to 92 senators}$$

9.77 c/a/m To find the number of households surveyed to predict the population proportion of a finite population, we are given N = 2000, level of confidence = 95%, and e = 0.04. Since the population proportion is not known, we shall use the conservative estimate of p = 0.5.

$$n = \frac{p(1-p)}{\dfrac{e^2}{z^2} + \dfrac{p(1-p)}{N}} = \frac{0.5(1-0.5)}{\dfrac{0.04^2}{1.96^2} + \dfrac{0.5(0.5)}{2,000}} = 461.6864 \text{, rounded up to 462}$$

9.79 p/a/m For the 99% confidence level, z = 2.58. The maximum likely error is e = 0.01, or 1 percentage point, and we will estimate the population proportion with p = 0.05. Applying the finite population formula with N = 2000, the necessary sample size is:

$$n = \frac{p(1-p)}{\dfrac{e^2}{z^2} + \dfrac{p(1-p)}{N}} = \frac{0.5(1-0.5)}{\dfrac{0.01^2}{2.58^2} + \dfrac{0.5(0.5)}{2000}} = 1225.08 \text{, rounded up to 1226}$$

CHAPTER EXERCISES

9.81 c/a/d Since the researchers are working independently,

P(neither of the confidence intervals include μ) =

   P(Researcher 1's interval does not contain μ) x P(Researcher 2's interval does not contain μ)

   = (1 - 0.90)(1 - 0.90) = 0.01

9.83 p/a/m

a. Since σ is unknown, we will use the t distribution. For a confidence level of 95%, the right-tail area of interest is (1 - 0.95)/2 = 0.025 with d.f. = n - 1 = 35 - 1 = 34. Referring to the 0.025 column and d.f. = 34 row of the t table, t = 2.032. The 95% confidence interval for μ is:

$$\bar{x} \pm t\frac{s}{\sqrt{n}} = 2400 \pm 2.032\frac{280}{\sqrt{35}} = 2400 \pm 96.17 \text{, or from \$2303.83 to \$2496.17}$$

b. No, the interval found in part (a) does not contain the actual value of μ = \$2298.

c. Approximately 95% of the intervals would include the actual value of μ while 100 - 95 = 5% of the intervals would not include the actual value of μ.

9.85 p/a/d Since σ is known, we will use the standard normal distribution. For a confidence level of 95%, z = 1.96. The 95% confidence interval for μ is:

$$\bar{x} \pm z\frac{\sigma}{\sqrt{n}} = 137 \pm 1.96\frac{5}{\sqrt{30}} = 137 \pm 1.789 \text{, or from 135.211 to 138.789}$$

We have 95% confidence that the current process mean is between 135.211 and 138.789 lbs.-ft. Since the desired process average of 135 lbs.-ft. is not in the 95% confidence interval found above, the machine may be in need of adjustment.

9.87 p/a/m From exercise 9.86, z = 1.96 and e = 0.05. We will estimate the population proportion with p = 0.20. The number of TV households needed in the sample now is:

$$n = \frac{z^2 p(1-p)}{e^2} = \frac{1.96^2 (0.2)(1-0.2)}{0.05^2} = 245.86 \text{, rounded up to 246.}$$

The sample sizes in exercises 9.86 and 9.87 can also be obtained using Excel worksheet template tmnforpi.xls, on the disk that came with the text. Enter the estimate for π, the maximum likely error desired, and the confidence level (0.95 for 95%), and Excel computes the necessary sample size:

	A	B	C	D
1	Sample size required for estimating a			
2	population proportion:			
3				
4	Estimate for pi:			0.50
5	Maximum likely error, e:			0.05
6				
7	Confidence level desired:			0.95
8	alpha = (1 - conf. level desired):			0.05
9	The corresponding z value is:			1.960
10				
11	The required sample size is n =			384.1

	A	B	C	D
1	Sample size required for estimating a			
2	population proportion:			
3				
4	Estimate for pi:			0.20
5	Maximum likely error, e:			0.05
6				
7	Confidence level desired:			0.95
8	alpha = (1 - conf. level desired):			0.05
9	The corresponding z value is:			1.960
10				
11	The required sample size is n =			245.9

9.89 c/c/m Using Minitab, we find the confidence intervals shown below:

```
One-Sample T: Income
Variable     N       Mean     StDev   SE Mean        90.0 % CI
Income       30      47.43    8.14    1.49     (  44.91,    49.96)

One-Sample T: Income
Variable     N       Mean     StDev   SE Mean        95.0 % CI
Income       30      47.43    8.14    1.49     (  44.39,    50.47)
```

9.91 p/a/m For the 90% confidence level, z = 1.645. The maximum likely error is e = 0.03, or 3 percentage points. Since no estimate has been made regarding the actual population proportion, we will be conservative and use p = 0.5. Applying the finite population formula with N = 904, the number of franchises needed in the sample is:

$$n = \frac{p(1-p)}{\frac{e^2}{z^2} + \frac{p(1-p)}{N}} = \frac{0.5(1-0.5)}{\frac{0.03^2}{1.645^2} + \frac{0.5(0.5)}{904}} = 410.41 \text{, rounded up to 411}$$

9.93 p/a/d For the 90% level of confidence, z = 1.645. The maximum likely error is e = 0.03.

Using p = 0.5, $n = \dfrac{z^2 p(1-p)}{e^2} = \dfrac{1.645^2 (0.5)(1-0.5)}{0.03^2} = 751.67$, rounded up to 752

Using p = 0.3, $n = \dfrac{z^2 p(1-p)}{e^2} = \dfrac{1.645^2 (0.3)(1-0.3)}{0.03^2} = 631.41$, rounded up to 632

The new graduate just saved the company (752 - 632) x 10 = $1200 in interview costs.

9.95 p/a/m

For a confidence level of 95%, z = 1.96. The 95% confidence interval for π is:

$$p \pm z\sqrt{\frac{p(1-p)}{n}} = 0.39 \pm 1.96\sqrt{\frac{0.39(1-0.39)}{500}} = 0.39 \pm 0.043 \text{, or from 0.347 to 0.433}$$

The maximum likely error is $e = z\sqrt{\frac{p(1-p)}{n}} = 0.043$. If we use p = 0.39 to estimate π, we will have

95% confidence that we are within 0.043 of the true population proportion.

For a confidence level of 99%, z = 2.58. The 99% confidence interval for π is:

$$p \pm z\sqrt{\frac{p(1-p)}{n}} = 0.39 \pm 2.58\sqrt{\frac{0.39(1-0.39)}{500}} = 0.39 \pm 0.056 \text{, or from 0.334 to 0.446}$$

The maximum likely error is $e = z\sqrt{\frac{p(1-p)}{n}} = 0.056$. If we use p = 0.39 to estimate π, we will have

99% confidence that we are within 0.056 of the true population proportion.

9.97 p/a/d For the 99% level of confidence, z = 2.58. The maximum likely error is e = 0.02
(2 percentage points). If we make no estimate regarding the actual population proportion, we can be
conservative and use p = 0.5. The recommended sample size would be:

$$n = \frac{z^2 p(1-p)}{e^2} = \frac{2.58^2(0.5)(1-0.5)}{0.02^2} = 4160.25 \text{, rounded up to 4161}$$

Persons who are aware that Count Chocula is a kid's cereal, and that senior citizens don't tend to consume
the product, might want to use a lower estimate, such as p = 0.10. In this case, we would end up with a
recommended sample size of just 1498.

9.99 p/a/d For the 90% level of confidence, z = 1.645. The maximum likely error is e = 0.03, or 3
percentage points. We will estimate the population proportion with p = 0.4, the part of our range that is
closest to the most conservative possible estimate, 0.5. The needed sample size is:

$$n = \frac{z^2 p(1-p)}{e^2} = \frac{1.645^2(0.4)(1-0.4)}{0.03^2} = 721.61 \text{, rounded up to 722.}$$

9.101 p/a/m For the 95% level of confidence, z = 1.96. The maximum acceptable error is e = 0.03, or 3 percentage points. Since no estimate has been made regarding the actual population proportion, we will be conservative and use p = 0.5. The recommended sample size is:

$$n = \frac{z^2 p(1-p)}{e^2} = \frac{1.96^2(0.5)(1-0.5)}{0.03^2} = 1067.11 \text{, rounded up to 1068.}$$

An alternative approach is to use the Excel worksheet template tmnforpi.xls, on the disk that came with the text. Just enter the estimate (in this case, 0.5), the maximum likely error desired (0.03), and the desired confidence level (0.95 for 95%):

	A	B	C	D
1	Sample size required for estimating a			
2	population proportion:			
3				
4	Estimate for pi:			0.50
5	Maximum likely error, e:			0.03
6				
7	Confidence level desired:			0.95
8	alpha = (1 - conf. level desired):			0.05
9	The corresponding z value is:			1.960
10				
11	The required sample size is n =			1067.1

9.103 p/a/m For the 95% confidence level, z = 1.96. The maximum likely error is e = 0.01. Since no estimate has been made regarding the actual population proportion, we will be conservative and use p = 0.5. Applying the finite population formula with a population size of N = 1733, the number of companies that must be sampled is:

$$n = \frac{p(1-p)}{\dfrac{e^2}{z^2} + \dfrac{p(1-p)}{N}} = \frac{0.5(1-0.5)}{\dfrac{0.01^2}{1.96^2} + \dfrac{0.5(1-0.5)}{1733}} = 1468.09 \text{, rounded up to 1469}$$

9.105 p/a/m For the 99% confidence level, z = 2.58. The maximum likely error is e = 0.02. Since no estimate has been made regarding the actual population proportion, we will be conservative and use p = 0.5. Applying the finite population formula with N = 37,700, the necessary sample size is:

$$n = \frac{p(1-p)}{\dfrac{e^2}{z^2} + \dfrac{p(1-p)}{N}} = \frac{0.5(1-0.5)}{\dfrac{0.02^2}{2.58^2} + \dfrac{(0.5)(1-0.5)}{37,700}} = 3746.79 \text{, rounded up to 3745}$$

9.107 p/a/m With n = 1000 and p = 0.40, the 90% and 95% confidence intervals can be determined by computing $p \pm z\sqrt{\dfrac{p(1-p)}{n}}$ with z = 1.645 and z = 1.96, respectively.

Using the Estimators.xls workbook that accompanies Data Analysis Plus, these confidence intervals are shown below as 0.375 to 0.425 and 0.370 to 0.430, respectively.

	A	B	C	D	E
1	z-Estimate of a Proportion				
2					
3	Sample proportion	0.40	Confidence Interval Estimate		
4	Sample size	1000	0.40	plus/minus	0.025
5	Confidence level	0.90	Lower confidence limit		0.375
6			Upper confidence limit		0.425

	A	B	C	D	E
1	z-Estimate of a Proportion				
2					
3	Sample proportion	0.40	Confidence Interval Estimate		
4	Sample size	1000	0.40	plus/minus	0.030
5	Confidence level	0.95	Lower confidence limit		0.370
6			Upper confidence limit		0.430

9.109 c/c/d Let x = number of confidence intervals that contain the population mean; x is binomial distributed with n = 20 and $\pi = 0.95$. Using Minitab:

```
Cumulative Distribution Function
Binomial with n = 20 and p = 0.950000
      x       P( X <= x)
   15.00        0.0026
```

and $P(x \geq 16) = 1 - P(x \leq 15) = 1 - 0.0026 = 0.9974$

9.111 p/a/d For the 95% confidence level, z = 1.96. The sample proportion is p = 12/300 = 0.04. Since the sample is less than 5% as large as the population of N = 8000, we do not need to use the finite population correction. The 95% confidence interval for π is:

$$p \pm z\sqrt{\frac{p(1-p)}{n}} = 0.04 \pm 1.96\sqrt{\frac{0.04(1-0.04)}{300}} = 0.04 \pm 0.022 \text{, or from } 0.018 \text{ to } 0.062$$

We can also obtain the confidence interval by using the Estimators.xls workbook that accompanies Data Analysis Plus.

	A	B	C	D	E
1	z-Estimate of a Proportion				
2					
3	Sample proportion	0.04	Confidence Interval Estimate		
4	Sample size	300	0.04	plus/minus	0.022
5	Confidence level	0.95	Lower confidence limit		0.018
6			Upper confidence limit		0.062

We have 95% confidence that the proportion of the boards that fall outside the specifications is between 0.018 and 0.062. Although 0.03 is in the interval, there are also values in the interval that are more than 0.03. It is possible that the supplier's claim is correct but it is also possible that the supplier's claim is not correct.

9.113 p/c/m Using Data Analysis Plus, the 95% confidence interval for the population mean is from $24.33 to $25.67. The Minitab counterpart is shown below the Excel printout.

	A	B	C	D
1	t-Estimate: Mean			
2				check
3	Mean			25
4	Standard Deviation			9.63
5	LCL			24.33
6	UCL			25.67

```
One-Sample T: check
Variable          N      Mean    StDev   SE Mean        95.0% CI
check           800    25.000    9.626     0.340  (  24.332,   25.668)
```

9.115 p/c/m Using Data Analysis Plus, the 95% confidence interval for the population mean is from 64.719 to 68.301 mph. The Minitab counterpart is shown below the Excel printout.

	A	B	C	D
1	t-Estimate: Mean			
2				
3				mph
4	Mean			66.510
5	Standard Deviation			9.028
6	LCL			64.719
7	UCL			68.301

```
One-Sample T: mph
Variable          N      Mean    StDev   SE Mean        95.0% CI
mph             100    66.510    9.028     0.903  (  64.719,   68.301)
```

We find that 70.0 mph exceeds the range of values described by the confidence interval. This suggests that the population mean mph on this section of highway is some value less than 70 mph. The Federal highway funds are not in danger.

CHAPTER 10
HYPOTHESIS TESTS INVOLVING
A SAMPLE MEAN OR PROPORTION

SECTION EXERCISES

10.1 d/p/m The null hypothesis is a statement about the value of a population parameter. It is assumed to be true unless we have evidence to the contrary. The alternative hypothesis is an assertion that holds if the null hypothesis is false. The null hypothesis is not always the same as the verbal claim or assertion that led to the test. The null hypothesis must always contain the equal sign. If the directional claim does not contain an equal sign, then the claim is put in the alternative hypothesis and the opposite is put in the null hypothesis.

10.3 d/p/m
a. Inappropriate. The value given is different in the two hypotheses.
b. Appropriate
c. Appropriate.
d. Inappropriate. The null and alternative hypotheses do not include all possible values of the population parameter.
e. Appropriate.
f. Inappropriate. A hypothesis is a statement about a population parameter, not a sample statistic like p.

10.5 d/p/m If the scientist's null hypothesis that "global warming is taking place" is correct, but people do not take her seriously, they would be making a Type II error by rejecting a true null hypotheis.

10.7 d/p/m The engineer would least like to commit a Type II error since a lot of people could be killed if this occurred. A Type II error would be committed if he decided that the stadium was structurally sound when it was not.

10.9 d/p/m Since this is a directional claim, we will use a one-tail test. $H_0: \pi \leq 0.10$; $H_1: \pi > 0.10$

10.11 d/p/m H_0: Person is not drunk; H_1: Person is drunk
A Type I error would be committed if the officer decides the person is drunk since he can't walk a straight line or close his eyes and touch his nose, when he really was not drunk. This could occur if a person is tired, frightened, or has a physical disability.
A Type II error would be committed if the officer decides the person is not drunk since he can walk a straight line or close his eyes and touch his nose, but he really is drunk. This could occur because a person drinks quite often and therefore can withstand more.

10.13 d/p/m She appears to favor Type I error. In this case, Type I error would be deciding that a drug is harmful when it really isn't.

10.15 d/p/m Let π = population proportion of women aged 40 - 49 in NYC who save in a 401(k) or individual retirement account. $H_0: \pi = 0.62$ $H_1: \pi \neq 0.62$
This test would be a z-test since $n\pi = 300(0.62) = 186$ and $n(1-\pi) = 300(1 - 0.62) = 114$ are ≥ 5.

10.17 d/p/m The larger the value of α, the greater the likelihood of committing a Type I error. For this exercise, a Type I error would be deciding the mean tensile strength of the rivets is below 3000 pounds when it is really 3000 pounds or above. With the null and alternative hypotheses, H_0: $\mu \geq 3000$; H_1: $\mu < 3000$:

a. The marketing director for a major competitor would prefer a numerically high level of significance (e.g., $\alpha = 0.20$) to be used in reaching a conclusion. This would make it easier to conclude that the mean tensile strength is below 3000 pounds when it really is 3000 or above.

b. The rivet manufacturer's advertising agency would prefer a numerically low level of significance (e.g., $\alpha = 0.01$) to be used in reaching a conclusion. This would make it more difficult to conclude that the mean tensile strength is below 3000 pounds when it really is 3000 or above. They have already claimed the mean tensile strength to be at least 3000 pounds, so they don't want a test result to suggest otherwise.

10.19 d/p/m If the sample size is large ($n \geq 30$), the central limit theorem assures us that the distribution of sample means will be approximately normally distributed regardless of the shape of the underlying population. The larger the sample size, the better this approximation becomes. When the central limit theorem applies, we may use the standard normal distribution to identify the critical values for the test statistic when σ is known.

10.21 d/p/m A p-value is the exact level of significance associated with the calculated value of the test statistic. It is the most extreme critical value that the test statistic would be capable of exceeding. If p-value $< \alpha$, reject H_0 and if p-value $\geq \alpha$, do not reject H_0.

10.23 d/p/m Since p-value = 0.04 is not less than $\alpha = 0.01$, the null hypothesis would be not be rejected. The sample result is not more extreme than you would have been willing to attribute to chance.

10.25 c/a/m Using the standard normal table,
a. p-value = $P(z \geq 1.54) = 0.5000 - 0.4382 = 0.0618$
b. p-value = $P(z \leq -1.03) = 0.5000 - 0.3485 = 0.1515$
c. p-value = $2P(z \leq -1.83) = 2(0.5000 - 0.4664) = 0.0672$

10.27 c/a/m Null and alternative hypotheses:
H_0: $\mu = 450$ H_1: $\mu \neq 450$ Level of significance: $\alpha = 0.05$
Test results: $\bar{x} = 458$, n = 35 (known: $\sigma = 20.5$)

Calculated value of test statistic: $z = \dfrac{\bar{x} - \mu_0}{\sigma_{\bar{x}}} = \dfrac{458 - 450}{20.5/\sqrt{35}} = 2.31$

Critical values: $z = -1.96$ and $z = 1.96$ (look up $0.5000 - (0.0500/2) = 0.4750$ in the body of the standard normal table.)
Decision rule: Reject H_0 if the calculated $z < -1.96$ or > 1.96, otherwise do not reject.
Conclusion: Since calculated test statistic falls in rejection region ($z = 2.31 > 1.96$), reject H_0.
Decision: At the 0.05 level, the results suggest that the population mean is not 450.
Using the standard normal distribution table, we can find the approximate p-value as twice the area to the right of $z = 2.31$. This is $2(0.5000 - 0.4896) = 0.0208$.

Given the summary data, we can also carry out this z-test using the Test Statistics.xls workbook that accompanies Data Analysis Plus. It is on the disk that comes with the textbook. The results are shown below. For this two-tail test, the p-value (0.0210) is less than the 0.05 level of significance being used to reach a conclusion, so the null hypothesis is rejected. For a true null hypothesis, there is only a 0.0210 probability that a sample mean this far away from 450 would occur by chance.

	A	B	C	D
1	z-Test of a Mean			
2				
3	Sample mean	458.0	z Stat	2.31
4	Population standard deviation	20.5	P(Z<=z) one-tail	0.0105
5	Sample size	35	z Critical one-tail	1.645
6	Hypothesized mean	450	P(Z<=z) two-tail	0.0210
7	Alpha	0.05	z Critical two-tail	1.960

10.29 p/a/m Null and alternative hypotheses:

H_0: $\mu = 2$ (machine in adjustment) H_1: $\mu \neq 2$ (machine out of adjustment)

Level of significance: $\alpha = 0.01$

Test results: $\bar{x} = 2.025$, n = 35 (known: $\sigma = 0.07$)

Calculated value of test statistic: $z = \dfrac{\bar{x} - \mu_0}{\sigma_{\bar{x}}} = \dfrac{2.025 - 2}{0.07 / \sqrt{35}} = 2.11$

Critical values: z = -2.58 and z = 2.58 (look up 0.5000 - (0.0100/2) = 0.4950 in the standard normal table.)

Decision rule: Reject H_0 if the calculated z < -2.58 or > 2.58, otherwise do not reject.

Conclusion: Since calculated test statistic falls in nonrejection region (-2.58 < z = 2.11 < 2.58) do not reject H_0.

Decision: At the 0.01 level, results suggest the machine is properly adjusted. It appears the mean length of nails produced by the machine could be 2 inches. The difference between the hypothesized population mean and the sample mean is judged to have been merely the result of chance variation.

Using the standard normal distribution table, we can find the approximate p-value as twice the area to the right of z = 2.11. This is 2(0.5000 - 0.4826) = 0.0348.

Given the summary data, we can also carry out this z-test using the Test Statistics.xls workbook that accompanies Data Analysis Plus. For this two-tail test, the p-value (0.0346) is not less than the 0.01 level of significance being used to reach a conclusion, so the null hypothesis is not rejected. For a true null hypothesis, there is a 0.0346 probability that a sample mean this far away from 2.000 inches would occur by chance.

	A	B	C	D
1	z-Test of a Mean			
2				
3	Sample mean	2.025	z Stat	2.11
4	Population standard deviation	0.07	P(Z<=z) one-tail	0.0173
5	Sample size	35	z Critical one-tail	2.326
6	Hypothesized mean	2.000	P(Z<=z) two-tail	0.0346
7	Alpha	0.01	z Critical two-tail	2.576

10.31 p/a/m Null and alternative hypotheses:

H_0: $\mu = 2.5$ (machine doesn't need maintenance) H_1: $\mu \neq 2.5$ (needs maintenance)

Level of significance: $\alpha = 0.01$

Test results: $\bar{x} = 2.509$, n = 34 (known: $\sigma = 0.027$)

Calculated value of test statistic: $z = \dfrac{\bar{x} - \mu_0}{\sigma_{\bar{x}}} = \dfrac{2.509 - 2.50}{0.027 / \sqrt{34}} = 1.94$

Critical values: $z = -2.58$ and $z = 2.58$

Decision rule: Reject H_0 if the calculated $z < -2.58$ or > 2.58, otherwise do not reject.

Conclusion: Calculated test statistic falls in nonrejection region, do not reject H_0.

Decision: At the 0.01 level, the results suggest that the machine is not in need of maintenance and calibration. The mean diameter of the tubing appears to still be 2.5 inches. The difference between the hypothesized population mean and the sample mean is judged to have been merely the result of chance variation.

Using the standard normal distribution table, we can find the approximate p-value as twice the area to the right of z = 1.94. This is 2(0.5000 - 0.4738) = 0.0524.

Given the summary data, we can also carry out this z-test using the Test Statistics.xls workbook that accompanies Data Analysis Plus. For this two-tail test, the p-value (0.0519) is not less than the 0.01 level of significance being used to reach a conclusion, so the null hypothesis is not rejected. For a true null hypothesis, there is a 0.0519 probability that a sample mean this far away from 2.500 would occur by chance.

	A	B	C	D
1	z-Test of a Mean			
2				
3	Sample mean	2.509	z Stat	1.94
4	Population standard deviation	0.027	P(Z<=z) one-tail	0.0260
5	Sample size	34	z Critical one-tail	2.326
6	Hypothesized mean	2.500	P(Z<=z) two-tail	0.0519
7	Alpha	0.01	z Critical two-tail	2.576

10.33 p/c/m The null and alternative hypotheses are H_0: $\mu = \$5976$ and H_1: $\mu \neq \$5976$. The Data Analysis Plus and Minitab results are shown below.

	A	B	C	D
1	Z-Test: Mean			
2				
3				price
4	Mean			6310
5	Standard Deviation			1020.80
6	Observations			40
7	Hypothesized Mean			5976
8	SIGMA			1000
9	z Stat			2.1124
10	P(Z<=z) one-tail			0.017
11	z Critical one-tail			1.96
12	P(Z<=z) two-tail			0.035
13	z Critical two-tail			2.241

```
Test of mu = 5976 vs mu not = 5976
The assumed sigma = 1000

Variable            N       Mean    StDev   SE Mean
Price              40       6310     1021       158

Variable             97.5% CI              Z       P
Price          (   5956,    6664)        2.11   0.035
```

For this two-tail test, the p-value (0.035) is not less than the 0.025 level of significance being used to reach a conclusion, so the null hypothesis is not rejected. At this level of significance, we conclude that the mean price for home office conversions in this region could be the same as the mean price for the nation as a whole.

10.35 d/p/m

a. Do not reject H_0 since 170 is in the 90% confidence interval given.
b. Reject H_0 since 110 is not in the 90% confidence interval given.
c. Do not reject H_0 since 130 is in the 90% confidence interval given.
d. Reject H_0 since 200 is not in the 90% confidence interval given.

10.37 c/a/m From exercise 10.29, $\bar{x} = 2.025$ inches, $\sigma = 0.070$ inches, $n = 35$, the critical z for $\alpha = 0.01$ is $z = 2.58$, and the hypothesis test is $H_0: \mu = 2.000$ versus $H_1: \mu \neq 2.000$. The 99% confidence interval for μ is:

$$\bar{x} \pm z\frac{\sigma}{\sqrt{n}} = 2.025 \pm 2.58\frac{0.070}{\sqrt{35}} = 2.025 \pm 0.031, \text{ or from } 1.994 \text{ to } 2.056$$

Since 2.000 is within the 99% confidence interval for μ found above, the population mean could be equal to 2.000. In exercise 10.29, the null hypothesis was not rejected and we concluded that the population mean could be 2.000. Therefore, the conclusion using the confidence interval is the same as the conclusion from the hypothesis test. The confidence interval can also be obtained using the Estimators.xls workbook that accompanies Data Analysis Plus, as shown below. Because it does not rely on the printed standard normal table (with its gaps between listed values), this interval is more accurate, and has lower and upper limits of 1.995 inches and 2.055 inches, respectively.

	A	B	C	D	E
1	z-Estimate of a Mean				
2					
3	Sample mean	2.025	Confidence Interval Estimate		
4	Population standard deviation	0.07	2.025	plus/minus	0.030
5	Sample size	35	Lower confidence limit		1.995
6	Confidence level	0.99	Upper confidence limit		2.055

10.39 d/p/e The t statistic should be used in carrying out a hypothesis test for the mean when σ is unknown. When $n < 30$, we must assume the population is approximately normally distributed.

10.41 c/a/m Null and alternative hypotheses: H_0: $\mu \geq 90.0$ H_1: $\mu < 90.0$

Level of significance: $\alpha = 0.05$

Test results: $\bar{x} = 82.0$, $s = 20.5$, $n = 15$ (Note: population is approximately normally distributed.)

Calculated value of test statistic: $t = \dfrac{\bar{x} - \mu_0}{s_{\bar{x}}} = \dfrac{82.0 - 90.0}{20.5/\sqrt{15}} = -1.511$

Critical value: t = -1.761. For this test, $\alpha = 0.05$ and d.f. = (n - 1) = (15 - 1) = 14. Referring to the 0.05 column and the 14th row of the t table, the critical value is t = -1.761.

Decision rule: Reject H_0 if the calculated t < -1.761, otherwise do not reject.

Conclusion: Calculated test statistic falls in nonrejection region, do not reject H_0.

Decision: At the 0.05 level, the results suggest that the population mean could be at least 90.0. The sample mean could have been this low merely by chance.

Given the summary data, we can also carry out this test using the Test Statistics.xls workbook that accompanies Data Analysis Plus. In this left-tail test, the p-value (0.076) is not less than 0.05, so we do not reject the null hypothesis.

	A	B	C	D
1	t-Test of a Mean			
2				
3	Sample mean	82.0	t Stat	-1.511
4	Sample standard deviation	20.5	P(T<=t) one-tail	0.076
5	Sample size	15	t Critical one-tail	1.761
6	Hypothesized mean	90.0	P(T<=t) two-tail	0.153
7	Alpha	0.05	t Critical two-tail	2.145

10.43 p/a/m Null and alternative hypotheses:

H_0: $\mu = 299$ (the average flight is 299 miles, the value reported by the industry association) and

H_1: $\mu \neq 299$ (the average flight is not 299 miles)   Level of significance: $\alpha = 0.05$

Test results: $\bar{x} = 314.6$, $s = 42.8$, $n = 30$

Calculated value of test statistic: $t = \dfrac{\bar{x} - \mu_0}{s_{\bar{x}}} = \dfrac{314.6 - 299}{42.8/\sqrt{30}} = 1.996$

Critical values: t = -2.045 and t = 2.045 For this test, $\alpha = 0.05$ and d.f. = (n - 1) = (30 - 1) = 29. Referring to the 0.05/2 = 0.025 column and the 29th row of the t table, the critical values are t = -2.045 and t = 2.045.

Decision rule: Reject H_0 if the calculated t < -2.045 or > 2.045, otherwise do not reject.

Conclusion: Calculated test statistic falls in nonrejection region, do not reject H_0.

Decision: At the 0.05 level, the results do not cause us to doubt that the average length of a flight by regional airlines in the U.S. is the reported value, 299 miles. The difference between the hypothesized population mean and the sample mean is judged to have been merely the result of chance variation.

Given the summary data, we can also carry out this test using the Test Statistics.xls workbook that accompanies Data Analysis Plus. In this two-tail test, the p-value (0.055) is not less than 0.05, so we do not reject the null hypothesis.

	A	B	C	D
1	t-Test of a Mean			
2				
3	Sample mean	314.60	t Stat	1.996
4	Sample standard deviation	42.8	P(T<=t) one-tail	0.028
5	Sample size	30	t Critical one-tail	1.699
6	Hypothesized mean	299.00	P(T<=t) two-tail	0.055
7	Alpha	0.05	t Critical two-tail	2.045

10.45 p/a/m Null and alternative hypotheses:

H_0: $\mu = 150$ (Taxco's assertion is accurate) H_1: $\mu \neq 150$ (assertion is not accurate)

Level of significance: $\alpha = 0.10$

Test results: $\bar{x} = 125$, $s = 43$, $n = 12$ (assumed: population is approximately normally distributed)

Calculated value of test statistic: $t = \dfrac{\bar{x} - \mu_0}{s_{\bar{x}}} = \dfrac{125 - 150}{43 / \sqrt{12}} = -2.014$

Critical values: $t = -1.796$ and $t = 1.796$ For this test, $\alpha = 0.10$ and d.f. $= (n - 1) = (12 - 1) = 11$.
    Referring to the $0.10/2 = 0.05$ column and the 11th row of the t table, the critical values
    are $t = -1.796$ and $t = 1.796$.

Decision rule: Reject H_0 if the calculated $t < -1.796$ or > 1.796, otherwise do not reject.

Conclusion: Calculated test statistic falls in rejection region, reject H_0.

Decision: At the 0.10 level, the results suggest that Taxco's assertion that the mean refund for those
    customers who received refunds last year was $150 is not accurate.

Given the summary data, we can also carry out this test using the Test Statistics.xls workbook that accompanies Data Analysis Plus. In this two-tail test, the p-value (0.069) is less than 0.10, so we are able to reject the null hypothesis.

	A	B	C	D
1	t-Test of a Mean			
2				
3	Sample mean	125.0	t Stat	-2.014
4	Sample standard deviation	43	P(T<=t) one-tail	0.035
5	Sample size	12	t Critical one-tail	1.363
6	Hypothesized mean	150.0	P(T<=t) two-tail	0.069
7	Alpha	0.10	t Critical two-tail	1.796

10.47 p/a/d Null and alternative hypotheses:

H_0: $\mu \leq 80$ (the mean of cash sales is no more than $80)

H_1: $\mu > 80$ (the mean of cash sales is greater than $80)

Level of significance: $\alpha = 0.05$

Test results: $\bar{x} = 91$, $s = 21$, $n = 20$

Calculated value of test statistic: $t = \dfrac{\bar{x} - \mu_0}{s_{\bar{x}}} = \dfrac{91 - 80}{21 / \sqrt{20}} = 2.343$

Critical value: $t = 1.729$ For this test, $\alpha = 0.05$ and d.f. $= (n - 1) = (20 - 1) = 19$. Referring to the 0.05
    column and the 19th row of the t table, the critical value is $t = 1.729$.

Decision rule: Reject H_0 if the calculated $t > 1.729$, otherwise do not reject.

Conclusion: Calculated test statistic falls in rejection region, reject H_0.

Decision: At the 0.05 level, it appears that the agent's suspicion is confirmed. The mean of the scrap
    metal dealer's cash sales appears to exceed $80.

Given the summary data, we can also carry out this test using the Test Statistics.xls workbook that accompanies Data Analysis Plus. In this right-tail test, the p-value (0.015) is less than 0.05, so we are able to reject the null hypothesis.

	A	B	C	D
1	t-Test of a Mean			
2				
3	Sample mean	91.0	t Stat	2.343
4	Sample standard deviation	21.0	P(T<=t) one-tail	0.015
5	Sample size	20	t Critical one-tail	1.729
6	Hypothesized mean	80.0	P(T<=t) two-tail	0.030
7	Alpha	0.05	t Critical two-tail	2.093

10.49 p/a/m Null and alternative hypotheses: H_0: $\mu = \$183,000$ and H_1: $\mu \neq \$183,000$
The exercise can be solved by hand, but we will use the computer and the Test Statistics.xls workbook that accompanies Data Analysis Plus. In this two-tail test, the p-value (0.094) is not less than 0.05, so we do not reject the null hypothesis. At the 0.05 level of significance, the mean for this region may be the same as the national mean.

	A	B	C	D
1	t-Test of a Mean			
2				
3	Sample mean	198700	t Stat	1.708
4	Sample standard deviation	65000	P(T<=t) one-tail	0.047
5	Sample size	50	t Critical one-tail	1.677
6	Hypothesized mean	183000	P(T<=t) two-tail	0.094
7	Alpha	0.05	t Critical two-tail	2.010

10.51 p/a/m Null and alternative hypotheses: H_0: $\mu \geq 4000$ hours and H_1: $\mu < 4000$ hours
The exercise can be solved by hand, but we will use the computer and the Test Statistics.xls workbook that accompanies Data Analysis Plus. In this left-tail test, the p-value (0.019) is less than 0.025, so we reject H_0 and conclude that the conditions may be having an adverse effect on bulb life.

	A	B	C	D
1	t-Test of a Mean			
2				
3	Sample mean	3882	t Stat	-2.285
4	Sample standard deviation	200	P(T<=t) one-tail	0.019
5	Sample size	15	t Critical one-tail	2.145
6	Hypothesized mean	4000	P(T<=t) two-tail	0.038
7	Alpha	0.025	t Critical two-tail	2.510

10.53 p/a/m Using the Estimators.xls workbook, we obtain the 95% confidence interval shown below. We are 95% confident the population mean is within the interval from 86.29 to 92.71 seconds. Because the hypothesized population mean (93 minutes) is not within this interval, we conclude that the actual population mean must be some value other than 93 minutes. This is the same conclusion reached in the hypothesis test of exercise 10.52.

	A	B	C	D	E
1	t-Estimate of a Mean				
2					
3	Sample mean	89.5	Confidence Interval Estimate		
4	Sample standard deviation	11.3	89.50	plus/minus	3.21
5	Sample size	50	Lower confidence limit		86.29
6	Confidence level	0.95	Upper confidence limit		92.71

10.55 p/a/m Using the Estimators.xls workbook, we obtain the 90% confidence interval shown below. We are 90% confident the population mean completion time is within the interval from 35.657 minutes to 37.943 minutes. Because the hypothesized population mean (38 minutes) is not within this interval, we conclude that the actual population mean must be some value other than 38 minutes. This is the same conclusion reached in the hypothesis test of exercise 10.54.

	A	B	C	D	E
1	t-Estimate of a Mean				
2					
3	Sample mean	36.8	Confidence Interval Estimate		
4	Sample standard deviation	4.0	36.800	plus/minus	1.143
5	Sample size	35	Lower confidence limit		35.657
6	Confidence level	0.90	Upper confidence limit		37.943

10.57 p/c/m The null and alternative hypotheses are H_0: $\mu = \$706$ and H_1: $\mu \neq \$706$. The Data Analysis Plus and Minitab results are shown below.

	A	B	C	D
1	t-Test: Mean			
2				
3				*Expense*
4	Mean			739.575
5	Standard Deviation			136.051
6	Hypothesized Mean			706
7	df			79
8	t Stat			2.207
9	P(T<=t) one-tail			0.015
10	t Critical one-tail			1.664
11	P(T<=t) two-tail			0.030
12	t Critical two-tail			1.991

```
One-Sample T: Expense

Test of mu = 706 vs mu not = 706

Variable          N      Mean     StDev   SE Mean
Expense          80     739.6     136.1      15.2

Variable          95.0% CI            T      P
Expense      (  709.3,    769.9)    2.21   0.030
```

For this two-tail test, the p-value (0.030) is less than the 0.05 level of significance being used to reach a conclusion, so the null hypothesis is rejected. If the North Carolina mean were really $706, there would be only a 0.030 probability of obtaining a sample mean this far away from $706. We conclude that the mean for North Carolina motorists is some value other than $706.

10.59 d/p/e The normal distribution is a good approximation for the binomial distribution when $n\pi$ and $n(1-\pi)$ are both ≥ 5.

10.61 c/a/m Null and alternative hypotheses: H_0: $\pi \geq 0.50$ H_1: $\pi < 0.50$
Level of significance: $\alpha = 0.05$
Test results: p = 0.47, n = 1000

Calculated value of test statistic: $z = \dfrac{p - \pi_0}{\sigma_p} = \dfrac{0.47 - 0.50}{\sqrt{0.5(1-0.5)/1000}} = -1.90$

Critical value: z = -1.645
Decision rule: Reject H_0 if the calculated z < -1.645, otherwise do not reject.
Conclusion: Calculated test statistic falls in rejection region, reject H_0.
Decision: At the 0.05 level, the results suggest that the population proportion is less than 0.50.
Given the summary data, we can also use the Test Statistics.xls workbook that accompanies Data Analysis Plus on the disk provided with the text. For this left-tail test, the p-value (0.029) is less than the 0.05 level of significance being used to reach a conclusion, so reject the null hypothesis.

	A	B	C	D
1	z-Test of a Proportion			
2				
3	Sample proportion	0.47	z Stat	-1.90
4	Sample size	1000	P(Z<=z) one-tail	0.029
5	Hypothesized proportion	0.50	z Critical one-tail	1.645
6	Alpha	0.05	P(Z<=z) two-tail	0.058
7			z Critical two-tail	1.960

10.63 p/a/m Null and alternative hypotheses:

H_0: $\pi \leq 0.02$ (the proportion of defectives is no more than 0.02)

H_1: $\pi > 0.02$ (the proportion of defectives is greater than 0.02)

Level of significance: We will use $\alpha = 0.05$ in carrying out this right-tail test.

Test results: p = 0.04, n = 300

Calculated value of test statistic: $z = \dfrac{p - \pi_0}{\sigma_p} = \dfrac{0.04 - 0.02}{\sqrt{0.02(1 - 0.02)/300}} = 2.47$

Critical value: $z = 1.645$

Decision rule: Reject H_0 if the calculated z > 1.645, otherwise do not reject.

Conclusion: Calculated test statistic falls in rejection region, reject H_0.

Decision: At the 0.05 level, the results suggest that the supplier's claim is not correct. The true percentage of defectives in the shipment appears to be greater than 2%.

Given the summary data, we can also use the Test Statistics.xls workbook that accompanies Data Analysis Plus on the disk provided with the text. For this right-tail test, the p-value (0.007) is less than the 0.05 level of significance being used to reach a conclusion, so reject the null hypothesis.

	A	B	C	D
1	z-Test of a Proportion			
2				
3	Sample proportion	0.04	z Stat	2.47
4	Sample size	300	P(Z<=z) one-tail	0.007
5	Hypothesized proportion	0.02	z Critical one-tail	1.645
6	Alpha	0.05	P(Z<=z) two-tail	0.013
7			z Critical two-tail	1.960

10.65 p/a/m Null and alternative hypotheses:

H_0: $\pi \leq 0.05$ (the proportion who violated the agreement is no more than 0.05)

H_1: $\pi > 0.05$ (the proportion who violated the agreement is more than 0.05)

Level of significance: $\alpha = 0.025$

Test results: p = 0.08, n = 400

Calculated value of test statistic: $z = \dfrac{p - \pi_0}{\sigma_p} = \dfrac{0.08 - 0.05}{\sqrt{0.05(1 - 0.05)/400}} = 2.75$

Critical value: $z = 1.96$

Decision rule: Reject H_0 if the calculated z > 1.96, otherwise do not reject.

Conclusion: Calculated test statistic falls in rejection region, reject H_0.

Decision: At the 0.025 level, the data do not support the human resource's director's claim that no more than 5% of employees hired in the past year have violated their pre-employment agreement not to use any of five illegal drugs.

Given the summary data, we can also use the Test Statistics.xls workbook that accompanies Data Analysis Plus on the disk provided with the text. For this right-tail test, the p-value (0.003) is less than the 0.025 level of significance being used to reach a conclusion, so reject the null hypothesis.

	A	B	C	D
1	z-Test of a Proportion			
2				
3	Sample proportion	0.08	z Stat	2.75
4	Sample size	400	P(Z<=z) one-tail	0.003
5	Hypothesized proportion	0.05	z Critical one-tail	1.960
6	Alpha	0.025	P(Z<=z) two-tail	0.006
7			z Critical two-tail	2.241

10.67 p/a/m Null and alternative hypotheses:

H_0: $\pi \le 0.44$ (the proportion passing on the first try has not increased)

H_1: $\pi > 0.44$ (the proportion passing on the first try has increased)

Level of significance: $\alpha = 0.05$

Test results: p = 130/250 = 0.52, n = 250

Calculated value of test statistic: $z = \dfrac{p - \pi_0}{\sigma_p} = \dfrac{0.52 - 0.44}{\sqrt{0.44(1 - 0.44)/250}} = 2.55$

Critical value: z = 1.645

Decision rule: Reject H_0 if the calculated z > 1.645, otherwise do not reject.

Conclusion: Calculated test statistic falls in rejection region, reject H_0.

Decision: At the 0.05 level, we can conclude that the proportion passing on the first try has increased from 0.44.

The p-value for this right-tail test is the area to the right of z = 2.55, or 0.5000 - 0.4946 = 0.0054. Because the p-value is less than 0.05, we reject the null hypothesis. Using the Test Statistics.xls workbook, the corresponding results are shown below.

	A	B	C	D
1	z-Test of a Proportion			
2				
3	Sample proportion	0.52	z Stat	2.55
4	Sample size	250	P(Z<=z) one-tail	0.005
5	Hypothesized proportion	0.44	z Critical one-tail	1.645
6	Alpha	0.05	P(Z<=z) two-tail	0.011
7			z Critical two-tail	1.960

10.69 p/a/m The null and alternative hypotheses are H_0: $\pi = 0.62$ and H_1: $\pi \ne 0.62$.

This solution can be obtained with a pocket calculator and formulas, but we will use the computer. As shown in the Test Statistics.xls printout for this two-tail test, the p-value (0.023) is less than the 0.05 level of significance being used to reach a conclusion, so reject the null hypothesis. If the population proportion for this builder were really 0.62, there would be only a 0.023 probability of obtaining a sample proportion this far away from 0.62.

	A	B	C	D
1	z-Test of a Proportion			
2				
3	Sample proportion	0.58	z Stat	-2.27
4	Sample size	600	P(Z<=z) one-tail	0.012
5	Hypothesized proportion	0.62	z Critical one-tail	1.645
6	Alpha	0.05	P(Z<=z) two-tail	0.023
7			z Critical two-tail	1.960

10.71 p/a/m The null and alternative hypotheses are H_0: $\pi = 0.09$ and H_1: $\pi \neq 0.09$.

This solution can be obtained with a pocket calculator and formulas, but we will use the computer.

As shown in the Test Statistics.xls printout for this two-tail test, the p-value (0.118) is not less than the 0.10 level of significance used to reach a conclusion, so we do not reject the null hypothesis.

The percentage of young women who are low-paid in this county might be the same as the percentage of young woman who are low-paid in the nation as a whole.

	A	B	C	D
1	z-Test of a Proportion			
2				
3	Sample proportion	0.11	z Stat	1.56
4	Sample size	500	P(Z<=z) one-tail	0.059
5	Hypothesized proportion	0.09	z Critical one-tail	1.282
6	Alpha	0.10	P(Z<=z) two-tail	0.118
7			z Critical two-tail	1.645

10.73 p/a/m The null and alternative hypotheses are H_0: $\pi \leq 0.50$ and H_1: $\pi > 0.50$.

This solution can be obtained with a pocket calculator and formulas, but we will use the computer.

As shown in the Test Statistics.xls printout for this right-tail test, the p-value (0.079) is not less than the 0.025 level of significance being used to reach a conclusion, so we do not reject the null hypothesis.

The sample proportion is not significantly greater than the 0.50 value we would expect simply by chance.

	A	B	C	D
1	z-Test of a Proportion			
2				
3	Sample proportion	0.55	z Stat	1.41
4	Sample size	200	P(Z<=z) one-tail	0.079
5	Hypothesized proportion	0.50	z Critical one-tail	1.960
6	Alpha	0.025	P(Z<=z) two-tail	0.157
7			z Critical two-tail	2.241

10.75 p/a/m This solution can be obtained with a pocket calculator and formulas, but we will use the computer. As shown in the Estimators.xls printout below, the 90% confidence interval for the population proportion for this auditor is from 0.735 to 0.805. The hypothesized proportion (0.80) is within the interval, so we conclude that the population proportion for this auditor could be 0.80. This is the same conclusion that was reached in exercise 10.74.

	A	B	C	D	E
1	z-Estimate of a Proportion				
2					
3	Sample proportion	0.77	Confidence Interval Estimate		
4	Sample size	400	0.770	plus/minus	0.035
5	Confidence level	0.90	Lower confidence limit		0.735
6			Upper confidence limit		0.805

10.77 p/c/m As shown in the Data Analysis Plus printout, the 90% confidence interval for the population graduation rate for male basketball players from this region is from 0.402 to 0.518. Because the hypothesized proportion (0.41) is within this interval, we conclude that the population proportion for this region could be 0.41. This is the same as the conclusion that was reached in exercise 10.76.

	A	B
1	**z-Estimate: Proportion**	
2		*Status*
3	Sample Proportion	0.460
4	Observations	200
5	LCL	0.402
6	UCL	0.518

10.79 d/p/m The power of a test is the probability that the test will respond correctly by rejecting a false null hypothesis. By calculating the power of the test $(1 - \beta)$ for several assumed values for the population mean and plotting the power versus the population mean, we arrive at the power curve. By looking at the power curve, you can get an idea of how powerful the hypothesis test is for different possible values of the population mean.

10.81 d/p/m Alpha has already been specified as 0.05 so, when the sample size is increased, α will stay the same and β will be decreased for the test.

10.83 p/a/d From exercise 10.32, $\sigma = 0.20$, $n = 15$, $\sigma_{\bar{x}} = 0.05164$, the hypothesis test is $H_0: \mu \geq 3$ versus $H_1: \mu < 3$, and the decision rule is "Reject H_0 if the calculated $z < -1.645$."
First, get the decision rule in terms of $\bar{x}$.
   Sample mean, $\bar{x}$, corresponding to critical $z = -1.645$ is $3 - 1.645(0.05164) = 2.915$
   The decision rule in terms of $\bar{x}$ is: "Reject H_0 if $\bar{x} < 2.915$"
Next, convert the sample mean into a z value using the true mean of $\mu = 2.80$.

$$\text{when } \bar{x} = 2.915, \quad z = \frac{\bar{x} - \mu}{\sigma_{\bar{x}}} = \frac{2.915 - 2.80}{0.05164} = 2.23$$

$\beta = P(z \geq 2.23) = 0.5000 - 0.4871 = 0.0129$
Power of the test $= 1 - \beta = 1 - 0.0129 = 0.9871$

Using the Beta-means.xls workbook that accompanies Data Analysis Plus, we get a comparable result. Refer to the "Left-tail Test" worksheet and enter the requisite information into cells B3:B7. The power of the test is shown in D5.

	A	B	C	D
1	**Type II Error**			
2				
3	**H0: MU**	3.00	**Critical value**	2.92
4	**SIGMA**	0.20	**Prob(Type II error)**	0.0129
5	**Sample size**	15	**Power of the test**	0.9871
6	**ALPHA**	0.05		
7	**H1: MU**	2.80		

10.85 p/a/d From exercise 10.32, $\sigma = 0.20$, $n = 15$, and $\sigma_{\bar{x}} = 0.05164$. From exercise 10.83, the decision rule in terms of $\bar{x}$ is "Reject H_0 if $\bar{x} < 2.915$."

Now, find $1 - \beta$ for each assumed true population mean given.

Given $\mu = 2.80$:

$$\text{when } \bar{x} = 2.915, \quad z = \frac{\bar{x} - \mu}{\sigma_{\bar{x}}} = \frac{2.915 - 2.80}{0.05164} = 2.23$$

$$\text{and } 1 - \beta = 1 - P(z \geq 2.23) = 1 - (0.5000 - 0.4871) = 0.9871$$

Given $\mu = 2.85$:

$$\text{when } \bar{x} = 2.915, \quad z = \frac{\bar{x} - \mu}{\sigma_{\bar{x}}} = \frac{2.915 - 2.85}{0.05164} = 1.26$$

$$\text{and } 1 - \beta = 1 - P(z \geq 1.26) = 1 - (0.5000 - 0.3962) = 0.8962$$

Given $\mu = 2.90$:

$$\text{when } \bar{x} = 2.915, \quad z = \frac{\bar{x} - \mu}{\sigma_{\bar{x}}} = \frac{2.915 - 2.90}{0.05164} = 0.29$$

$$\text{and } 1 - \beta = 1 - P(z \geq 0.29) = 1 - (0.5000 - 0.1141) = 0.6141$$

Given $\mu = 2.95$:

$$\text{when } \bar{x} = 2.915, \quad z = \frac{\bar{x} - \mu}{\sigma_{\bar{x}}} = \frac{2.915 - 2.95}{0.05164} = -0.68$$

$$\text{and } 1 - \beta = 1 - P(z \geq -0.68) = 1 - (0.5000 + 0.2517) = 0.2483$$

Given $\mu = 3.00$:

$$\text{when } \bar{x} = 2.915, \quad z = \frac{\bar{x} - \mu}{\sigma_{\bar{x}}} = \frac{2.915 - 3.00}{0.05164} = -1.65$$

$$\text{and } 1 - \beta = 1 - P(z \geq -1.65) = 1 - (0.5000 + 0.4505) = 0.0495$$

Using Minitab, the following results are obtained. The "Difference" column refers to the difference between the assumed actual μ and the value of μ in the null hypothesis.

For example, the Difference = -0.20 row shows the power of the test when the assumed actual mean is 2.80 hours compared to the hypothesized mean of 3.00 hours.

Power and Sample Size

1-Sample Z Test

Testing mean = null (versus < null)
Calculating power for mean = null + difference
Alpha = 0.05 Sigma = 0.2

Difference	Sample Size	Power
-0.20	15	0.9871
-0.15	15	0.8961
-0.10	15	0.6147
-0.05	15	0.2493
0.00	15	0.0500

Using Excel to chart the power curve.

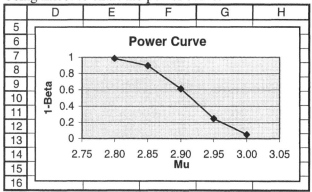

10.87 p/a/d H_0: $\pi \le 0.02$ versus H_1: $\pi > 0.02$, with $\alpha = 0.01$, n = 400, and the standard error of p calculated as:

$$\sigma_p = \sqrt{\frac{0.02(1-0.02)}{400}} = 0.007$$

a. z = 2.33 (Look up 0.5000 - 0.0100 = 0.4900 in the body of the standard normal table.)
b. Getting the decision rule in terms of p: p = 0.02 + 2.33(0.007) = 0.036
    and the decision rule is: "Reject H_0 if p > 0.036."
c. β = P(fail to reject a false H_0)

Given $\pi = 0.02$:

$\quad$ when p = 0.036, $z = \dfrac{p-\pi}{\sigma_p} = \dfrac{0.036-0.02}{0.007} = 2.29$

$\quad$ and β = P(z $\le$ 2.29) = 0.5000 + 0.4890 = 0.9890

Given $\pi = 0.03$:

$\quad$ when p = 0.036, $z = \dfrac{p-\pi}{\sigma_p} = \dfrac{0.036-0.03}{0.007} = 0.86$

$\quad$ and β = P(z $\le$ 0.86) = 0.5000 + 0.3051 = 0.8051

Given $\pi = 0.04$:

$\quad$ when p = 0.036, $z = \dfrac{p-\pi}{\sigma_p} = \dfrac{0.036-0.04}{0.007} = -0.57$

$\quad$ and β = P(z $\le$ -0.57) = 0.5000 - 0.2157 = 0.2843

Given $\pi = 0.05$:

$\quad$ when p = 0.036, $z = \dfrac{p-\pi}{\sigma_p} = \dfrac{0.036-0.05}{0.007} = -2.00$

$\quad$ and β = P(z $\le$ -2.00) = 0.5000 - 0.4772 = 0.0228

Given $\pi = 0.06$:

$\quad$ when p = 0.036, $z = \dfrac{p-\pi}{\sigma_p} = \dfrac{0.036-0.06}{0.007} = -3.43$

$\quad$ and β = P(z $\le$ -3.43) = 0.5000 - 0.5000 = 0.0000

d. Using the calculations carried out in part c,
   When $\pi = 0.02$, $1 - \beta = 1 - 0.9890 = 0.0110$
   When $\pi = 0.03$, $1 - \beta = 1 - 0.8051 = 0.1949$
   When $\pi = 0.04$, $1 - \beta = 1 - 0.2843 = 0.7157$
   When $\pi = 0.05$, $1 - \beta = 1 - 0.0228 = 0.9772$
   When $\pi = 0.06$, $1 - \beta = 1 - 0.0000 = 1.0000$

Using Minitab, note that the "Alternative Proportion" column refers to the assumed actual π and the entries are in scientific notation -- e.g., "2.00E-02" represents 2.00×10^{-2}, or 0.02. The Minitab results are much more accurate than the ones calculated above, largely due to our rounding in the quantities either leading to the calculation or resulting from it, including p, σ_p, and z.

Power and Sample Size

Test for One Proportion

Testing proportion = 0.02 (versus > 0.02)
Alpha = 0.01

Alternative Proportion	Sample Size	Power
2.00E-02	400	0.0100
3.00E-02	400	0.2306
4.00E-02	400	0.6477
5.00E-02	400	0.8959
6.00E-02	400	0.9771

Using Excel to chart the power curve.

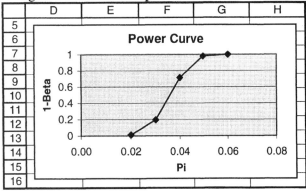

The operating characteristic curve.

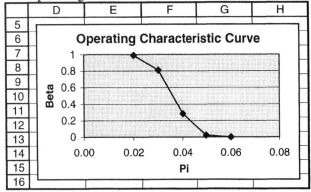

10.89 p/a/m From exercise 10.86,

When $\pi = 0.02$, $\beta = 0.9463$ When $\pi = 0.05$, $\beta = 0.0179$
When $\pi = 0.03$, $\beta = 0.6443$ When $\pi = 0.06$, $\beta = 0.0000$
When $\pi = 0.04$, $\beta = 0.1922$ When $\pi = 0.07$, $\beta = 0.0000$

The operating characteristic curve is shown below:

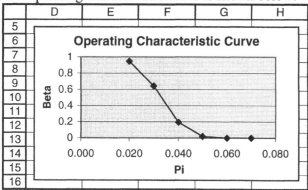

CHAPTER EXERCISES

10.91 d/p/m H_0: The employee has not taken drugs, and H_1: The employee has taken drugs
A Type I error will occur if we decide the employee has taken drugs but he really has not taken drugs.
A Type II error will occur if we decide the employee has not taken drugs but he really has taken drugs.

10.93 p/a/m Null and alternative hypotheses:
H_0: $\mu \le 45.4$ (mean productivity has not increased) and H_1: $\mu > 45.4$ (productivity has increased)
Level of significance: $\alpha = 0.01$
Test results: $\bar{x} = 47.5$, n = 30 (known: $\sigma = 4.5$)

Calculated value of test statistic: $z = \dfrac{\bar{x} - \mu_0}{\sigma_{\bar{x}}} = \dfrac{47.5 - 45.4}{4.5 / \sqrt{30}} = 2.56$

Critical value: z = 2.33 (look up 0.5000 - 0.0100 = 0.4900 in the standard normal table)
Decision rule: Reject H_0 if the calculated z > 2.33, otherwise do not reject.
Conclusion: Calculated test statistic falls in rejection region, reject H_0.
Decision: At the 0.01 level, it appears the efficiency expert's efforts have been successful, since the mean productivity is now significantly more than 45.4 units per hour.
Given the information in this exercise, we could also use Excel worksheet template tmztest.xls to obtain a solution. The template is on the disk supplied with the text, and you need only enter the values for n, $\bar{x}$, σ, and the hypothesized value for μ. The Excel printout is shown below, and the right-tail portion of the p-value section is in bold type for emphasis. The p-value for the test is 0.0053.

	A	B	C	D	E
1	1-Sample Z-Test, Known Sigma				
2					
3	Sample Summary and Assumed Values:		Calculated Values:		
4	Observed Sample Mean	47.5000	Std. Error	0.8216	
5	Sample Size	30	z =	2.5560	
6	Hypothesized Pop. Mean	45.4000	p-Value If the Test Is:		
7	Assumed Pop. Std. Deviation	4.5000	Left-Tail	Two-Tail	**Right-Tail**
8			0.9947	0.0106	**0.0053**

10.95 p/a/m

a. Null and alternative hypotheses:

H_0: μ = 178,600 (The average life insurance in this city is the same as the national average.)

H_1: $\mu \neq$ 178,600 (The average life insurance in this city is not the same as the national average.)

Level of significance: $\alpha = 0.05$

Test results: $\overline{x}$ = 188,800, s = 25,500, n = 30

Calculated value of test statistic: $t = \dfrac{\overline{x} - \mu_0}{s_{\overline{x}}} = \dfrac{188,800 - 178,600}{25,500 / \sqrt{30}} = 2.191$

Critical values: t = -2.045 and t = 2.045   For this test, $\alpha = 0.05$ and d.f. = (n - 1) = (30 - 1) = 29. Referring to the 0.05/2 = 0.025 column and the 29th row of the t table, the critical values are t = -2.045 and t = 2.045.

Decision rule: Reject H_0 if the calculated t < -2.045 or > 2.045, otherwise do not reject.

Conclusion: Calculated test statistic falls in rejection region, reject H_0.

Decision: At the 0.05 level, the average amount of life insurance in this city appears to be different from the national average of $178,600.

b. The 95% confidence interval for μ is: $\overline{x} \pm t \dfrac{s}{\sqrt{n}} = 188,800 \pm 2.045 \dfrac{25,500}{\sqrt{30}} = 188,800 \pm 9520.79$,

or from $179,279.21 to $198,320.79. Since $178,600 is not in the 95% confidence interval for μ found above, reject H_0. This is the same conclusion that was reached in part a.

Given the information in this exercise, we can also use the Test Statistics.xls and Estimators.xls workbooks that accompany Data Analysis Plus on the disk supplied with the text. The t-test result and the 95% confidence interval are shown in the printouts below. The p-value for this two-tail test is 0.037. The lower and upper limits of the 95% confidence interval are $179,278 and $198,322. Because they are not dependent on the printed t table and its inherent gaps between listed values, these confidence limits are more accurate than the ones calculated above.

	A	B	C	D
1	t-Test of a Mean			
2				
3	Sample mean	188800	t Stat	2.19
4	Sample standard deviation	25500.00	P(T<=t) one-tail	0.018
5	Sample size	30	t Critical one-tail	1.699
6	Hypothesized mean	178600	P(T<=t) two-tail	0.037
7	Alpha	0.05	t Critical two-tail	2.045

	A	B	C	D	E
1	t-Estimate of a Mean				
2					
3	Sample mean	188800	Confidence Interval Estimate		
4	Sample standard deviation	25500	188800.0000	plus/minus	9521.862
5	Sample size	30	Lower confidence limit		179278
6	Confidence level	0.95	Upper confidence limit		198322

10.97 p/a/d Null and alternative hypotheses:

H_0: $\mu \geq 356$ (Mr. Jones is not too lenient with audits)

H_1: $\mu < 356$ (Mr. Jones is too lenient with audits)

Level of significance: $\alpha = 0.05$

Test results: $\bar{x} = 322$, $s = 90$, $n = 30$

Calculated value of test statistic: $\quad t = \dfrac{\bar{x} - \mu_0}{s_{\bar{x}}} = \dfrac{322 - 356}{90 / \sqrt{30}} = -2.069$

Critical value: $t = -1.699$ $\quad$ For this test, $\alpha = 0.05$ and d.f. = (n - 1) = (30 - 1) = 29. Referring to the 0.05 column and the 29th row of the t table, the critical value is t = -1.699.

Decision rule: Reject H_0 if the calculated t < -1.699, otherwise do not reject.

Conclusion: Calculated test statistic falls in rejection region, reject H_0.

Decision: At the 0.05 level, the suspicions regarding Mr. Jones appear to be justified. The average amount of extra taxes collected by Mr. Jones appears to be less than \$356.

Using the Test Statistics.xls workbook that accompanies Data Analysis Plus, the corresponding printout is shown below. For this left-tail test, the p-value is 0.024.

	A	B	C	D
1	t-Test of a Mean			
2				
3	Sample mean	322	t Stat	-2.07
4	Sample standard deviation	90.00	P(T<=t) one-tail	0.024
5	Sample size	30	t Critical one-tail	1.699
6	Hypothesized mean	356	P(T<=t) two-tail	0.048
7	Alpha	0.05	t Critical two-tail	2.045

10.99 p/a/m Null and alternative hypotheses:

H_0: $\pi \leq 0.032$ (This region does not have more of a burglary problem than the nation.)

H_1: $\pi > 0.032$ (This region does have more of a burglary problem than the nation.)

Level of significance: $\alpha = 0.05$

Test results: $p = 18/300 = 0.06$, $n = 300$

Calculated value of test statistic: $\quad z = \dfrac{p - \pi_0}{\sigma_p} = \dfrac{0.06 - 0.032}{\sqrt{0.032(1 - 0.032)/300}} = 2.76$

Critical value: $z = 1.645$

Decision rule: Reject H_0 if the calculated z > 1.645, otherwise do not reject.

Conclusion: Calculated test statistic falls in rejection region, reject H_0.

Decision: At the 0.05 level, this region should be considered as having a burglary problem greater than the nation as a whole, since the percentage of households burglarized in this region is significantly larger than 3.2%. Using the standard normal table, p-value = P(z > 2.76) = 0.5000 - 0.4971 = 0.0029. From the p-value perspective, we reject H_0 since p-value = 0.0029 is less than $\alpha = 0.05$.

Using the Test Statistics.xls workbook that accompanies Data Analysis Plus, the corresponding printout is shown below. For this right-tail test, the p-value is listed as 0.0029.

	A	B	C	D
1	z-Test of a Proportion			
2				
3	Sample proportion	0.06	z Stat	2.76
4	Sample size	300	P(Z<=z) one-tail	0.0029
5	Hypothesized proportion	0.032	z Critical one-tail	1.6449
6	Alpha	0.05	P(Z<=z) two-tail	0.0059
7			z Critical two-tail	1.9600

10.101 p/a/m Null and alternative hypotheses:

H_0: $\pi = 0.30$ (the administrator's statement is correct) and H_1: $\pi \neq 0.30$ (statement is not correct)

Level of significance: $\alpha = 0.05$

Test results: $p = 0.35$, $n = 400$

Calculated value of test statistic: $z = \dfrac{p - \pi_0}{\sigma_p} = \dfrac{0.35 - 0.30}{\sqrt{0.30(1 - 0.30)/400}} = 2.18$

Critical values: $z = -1.96$ and $z = 1.96$

Decision rule: Reject H_0 if the calculated $z < -1.96$ or > 1.96, otherwise do not reject.

Conclusion: Calculated test statistic falls in rejection region, reject H_0.

Decision: At the 0.05 level, the administrator's statement does not appear to be correct. Based on these results, the true proportion of emergency room patients that are not really in need of emergency treatment is not 0.30.

Using the Test Statistics.xls workbook that accompanies Data Analysis Plus, the corresponding printout is shown below. For this two-tail test, the p-value is listed as 0.029.

	A	B	C	D
1	z-Test of a Proportion			
2				
3	Sample proportion	0.35	z Stat	2.18
4	Sample size	400	P(Z<=z) one-tail	0.015
5	Hypothesized proportion	0.30	z Critical one-tail	1.645
6	Alpha	0.05	P(Z<=z) two-tail	0.029
7			z Critical two-tail	1.960

10.103 p/a/m Null and alternative hypotheses:

H_0: $\pi \leq 0.10$ (exterminator's claim is correct) and H_1: $\pi > 0.10$ (claim is not correct)

Level of significance: $\alpha = 0.05$

Test results: $p = 14/100 = 0.14$, $n = 100$

Calculated value of test statistic: $z = \dfrac{p - \pi_0}{\sigma_p} = \dfrac{0.14 - 0.10}{\sqrt{0.10(1 - 0.10)/100}} = 1.33$

Critical value: $z = 1.645$

Decision rule: Reject H_0 if the calculated $z > 1.645$, otherwise do not reject.

Conclusion: Calculated test statistic falls in nonrejection region, do not reject H_0.

Decision: At the 0.05 level, we have no reason to doubt the exterminator's claim. The proportion of homes the exterminator treats that have termite problems within one year after treatment is not significantly larger than 0.10.

Using the Test Statistics.xls workbook that accompanies Data Analysis Plus, the corresponding printout is shown below. For this right-tail test, the p-value is listed as 0.091.

	A	B	C	D
1	z-Test of a Proportion			
2				
3	Sample proportion	0.14	z Stat	1.33
4	Sample size	100	P(Z<=z) one-tail	0.091
5	Hypothesized proportion	0.10	z Critical one-tail	1.645
6	Alpha	0.05	P(Z<=z) two-tail	0.182
7			z Critical two-tail	1.960

10.105 p/a/d Null and alternative hypotheses: H_0: $\pi \geq 0.75$ and H_1: $\pi < 0.75$

The standard error of p can be calculated as: $\sigma_p = \sqrt{\dfrac{0.75(1-0.75)}{40}} = 0.0685$

We must first express the decision rule "Reject H_0 if z < -1.645" in terms of p:

The sample proportion, p corresponding to z = -1.645 is p = 0.75 - 1.645(0.0685) = 0.637

The decision rule in terms of p will be: "Reject H_0 if p < 0.637."

a. Let $\pi = 0.75$:

when p = 0.637, $\quad z = \dfrac{p - \pi}{\sigma_p} = \dfrac{0.637 - 0.75}{0.0685} = -1.65$

and $\beta = P(z \geq -1.65) = 0.5000 + 0.4505 = 0.9505$

b. Let $\pi = 0.70$:

when p = 0.637, $\quad z = \dfrac{p - \pi}{\sigma_p} = \dfrac{0.637 - 0.70}{0.0685} = -0.92$

and $\beta = P(z \geq -0.92) = 0.5000 + 0.3212 = 0.8212$

c. Let $\pi = 0.65$:

when p = 0.637, $\quad z = \dfrac{p - \pi}{\sigma_p} = \dfrac{0.637 - 0.65}{0.0685} = -0.19$

and $\beta = P(z \geq -0.19) = 0.5000 + 0.0753 = 0.5753$

d. Let $\pi = 0.60$:

when p = 0.637, $\quad z = \dfrac{p - \pi}{\sigma_p} = \dfrac{0.637 - 0.60}{0.0685} = 0.54$

and $\beta = P(z \geq 0.54) = 0.5000 - 0.2054 = 0.2946$

e. Let $\pi = 0.55$:

when p = 0.637, $\quad z = \dfrac{p - \pi}{\sigma_p} = \dfrac{0.637 - 0.55}{0.0685} = 1.27$

and $\beta = P(z \geq 1.27) = 0.5000 - 0.3980 = 0.1020$

f. When $\pi = 0.75$, $1 - \beta = 1 - 0.9505 = 0.0495$ When $\pi = 0.70$, $1 - \beta = 1 - 0.8212 = 0.1788$
When $\pi = 0.65$, $1 - \beta = 1 - 0.5753 = 0.4247$ When $\pi = 0.60$, $1 - \beta = 1 - 0.2946 = 0.7054$
When $\pi = 0.55$, $1 - \beta = 1 - 0.1020 = 0.8980$

Using Minitab, note that the "Alternative Proportion" column refers to the assumed actual π. The Minitab results are much more accurate than the ones calculated above, largely due to our rounding in the quantities either leading to the calculation or resulting from it, including p, σ_p, and z.

Power and Sample Size

Test for One Proportion

Testing proportion = 0.75 (versus < 0.75)
Alpha = 0.05

Alternative Proportion	Sample Size	Power
0.750000	40	0.0500
0.700000	40	0.1937
0.650000	40	0.4336
0.600000	40	0.6853
0.550000	40	0.8667

Using Excel to chart the power curve. This plot graphs the power of the test = $1 - \beta$ = probability that the hypothesis test will correctly reject a false null hypothesis against the assumed value of π.

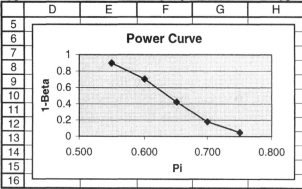

10.107 p/a/d

a. Shop-Mart should consider switching to Phipps bulbs. If the Phipps bulbs were really no better than the G&E, they would have had only a 0.012 probability of having this great an advantage in our tests just by chance.

b. G&E might like to use the 0.005 level of significance in reaching a conclusion. Since the p-value is not less than 0.005, using the $\alpha = 0.005$ level would lead to the conclusion that the Phipps advantage in the test could have been merely due to chance.

c. Phipps might like to use the 0.02 level of significance in reaching a conclusion. Since the p-value is less than 0.02, using the $\alpha = 0.02$ level would lead to the conclusion that the Phipps advantage in the test was not merely due to chance, and that the Phipps bulbs really are better.

d. If the test had been two-tail instead of one-tail, the p-value would have been 0.024. We would have had to consider two tail areas instead of just one, and the mirror-image area on the other side would also have been 0.012. In this case, the p-value would have been 2(0.012) = 0.024.

10.109 p/c/m The null and alternative hypotheses are H_0: $\pi = 0.27$ and H_1: $\pi \neq 0.27$.
Printout results are shown below for Data Analysis Plus. Of the 400 crimes in the sample, 33.0% involved a weapon. In this two-tail test, the p-value is 0.007, which is less than the 0.05 level of significance being used to reach a conclusion. We reject the null hypothesis and conclude that this city's experience is different from the nation as a whole in terms of the percent of crimes that involve a weapon. If the city were really the same as the rest of the nation, there would be only a 0.007 probability of obtaining a sample proportion this far away from 0.27.

	A	B	C	D
1	z-Test: Proportion			
2				
3				*Weapon*
4	Sample Proportion			0.33
5	Observations			400
6	Hypothesized Proportion			0.27
7	z Stat			2.703
8	P(Z<=z) one-tail			0.003
9	z Critical one-tail			1.645
10	P(Z<=z) two-tail			0.007
11	z Critical two-tail			1.96

10.111 p/c/m The null and alternative hypotheses are $H_0: \mu \geq 3.5$ and $H_1: \mu < 3.5$.
Printout results are shown for Data Analysis Plus and Minitab. In this left-tail test, the sample mean of 3.293 ounces is less than the hypothesized mean (3.5000), but the p-value (0.059) is not less than the level of significance used to reach a conclusion (0.01), so we do not reject the null hypothesis. The new procedure has not significantly reduced the average amount of aluminum trimmed and recycled. At this level of significance, a sample mean this small could have happened by chance.

	A	B	C	D
1	t-Test: Mean			
2				
3				Ounces
4	Mean			3.293
5	Standard Deviation			0.764
6	Hypothesized Mean			3.5
7	df			34
8	t Stat			-1.605
9	P(T<=t) one-tail			0.059
10	t Critical one-tail			2.441
11	P(T<=t) two-tail			0.118
12	t Critical two-tail			2.728

```
One-Sample T: Ounces

Test of mu = 3.5 vs mu < 3.5

Variable          N        Mean     StDev    SE Mean
Ounces           35       3.293     0.764      0.129

Variable     95.0% Upper Bound        T        P
Ounces                     3.511    -1.61    0.059
```

10.113 p/c/m The null and alternative hypotheses are $H_0: \mu \leq 12,000$ hours and $H_1: \mu > 12,000$ hours
Printout results are shown below for Data Analysis Plus and Minitab. In this right-tail test, the sample mean of 12,070.38 hours exceeds the hypothesized mean (12,000) and the p-value is 0.282.
Since the p-value is not less than the level of significance specified (0.025), we do not reject the null hypothesis. The new injection pumps may be no better than the ones already in use.

	A	B	C	D
1	t-Test: Mean			
2				
3				Hours
4	Mean			12070.38
5	Standard Deviation			856.20
6	Hypothesized Mean			12000
7	df			49
8	t Stat			0.581
9	P(T<=t) one-tail			0.282
10	t Critical one-tail			2.010
11	P(T<=t) two-tail			0.564
12	t Critical two-tail			2.312

```
One-Sample T: Hours

Test of mu = 12000 vs mu > 12000

Variable          N        Mean     StDev    SE Mean
Hours            50       12070      856        121

Variable     95.0% Lower Bound        T        P
Hours                     11867     0.58    0.282
```

CHAPTER 11
HYPOTHESIS TESTS INVOLVING
TWO SAMPLE MEANS OR PROPORTIONS

SECTION EXERCISES

11.1 d/p/m The t distribution is often associated with small sample tests. However, it is appropriate when the population standard deviations are unknown, regardless of how large or small the samples happen to be.

11.3 p/a/m The null and alternative hypotheses are:

$H_0: \mu_1 = \mu_2$, or $\mu_1 - \mu_2 = 0$ (average number of TV hours for seniors and sophomores are equal)

$H_1: \mu_1 \neq \mu_2$, or $\mu_1 - \mu_2 \neq 0$ (average number of TV hours for seniors and sophomores are not equal)

$\overline{x}_1$ and $\overline{x}_2$ are the mean number of TV hours for seniors and sophomores.

The pooled estimate of the common variance, and the test statistic, t, are:

$$s_p^2 = \frac{(32-1)(1.2)^2 + (30-1)(1.4)^2}{32+30-2} = 1.691 \quad \text{and} \quad t = \frac{(3.9-3.5)-0}{\sqrt{1.691(\frac{1}{32}+\frac{1}{30})}} = 1.210$$

We will use the $\alpha = 0.05$ level of significance in carrying out the test. For a right-tail area of $0.05/2 = 0.025$ and d.f. = $(n_1 + n_2 - 2) = (32 + 30 - 2) = 60$, the critical values are t = -2.000 and t = 2.000. The decision rule is "Reject H_0 if the calculated t < -2.000 or > 2.000, otherwise do not reject." Since the calculated test statistic falls in the nonrejection region, do not reject H_0.

At the 0.05 level, it appears that the average number of TV hours per week could be equal for high school seniors and sophomores. The observed difference between the groups' means is judged to have been due to chance. Using the t table with d.f. = 60, $P(t \geq 1.210)$ is greater than 0.10. Therefore, p-value is greater than 0.20.

Given the summary data, we can also carry out this pooled-variances t-test using the Test Statistics.xls workbook that accompanies Data Analysis Plus. It is on the disk that comes with the textbook. This procedure requires that we first square the two sample standard deviations (1.2 and 1.4) to obtain the sample variances (1.44 and 1.96). The results are shown below. For this two-tail test, the p-value (0.2309) is not less than the 0.05 level of significance being used to reach a conclusion, so the null hypothesis is not rejected. For a true null hypothesis, there is a 0.2309 probability that this much of a difference between the sample means could occur by chance.

	A	B	C	D	E
1	t-Test of the Difference Between Two Means (Equal-Variances)				
2					
3		Sample 1	Sample 2	t Stat	1.210
4	Mean	3.9	3.5	P(T<=t) one-tail	0.1155
5	Variance	1.44	1.96	t Critical one-tail	1.6706
6	Observations	32	30	P(T<=t) two-tail	0.2309
7	Pooled Variance estimate	1.691		t Critical two-tail	2.0003
8	Hypothesized difference	0			
9	Alpha	0.05			

11.5 p/a/m The null and alternative hypotheses are:

H_0: $\mu_1 \geq \mu_2$, or $\mu_1 - \mu_2 \geq 0$ (the experimental booklet is not better than the standard one)

H_1: $\mu_1 < \mu_2$, or $\mu_1 - \mu_2 < 0$ (the experimental booklet is better than the standard one)

$\bar{x}_1$ and $\bar{x}_2$ are the mean exam scores for the standard version and the one claimed to be superior. The pooled estimate of the common variance, and the test statistic, t, are:

$$s_p^2 = \frac{(13-1)(9.3)^2 + (13-1)(10.1)^2}{13+13-2} = 94.25 \quad \text{and} \quad t = \frac{(72.0 - 80.2) - 0}{\sqrt{94.25(\frac{1}{13} + \frac{1}{13})}} = -2.153$$

We will use the $\alpha = 0.05$ level of significance. For a left-tail area of 0.05 and d.f. = $(n_1 + n_2 - 2) = (13 + 13 - 2) = 24$, the critical value is t = -1.711. The decision rule is: "Reject H_0 if the calculated t < -1.711, otherwise do not reject." Since the calculated test statistic falls in the rejection region, reject H_0. At the 0.05 level, the experimental booklet appears to be better than the standard version.

Given the summary data, we can also carry out this pooled-variances t-test using the Test Statistics.xls workbook that accompanies Data Analysis Plus. It is on the disk that comes with the textbook.
This procedure requires that we first square the two sample standard deviations (9.3 and 10.1) to obtain the sample variances (86.49 and 102.01). For this left-tail test, the p-value (0.0208) is less than the 0.05 level of significance being used to reach a conclusion, so the null hypothesis is rejected. For a true null hypothesis, there is only a 0.0208 probability that the difference between the sample means (72.0 - 80.2) would be this much less than 0 just by chance.

	A	B	C	D	E
1	t-Test of the Difference Between Two Means (Equal-Variances)				
2					
3		Sample 1	Sample 2	t Stat	-2.153
4	Mean	72.0	80.2	P(T<=t) one-tail	0.0208
5	Variance	86.49	102.01	t Critical one-tail	1.7109
6	Observations	13	13	P(T<=t) two-tail	0.0415
7	Pooled Variance estimate	94.250		t Critical two-tail	2.0639
8	Hypothesized difference	0			
9	Alpha	0.05			

11.7 p/a/m The null and alternative hypotheses are:

H_0: $\mu_1 = \mu_2$, or $\mu_1 - \mu_2 = 0$ (the average dexterity scores are equal for the 2 shifts)

H_1: $\mu_1 \neq \mu_2$, or $\mu_1 - \mu_2 \neq 0$ (the average dexterity scores are not equal for the 2 shifts)

$\bar{x}_1$ and $\bar{x}_2$ are the average dexterity scores for workers on the day and night shifts. The pooled estimate of the common variance, and the test statistic, t, are:

$$s_p^2 = \frac{(37-1)(12.3)^2 + (42-1)(8.4)^2}{37+42-2} = 108.304 \quad \text{and} \quad t = \frac{(73.1 - 77.3) - 0}{\sqrt{108.304(\frac{1}{37} + \frac{1}{42})}} = -1.790$$

We will use the $\alpha = 0.05$ level of significance. For a right-tail area of 0.05/2 = 0.025 and d.f. = $(n_1 + n_2 - 2) = (37 + 42 - 2) = 77$, the critical values are t = -1.991 and t = 1.991. The decision rule is: "Reject H_0 if the calculated t < -1.991 or > 1.991, otherwise do not reject." Since the calculated test statistic falls in the nonrejection region, do not reject H_0. At the 0.05 level, it appears that the average dexterity scores for the two shifts are equal. The observed difference between the shifts' means is judged to have been due to chance. Using the t table with d.f. = 77, P(t ≤ -1.790) is between 0.025 and 0.05. Therefore, p-value = 2P(t ≤ -1.790) is between 0.05 and 0.10.

Given the summary data, we can also carry out this pooled-variances t-test using the Test Statistics.xls workbook that accompanies Data Analysis Plus. This procedure requires that we first square the two sample standard deviations (12.3 and 8.4) to obtain the sample variances (151.29 and 70.56).

The Estimators.xls workbook that accompanies Data Analysis Plus has been used to obtain the 95% confidence interval for $(\mu_1 - \mu_2)$. For this two-tail test, the p-value (0.0774) is not less than the 0.05 level of significance being used to reach a conclusion, so H_0 is not rejected. For a true H_0, there is a 0.0774 probability that this much of a difference between the sample means could occur by chance.

The 95% confidence interval for $(\mu_1 - \mu_2)$ is from -8.872 to 0.472. This interval includes 0, so we conclude that $(\mu_1 - \mu_2)$ could be 0. The conclusion using the 95% confidence interval is consistent with the results of the two-tail hypothesis test at the 0.05 level of significance

	A	B	C	D	E
1	t-Test of the Difference Between Two Means (Equal-Variances)				
2					
3		Sample 1	Sample 2	t Stat	-1.790
4	Mean	73.1	77.3	P(T<=t) one-tail	0.0387
5	Variance	151.29	70.56	t Critical one-tail	1.6649
6	Observations	37	42	P(T<=t) two-tail	0.0774
7	Pooled Variance estimate	108.304		t Critical two-tail	1.9913
8	Hypothesized difference	0			
9	Alpha	0.05			

	A	B	C	D	E	F
1	t-Estimate of the Difference Between Two Means (Equal-Variances)					
2						
3		Sample 1	Sample 2	Confidence Interval Estimate		
4	Mean	73.1	77.3	-4.200	plus/minus	4.672
5	Variance	151.29	70.56	Lower confidence limit		-8.872
6	Sample size	37	42	Upper confidence limit		0.472
7	Pooled Variance	108.304				
8	Confidence level	0.95				

11.9 p/a/m The null and alternative hypotheses are:

H_0: $\mu_1 = \mu_2$, or $\mu_1 - \mu_2 = 0$ (the mean times for business/financial news are the same)
H_1: $\mu_1 \neq \mu_2$, or $\mu_1 - \mu_2 \neq 0$ (the mean times are not the same for the two channels.)
$\bar{x}_1$ and $\bar{x}_2$ are the average times for channel 2 and channel 4, respectively.

After computing the mean and standard deviation for each sample, the pooled estimate of the common variance and the test statistic, t, are:

$$s_p^2 = \frac{(10-1)(0.743)^2 + (10-1)(0.846)^2}{10+10-2} = 0.796 \quad \text{and} \quad t = \frac{(3.81-4.51)-0}{\sqrt{0.796(\frac{1}{10}+\frac{1}{10})}} = -1.97$$

We will use the $\alpha = 0.10$ level of significance. For a right-tail area of 0.10/2 = 0.05 and d.f. = $(n_1 + n_2 - 2) = (10 + 10 - 2) = 18$, the critical values are t = -1.734 and t = 1.734. The decision rule is: "Reject H_0 if the calculated t < -1.734 or > 1.734, otherwise do not reject." Since the calculated test statistic falls in the rejection region, we reject H_0. At the 0.10 level, it appears that the mean times devoted to business/financial news are not equal for the two channels. Using the t table with d.f. = 18, $P(t \leq -1.97)$ is between 0.05 and 0.025. Therefore, p-value = $2P(t \leq -1.97)$ is between 0.10 and 0.05. These results can also be obtained using Minitab. The 90% confidence interval for $(\mu_1 - \mu_2)$ is shown as (-1.318 to -0.082), an interval that does not include a difference of zero. This is consistent with the results of the two-tail hypothesis test.

```
Two-Sample T-Test and CI: chan2, chan4

Two-sample T for chan2 vs chan4
          N     Mean    StDev   SE Mean
chan2    10    3.810    0.743     0.23
chan4    10    4.510    0.846     0.27


Difference = mu chan2 - mu chan4
Estimate for difference:  -0.700
90% CI for difference: (-1.318, -0.082)
T-Test of difference = 0 (vs not =): T-Value = -1.97   P-Value = 0.065   DF = 18
Both use Pooled StDev = 0.796
```

11.11 p/a/m The null and alternative hypotheses are:

H_0: $\mu_1 \geq \mu_2$, or $\mu_1 - \mu_2 \geq 0$ (FoodFarm's peanut butter does not contain less fat)

H_1: $\mu_1 < \mu_2$, or $\mu_1 - \mu_2 < 0$ (FoodFarm's peanut butter does contain less fat)

$\overline{x}_1$ and $\overline{x}_2$ are the mean fat contents for FoodFarm's and a major competitor's peanut butter. The pooled estimate of the common variance, and the test statistic, t, are:

$$s_p^2 = \frac{(11-1)(2.1)^2 + (11-1)(1.8)^2}{11+11-2} = 3.825 \quad \text{and} \quad t = \frac{(31.3-33.2)-0}{\sqrt{3.825(\frac{1}{11}+\frac{1}{11})}} = -2.278$$

We will use the $\alpha = 0.05$ level of significance in carrying out the test. For a left-tail area of 0.05 and d.f. = $(n_1 + n_2 - 2) = (11+ 11 - 2) = 20$, the critical value is t = -1.725. The decision rule is "Reject H_0 if the calculated t < -1.725, otherwise do not reject." Since the calculated test statistic falls in the rejection region, reject H_0. At the 0.05 level, it appears that FoodFarm's claim their peanut butter contains less fat on the average could be valid. Using the t table with d.f. = 20, p-value = $P(t \leq -2.278)$ is between 0.01 and 0.025. Based on the summary statistics, Excel worksheet template tmt2pool.xls, on the disk that accompanies the text, was used in generating the following printout corresponding to this solution:

	A	B	C	D	E
1	2-Sample t-Test Comparing Means, Pooled Variances (using summary stats)				
2	Hypothesized Difference (mu1 - mu2)	0.000			
3			Calculated Values:		
4	Summary of Sample Data:		pooled var =	3.825	
5	Mean for Sample 1, xbar1	31.300	d.f. =	20	
6	Std. Dev. For Sample 1, s1	2.100	Std. Error =	0.834	
7	Size of Sample 1, n1	11	t =	-2.278	
8	Mean for Sample 2, xbar2	33.200	p-Value If the Test Is:		
9	Std. Dev. For Sample 2, s2	1.800	Left-Tail	Two-Tail	Right-Tail
10	Size of Sample 2, n2	11	0.0169	0.0338	0.9831

11.13 p/c/m The null and alternative hypotheses are:

H_0: $\mu_1 \leq \mu_2$, or $\mu_1 - \mu_2 \leq 0$ and H_1: $\mu_1 > \mu_2$, or $\mu_1 - \mu_2 > 0$

The Minitab printout is shown below. For this right-tail test, the p-value (to three decimal places, 0.000) is less than the 0.05 level of significance being used to reach a conclusion, so H_0 is rejected. We conclude that the photograph works as intended. For a true H_0, the probability is virtually zero that the sample mean for the photo group would be this much greater than that for the no-photo group just by chance.

```
Two-Sample T-Test and CI: Photo, NoPhoto

Two-sample T for Photo vs NoPhoto

            N      Mean    StDev   SE Mean
Photo     105     17.60     5.71      0.56
NoPhoto   105     11.00     5.81      0.57


Difference = mu Photo - mu NoPhoto
Estimate for difference:  6.600
95% lower bound for difference: 5.286
T-Test of difference = 0 (vs >): T-Value = 8.30   P-Value = 0.000   DF = 208
Both use Pooled StDev = 5.76
```

11.15 p/c/m As shown in the solution to exercise 11.14, the 90% confidence interval for $(\mu_1 - \mu_2)$ is from 0.639 to 3.961. This interval does not include 0, so we conclude that $(\mu_1 - \mu_2)$ could not be 0. The conclusion using the 90% confidence interval is consistent with the result of the two-tail hypothesis test at the 0.10 level of significance.

11.17 p/c/m As shown in the solution to exercise 11.16, the 95% confidence interval for $(\mu_1 - \mu_2)$ is from -3.33 to 31.20. This interval includes 0, so we conclude that $(\mu_1 - \mu_2)$ could be 0. The conclusion using the 95% confidence interval is consistent with the result of the two-tail hypothesis test at the 0.05 level.

11.19 c/a/m The null and alternative hypotheses are:

H_0: $\mu_1 \geq \mu_2$, or $\mu_1 - \mu_2 \geq 0$ and H_1: $\mu_1 < \mu_2$, or $\mu_1 - \mu_2 < 0$

Assuming $\sigma_1 \neq \sigma_2$, we will use the unequal-variances t-test and the computer. Hand calculations for the t statistic and d.f. can be carried out as shown below.

$$t = \frac{(\bar{x}_1 - \bar{x}_2) - (\mu_1 - \mu_2)_0}{\sqrt{\dfrac{s_1^2}{n_1} + \dfrac{s_2^2}{n_2}}} = \frac{(165.0 - 172.9) - 0}{\sqrt{\dfrac{21.5^2}{40} + \dfrac{31.3^2}{32}}} = -1.217$$

$$d.f. = \frac{[s_1^2/n_1 + s_2^2/n_2]^2}{\dfrac{[s_1^2/n_1]^2}{(n_1 - 1)} + \dfrac{[s_2^2/n_2]^2}{(n_2 - 1)}} = \frac{[21.5^2/40 + 31.3^2/32]^2}{\dfrac{[21.5^2/40]^2}{(40-1)} + \dfrac{[31.3^2/32]^2}{(32-1)}} = 52.84$$

We will use the Test Statistics.xls workbook that accompanies Data Analysis Plus. This procedure requires that we first square the two sample standard deviations (21.5 and 31.3) to obtain the sample variances (462.25 and 979.69). The p-value for this left-tail test (0.1146) is not less than the 0.10 level of significance used to reach a conclusion, so H_0 is not rejected. We conclude that the mean for population one could be at least as large as that for population two. For a true H_0, the probability is 0.1146 that the mean for sample one would be this much less than that for sample two just by chance.

	A	B	C	D	E
1	t-Test of the Difference Between Two Means (Unequal-Variances)				
2					
3		Sample 1	Sample 2		
4	Mean	165	172.9	t Stat	-1.217
5	Variance	462.25	979.69	P(T<=t) one-tail	0.1146
6	Sample size	40	32	t Critical one-tail	1.2980
7	Degrees of freedom	52.84		P(T<=t) two-tail	0.2293
8	Hypothesized difference	0		t Critical two-tail	1.6747
9	Alpha	0.10			

11.21 p/a/m The null and alternative hypotheses are:

H_0: $\mu_1 = \mu_2$, or $\mu_1 - \mu_2 = 0$ and H_1: $\mu_1 \neq \mu_2$, or $\mu_1 - \mu_2 \neq 0$

Assuming $\sigma_1 \neq \sigma_2$, we will use the unequal-variances t-test and the computer.

Hand calculations for the t statistic and d.f. can be carried out as shown below.

$$t = \frac{(\overline{x}_1 - \overline{x}_2) - (\mu_1 - \mu_2)_0}{\sqrt{\dfrac{s_1^2}{n_1} + \dfrac{s_2^2}{n_2}}} = \frac{(20.4 - 133.0) - 0}{\sqrt{\dfrac{20.4^2}{30} + \dfrac{38.7^2}{36}}} = -1.141$$

$$d.f. = \frac{[s_1^2/n_1 + s_2^2/n_2]^2}{\dfrac{[s_1^2/n_1]^2}{(n_1-1)} + \dfrac{[s_2^2/n_2]^2}{(n_2-1)}} = \frac{[20.4^2/30 + 38.7^2/36]^2}{\dfrac{[20.4^2/30]^2}{(30-1)} + \dfrac{[38.7^2/36]^2}{(36-1)}} = 54.87$$

With df = 54.87 (rounded to df = 55 for reference to our t table) and a two-tail test at the 0.05 level of significance, the critical values of t are t = -2.004 and t = +2.004. The calculated t = -1.141 lies within these critical values and we are not able to reject the null hypothesis of equal population means.

Using the t table, we see that the calculated t = -1.141 is even within the critical values of t for a two-tail test at the 0.20 level of significance (-1.297 and +1.297), so p-value is > 0.20.

Using the tmt2uneq.xls Excel worksheet template on the disk accompanying the text, the p-value for this two-tail test (0.2588) is not less than the 0.05 level of significance used to reach a conclusion, so H_0 is not rejected. We conclude that the population means could be equal. The 95% confidence interval for the difference between the population means is shown below as (-23.43, 6.43). The interval includes a difference of 0, and this is consistent with the findings of our two-tail test at the 0.05 level.

	A	B	C	D	E	F
1	2-Sample t-Test Comparing Means, Unequal Variances (based on summary statistics)					
2						
3	Hypothesized Difference (mu1 - mu2)	0.000				
4						
5	Summary of Sample Data:		Calculated Values:			
6	Mean for Sample 1, xbar1	124.5	d.f. =	54.87		
7	Std. Dev. For Sample 1, s1	20.4	Std. Error =	7.44812		
8	Size of Sample 1, n1	30	t =	-1.141		
9	Mean for Sample 2, xbar2	133.0	p-value if the test is:			
10	Std. Dev. For Sample 2, s2	38.7	Left-Tail	**Two-Tail**	Right-Tail	
11	Size of Sample 2, n2	36	0.1294	**0.2588**	0.8706	
12						
13	Confidence level desired	0.95				
14	Lower confidence limit	-23.43				
15	Upper confidence limit	6.43				

11.23 p/a/m The null and alternative hypotheses are:

H_0: $\mu_1 = \mu_2$, or $\mu_1 - \mu_2 = 0$ and H_1: $\mu_1 \neq \mu_2$, or $\mu_1 - \mu_2 \neq 0$

Assuming $\sigma_1 \neq \sigma_2$, we use the unequal-variances t-test and the computer. Hand calculations for the t statistic and d.f. can be carried out as shown below.

$$t = \frac{(\bar{x}_1 - \bar{x}_2) - (\mu_1 - \mu_2)_0}{\sqrt{\dfrac{s_1^2}{n_1} + \dfrac{s_2^2}{n_2}}} = \frac{(375 - 425) - 0}{\sqrt{\dfrac{75^2}{20} + \dfrac{143^2}{25}}} = -1.508$$

$$d.f. = \frac{[s_1^2 / n_1 + s_2^2 / n_2]^2}{\dfrac{[s_1^2 / n_1]^2}{(n_1 - 1)} + \dfrac{[s_2^2 / n_2]^2}{(n_2 - 1)}} = \frac{[75^2 / 20 + 143^2 / 25]^2}{\dfrac{[75^2 / 20]^2}{(20 - 1)} + \dfrac{[143^2 / 25]^2}{(25 - 1)}} = 37.71$$

With df = 37.71 (rounded to df = 38 for reference to our t table) and a two-tail test at the 0.05 level of significance, the critical values of t are t = -2.024 and t = +2.024. The calculated t = -1.508 lies within these critical values and we are not able to reject the null hypothesis of equal population means. Using the t table, the most accurate statement we could make about the p-value for the test is that it is between 0.10 and 0.20.

Given the summary data, we can also carry out this unequal-variances t-test using the Test Statistics.xls workbook that accompanies Data Analysis Plus. This procedure requires that we first square the two sample standard deviations (75 and 143) to obtain the sample variances (5625 and 20,449).

The Estimators.xls workbook that accompanies Data Analysis Plus has been used to obtain the 95% confidence interval for $(\mu_1 - \mu_2)$. For this two-tail test, the p-value (0.1400) is not less than the 0.05 level of significance being used to reach a conclusion, so H_0 is not rejected. For a true H_0, there is a 0.1400 probability that this much of a difference between the sample means could occur by chance. The 95% confidence interval for $(\mu_1 - \mu_2)$ is from -\$117.18 to \$17.18. This interval includes 0, so we conclude that $(\mu_1 - \mu_2)$ could be 0. The conclusion using the 95% confidence interval is consistent with the result of the two-tail hypothesis test at the 0.05 level of significance.

	A	B	C	D	E
1	t-Test of the Difference Between Two Means (Unequal-Variances)				
2					
3		Sample 1	Sample 2		
4	Mean	375	425	t Stat	-1.508
5	Variance	5625	20449	P(T<=t) one-tail	0.0700
6	Sample size	20	25	t Critical one-tail	1.6871
7	Degrees of freedom	37.71		P(T<=t) two-tail	0.1400
8	Hypothesized difference	0		t Critical two-tail	2.0262
9	Alpha	0.05			

	A	B	C	D	E	F
1	t-Estimate of the Difference Between Two Means (Unequal-Variances)					
2						
3		Sample 1	Sample 2	Confidence Interval Estimate		
4	Mean	375	425.0	-50.00	plus/minus	67.18
5	Variance	5625	20449	Lower confidence limit		-117.18
6	Sample size	20	25	Upper confidence limit		17.18
7	Degrees of freedom	37.71				
8	Confidence level	0.95				

11.25 p/a/d The null and alternative hypotheses are:

$H_0: \mu_1 - \mu_2 = -2.0$ and $H_1: \mu_1 - \mu_2 \neq -2.0$

Assuming $\sigma_1 \neq \sigma_2$, we will carry out this unequal-variances t-test using the Test Statistics.xls workbook that accompanies Data Analysis Plus. This procedure requires that we first square the two sample standard deviations (0.6 and 0.9) to obtain the sample variances (0.36 and 0.81). The Estimators.xls workbook that accompanies Data Analysis Plus has been used to obtain the 95% confidence interval for $(\mu_1 - \mu_2)$. Note that the hypothesized difference in this exercise is *not* zero. For this two-tail test, the p-value (0.0748) is not less than the 0.05 level of significance being used to reach a conclusion, so H_0 is not rejected. For a true H_0, there is a 0.0748 probability that the difference between the sample means could have been this far away from -0.2 visits per year by chance. The 95% confidence interval for $(\mu_1 - \mu_2)$ is from -2.03 to -1.37 visits per year. This interval includes -2.0, the hypothesized difference, so we conclude that $(\mu_1 - \mu_2)$ could be -2.0 visits per year. The conclusion using the 95% confidence interval is consistent with the result of the two-tail hypothesis test at the 0.05 level of significance.

	A	B	C	D	E
1	t-Test of the Difference Between Two Means (Unequal-Variances)				
2					
3		Sample 1	Sample 2		
4	Mean	2.2	3.9	t Stat	1.811
5	Variance	0.36	0.81	P(T<=t) one-tail	0.0374
6	Sample size	50	40	t Critical one-tail	1.6686
7	Degrees of freedom	65.11		P(T<=t) two-tail	0.0748
8	Hypothesized difference	-2		t Critical two-tail	1.9971
9	Alpha	0.05			

	A	B	C	D	E	F
1	t-Estimate of the Difference Between Two Means (Unequal-Variances)					
2						
3		Sample 1	Sample 2	Confidence Interval Estimate		
4	Mean	2.2	3.9	-1.70	plus/minus	0.33
5	Variance	0.36	0.81	Lower confidence limit		-2.03
6	Sample size	50	40	Upper confidence limit		-1.37
7	Degrees of freedom	65.11				
8	Confidence level	0.95				

11.27 p/c/m The null and alternative hypotheses are:

$H_0: \mu_1 \leq \mu_2$, or $\mu_1 - \mu_2 \leq 0$ and $H_1: \mu_1 > \mu_2$, or $\mu_1 - \mu_2 > 0$

Assuming $\sigma_1 \neq \sigma_2$, we use the unequal-variances t-test and Minitab. The printout is shown below. The p-value for this right-tail test is, to three decimal places, 0.000. This is less than the 0.01 level of significance used to reach a conclusion, so H_0 is rejected. We conclude that the mean time spent shopping by women is greater than that by men. For a true H_0, the probability is virtually zero that the sample mean for women would be this much greater than that for men just by chance.

```
Two-Sample T-Test and CI: women, men

Two-sample T for women vs men

          N      Mean    StDev   SE Mean
women    15     19.10     5.36       1.4
men      14     12.70     3.28      0.88

Difference = mu women - mu men
Estimate for difference:  6.40
95% lower bound for difference: 3.59
T-Test of difference = 0 (vs >): T-Value = 3.91  P-Value = 0.000  DF = 23
```

11.29 p/c/m The null and alternative hypotheses are:

H_0: $\mu_1 \leq \mu_2$, or $\mu_1 - \mu_2 \leq 0$ and H_1: $\mu_1 > \mu_2$, or $\mu_1 - \mu_2 > 0$

Assuming $\sigma_1 \neq \sigma_2$, we use the unequal-variances t-test and Minitab. The printout is shown below. The p-value for this right-tail test is 0.095. This is not less than the 0.05 level of significance used to reach a conclusion, so H_0 is not rejected. We conclude that the mean spending amount for Japanese visitors might be no higher than for British visitors. For a true H_0, the probability is 0.095 that the sample mean for Japanese visitors would be this much higher than that for British visitors just by chance.

```
Two-Sample T-Test and CI: Japanese, British

Two-sample T for Japanese vs British

            N    Mean    StDev   SE Mean
Japanese   31    1953     615       110
British    34    1783     378        65

Difference = mu Japanese - mu British
Estimate for difference:  170
95% lower bound for difference: -45
T-Test of difference = 0 (vs >): T-Value = 1.33   P-Value = 0.095   DF = 48
```

11.31 d/p/e When both sample sizes are greater than or equal to 30, it is appropriate to use the z-test as an approximation to the unequal-variances t-test when comparing the means of two independent samples.

11.33 c/a/m The null and alternative hypotheses are:

H_0: $\mu_1 \leq \mu_2$, or $\mu_1 - \mu_2 \leq 0$ and H_1: $\mu_1 > \mu_2$, or $\mu_1 - \mu_2 > 0$

The test statistic is: $z = \dfrac{(\overline{x}_1 - \overline{x}_2) - (\mu_1 - \mu_2)_0}{\sqrt{\dfrac{s_1^2}{n_1} + \dfrac{s_2^2}{n_2}}} = \dfrac{(85.2 - 81.7) - 0}{\sqrt{\dfrac{9.6^2}{40} + \dfrac{4.1^2}{32}}} = 2.08$

For a right-tail test at the 0.05 significance level, the critical value is $z = 1.645$. The decision rule is: "Reject H_0 if the calculated $z > 1.645$, otherwise do not reject." Since the calculated test statistic falls in the rejection region, reject H_0. At the 0.05 level, we can conclude that μ_1 is greater than μ_2.

11.35 p/a/m The null and alternative hypotheses are:

H_0: $\mu_1 = \mu_2$, or $\mu_1 - \mu_2 = 0$ and H_1: $\mu_1 \neq \mu_2$, or $\mu_1 - \mu_2 \neq 0$

Using the tmz2test.xls Excel worksheet template on the disk accompanying the text, the p-value for this two-tail test (0.2538) is not less than the 0.05 level of significance used to reach a conclusion, so H_0 is not rejected. We conclude that the population means could be equal. The 95% confidence interval for the difference between the population means is shown below as (-23.10, 6.10). The interval includes a difference of 0, and this is consistent with the findings of our two-tail test at the 0.05 level.

	A	B	C	D	E
1	2-Sample Z-Test Comparing Means (based on summary statistics)				
2					
3	Hypothesized Difference (mu1 - mu2)	0.000			
4					
5	Summary of Sample Data:		Calculated Values:		
6	Mean for Sample 1, xbar1	124.5	Std. Error =	7.448	
7	Std. Dev. For Sample 1, s1	20.4	z =	-1.141	
8	Size of Sample 1, n1	30			
9	Mean for Sample 2, xbar2	133.0	p-value if the test is:		
10	Std. Dev. For Sample 2, s2	38.7	Left-Tail	**Two-Tail**	Right-Tail
11	Size of Sample 2, n2	36	0.1269	**0.2538**	0.8731
12					
13	Confidence level desired	0.95			
14	Lower confidence limit	-23.10			
15	Upper confidence limit	6.10			

10.37 p/a/m The null and alternative hypotheses are:

H_0: $\mu_1 = \mu_2$, or $\mu_1 - \mu_2 = 0$ (the average number of cups is the same for both age groups)

H_1: $\mu_1 \neq \mu_2$, or $\mu_1 - \mu_2 \neq 0$ (the average number of cups is not the same for both age groups)

$\bar{x}_1$ and $\bar{x}_2$ are the average cups of coffee per day for the 30-59 and the 60-or-over age groups.

The test statistic is: $z = \dfrac{(\bar{x}_1 - \bar{x}_2) - (\mu_1 - \mu_2)_0}{\sqrt{\dfrac{s_1^2}{n_1} + \dfrac{s_2^2}{n_2}}} = \dfrac{(0.57 - 0.75) - 0}{\sqrt{\dfrac{0.43^2}{200} + \dfrac{0.68^2}{100}}} = -2.416$

For a two-tail test at the 0.05 significance level, the critical values are z = -1.96 and z = 1.96. The decision rule is "Reject H_0 if the calculated z < -1.96 or z > 1.96, otherwise do not reject." Since the calculated test statistic falls in the rejection region, reject H_0. At the 0.05 level, we can conclude the average number of cups of coffee per day for persons in the 30-59 age group is not equal to the amount consumed by those in the 60-or-over group.

Based on the summary statistics, Excel worksheet template tmz2test.xls was used in generating the printout shown below. For this two-tail test, the p-value (0.0157) is less than the 0.05 level of significance being used to reach a conclusion, so H_0 is rejected. For a true H_0, the probability is just 0.0157 that such a large difference between the sample means could occur by chance.

The 95% confidence interval for ($\mu_1 - \mu_2$) is from -0.326 to -0.034 cups per day. This interval does not include 0, so we conclude that ($\mu_1 - \mu_2$) could not be 0. The conclusion using the 95% confidence interval is consistent with the result of the two-tail hypothesis test at the 0.05 level of significance.

	A	B	C	D	E
1	2-Sample Z-Test Comparing Means (based on summary statistics)				
2					
3	Hypothesized Difference (mu1 - mu2)	0.000			
4					
5	Summary of Sample Data:		Calculated Values:		
6	Mean for Sample 1, xbar1	0.57	Std. Error =	0.074	
7	Std. Dev. For Sample 1, s1	0.43	z =	-2.416	
8	Size of Sample 1, n1	200			
9	Mean for Sample 2, xbar2	0.75	p-value If the Test Is:		
10	Std. Dev. For Sample 2, s2	0.68	Left-Tail	**Two-Tail**	Right-Tail
11	Size of Sample 2, n2	100	0.0078	**0.0157**	0.9922
12					
13	Confidence level desired	0.95			
14	Lower confidence limit	-0.326			
15	Upper confidence limit	-0.034			

11.39 p/a/m The null and alternative hypotheses are:

H_0: $\mu_1 \geq \mu_2$, or $\mu_1 - \mu_2 \geq 0$ (average no. of days missed is at least as high for males as for females)

H_1: $\mu_1 < \mu_2$, or $\mu_1 - \mu_2 < 0$ (average no. of days missed is lower for males)

$\bar{x}_1$ and $\bar{x}_2$ are the sample average number of sick days for males and females.

The test statistic is: $z = \dfrac{(\bar{x}_1 - \bar{x}_2) - (\mu_1 - \mu_2)_0}{\sqrt{\dfrac{s_1^2}{n_1} + \dfrac{s_2^2}{n_2}}} = \dfrac{(4.1 - 5.6) - 0}{\sqrt{\dfrac{1.2^2}{400} + \dfrac{1.8^2}{400}}} = -13.868$

For a left-tail test at the 0.01 significance level, the critical value is z = -2.33. The decision rule is: "Reject H_0 if the calculated z < -2.33, otherwise do not reject." Since the calculated test statistic falls in the rejection region, reject H_0. At the 0.01 level, we can conclude the average number of work days lost due to illness or injury is less for males than for females.

Based on the summary statistics, Excel worksheet template tmz2test.xls was used in generating the following printout. For this left-tail test, the p-value (0.0000) is less than the 0.01 level of significance being used to reach a conclusion, so H_0 is rejected. For a true H_0, the probability is virtually zero that such a large negative difference between the sample means could occur by chance.

	A	B	C	D	E
1	2-Sample Z-Test Comparing Means (based on summary statistics)				
2					
3	Hypothesized Difference (mu1 - mu2)	0.000			
4					
5	Summary of Sample Data:		Calculated Values:		
6	Mean for Sample 1, xbar1	4.10	Std. Error =	0.108	
7	Std. Dev. For Sample 1, s1	1.20	z =	-13.868	
8	Size of Sample 1, n1	400			
9	Mean for Sample 2, xbar2	5.60	p-value If the Test Is:		
10	Std. Dev. For Sample 2, s2	1.80	**Left-Tail**	Two-Tail	Right-Tail
11	Size of Sample 2, n2	400	**0.0000**	0.0000	1.0000

11.41 p/c/m The null and alternative hypotheses are:

H_0: $\mu_1 \le \mu_2$, or $\mu_1 - \mu_2 \le 0$ (average waiting time is the same or less for callers to Gateway)

H_1: $\mu_1 > \mu_2$, or $\mu_1 - \mu_2 > 0$ (average waiting time is greater for callers to Gateway)

To apply the z-test, we must first use Excel or Minitab to obtain the summary statistics for the two samples, since Excel and Minitab do not provide the two-sample z-test for means. These summary statistics can then be used with Excel worksheet template tmz2test.xls to produce the results shown below. For this right-tail test, the p-value (0.0300) is less than the 0.05 level of significance being used to reach a conclusion, so H_0 is rejected and we conclude that Gateway's mean technical support hold time is greater than that for IBM. For a true H_0, the probability is 0.0300 that such a large positive difference between the sample means could occur by chance.

	G	H	I	J	K
1	2-Sample Z-Test Comparing Means (based on summary statistics)				
2					
3	Hypothesized Difference (mu1 - mu2)	0.000			
4					
5	Summary of Sample Data:		Calculated Values:		
6	Mean for Sample 1, xbar1	16.50	Std. Error =	1.277	
7	Std. Dev. For Sample 1, s1	4.96	z =	1.880	
8	Size of Sample 1, n1	30			
9	Mean for Sample 2, xbar2	14.10	p-value If the Test Is:		
10	Std. Dev. For Sample 2, s2	4.93	Left-Tail	Two-Tail	**Right-Tail**
11	Size of Sample 2, n2	30	0.9700	0.0601	**0.0300**

The unequal-variances t-test would provide comparable results, as shown in the Minitab printout:

```
Two-Sample T-Test and CI: Gateway, IBM

Two-sample T for Gateway vs IBM
          N      Mean     StDev    SE Mean
Gateway   30     16.50    4.96     0.91
IBM       30     14.10    4.93     0.90

Difference = mu Gateway - mu IBM
Estimate for difference:  2.40
95% lower bound for difference: 0.26
T-Test of difference = 0 (vs >): T-Value = 1.88   P-Value = 0.033   DF = 57
```

125

11.43 p/c/m The null and alternative hypotheses are:

H_0: $\mu_1 \leq \mu_2$, or $\mu_1 - \mu_2 \leq 0$ (the experimental program is no better than the conventional program)

H_1: $\mu_1 > \mu_2$, or $\mu_1 - \mu_2 > 0$ (the experimental program is better)

In order to apply the z-test, we must first use Excel or Minitab to obtain the summary statistics for the two samples. These summary statistics can then be used with Excel worksheet template tmz2test.xls to produce the results shown below. For this right-tail test, the p-value (0.1178) is not less than the 0.025 level of significance being used to reach a conclusion, so H_0 is not rejected and we conclude that the experimental program may be no better than the conventional program. For a true H_0, the probability is 0.1178 that the sample mean for the experimental group could be this much larger than that for the conventional group simply by chance.

	G	H	I	J	K
1	2-Sample Z-Test Comparing Means (based on summary statistics)				
2					
3	Hypothesized Difference (mu1 - mu2)	0.000			
4					
5	Summary of Sample Data:		Calculated Values:		
6	Mean for Sample 1, xbar1	1041.10	Std. Error =	24.135	
7	Std. Dev. For Sample 1, s1	130.30	z =	1.186	
8	Size of Sample 1, n1	40			
9	Mean for Sample 2, xbar2	1012.48	p-value If the Test Is:		
10	Std. Dev. For Sample 2, s2	79.51	Left-Tail	Two-Tail	Right-Tail
11	Size of Sample 2, n2	40	0.8822	0.2357	0.1178

The unequal-variances t-test would provide comparable results, as shown in the Minitab printout:

```
Two-Sample T-Test and CI: NewProg, OldProg

Two-sample T for NewProg vs OldProg
           N      Mean    StDev   SE Mean
NewProg   40      1041      130        21
OldProg   40    1012.5     79.5        13

Difference = mu NewProg - mu OldProg
Estimate for difference:  28.6
95% lower bound for difference: -11.7
T-Test of difference = 0 (vs >): T-Value = 1.19   P-Value = 0.120   DF = 64
```

11.45 d/p/e This example involves dependent samples since all 20 consumers rate the taste of both brands of instant coffee.

11.47 p/a/m First, some preliminary calculations:

Truck number	x_1 = Current filter	x_2 = New filter	Difference, d = x_1 - x_2
1	7.6	7.3	0.3
2	5.1	7.2	-2.1
3	10.4	6.8	3.6
4	6.9	10.6	-3.7
5	5.6	8.8	-3.2
6	7.9	8.7	-0.8
7	5.4	5.7	-0.3
8	5.7	8.7	-3.0
9	5.5	8.9	-3.4
10	5.3	7.1	-1.8

The null and alternative hypotheses:

H_0: $\mu_d \geq 0$ (the new filter is not superior to the current filter)

H_1: $\mu_d < 0$ (the new filter is superior to the current filter)

For each truck, d = x_1 - x_2, with x_1 = mpg of trucks with current filter and x_2 = mpg of trucks with new filter. The sample mean and standard deviation for d are:

$$\bar{d} = \frac{0.3 - 2.1 + 3.6 - 3.7 - 3.2 - 0.8 - 0.3 - 3.0 - 3.4 - 1.8}{10} = -1.44$$

$$s_d = \sqrt{\frac{\sum d_i^2 - n(\overline{d})^2}{n-1}} = \sqrt{\frac{65.92 - 10(-1.44)^2}{10-1}} = 2.241$$

The test statistic is: $t = \dfrac{\overline{d}}{s_d / \sqrt{n}} = \dfrac{-1.44}{2.241/\sqrt{10}} = -2.032$

For a left-tail area of 0.05 and d.f. = (n - 1) = (10 - 1) = 9, the critical value is t = -1.833. The decision rule is: "Reject H_0 if the calculated t < -1.833, otherwise do not reject." Since the calculated test statistic falls in the rejection region, reject H_0. At the 0.05 level, it appears the new filtration system could be superior to the current one. The corresponding Minitab printout is shown below.

```
Paired T-Test and CI: Current, New

Paired T for Current - New
                    N       Mean      StDev    SE Mean
Current            10      6.540      1.686      0.533
New                10      7.980      1.409      0.445
Difference         10     -1.440      2.241      0.709

95% upper bound for mean difference: -0.141
T-Test of mean difference = 0 (vs < 0): T-Value = -2.03  P-Value = 0.036
```

11.49 p/a/m First, some preliminary calculations:

Person number	x_1 = Weight before	x_2 = Weight after	Difference, d = x_1 - x_2
1	198	194	4
2	154	151	3
3	124	126	-2
4	110	104	6
5	127	123	4
6	162	155	7
7	141	129	12
8	180	165	15

The null and alternative hypotheses:

H_0: $\mu_d \leq 0$ (the aerobics class was not effective)

H_1: $\mu_d > 0$ (the aerobics class was effective)

For each person the difference is d = x_1 - x_2, with x_1 = weight before the class and x_2 = weight after the class. The sample mean and standard deviation for d are:

$$\overline{d} = \frac{4+3-2+6+4+7+12+15}{8} = 6.125 \text{ and } s_d = \sqrt{\frac{\sum d_i^2 - n(\overline{d})^2}{n-1}} = \sqrt{\frac{499 - 8(6.125)^2}{8-1}} = 5.330$$

The test statistic is: $t = \dfrac{\overline{d}}{s_d / \sqrt{n}} = \dfrac{6.125}{5.33/\sqrt{8}} = 3.250$

For a right-tail area of 0.05 and d.f. = (n - 1) = (8 - 1) = 7, the critical value is t = 1.895. The decision rule is: "Reject H_0 if the calculated t > 1.895, otherwise do not reject." Since the calculated test statistic falls in the rejection region, reject H_0. At the 0.05 level, we can conclude that the average weight of aerobic students is less after participating in the class. Therefore, the program is effective.

Using the t table with d.f. = 7, p-value = P(t ≥ 3.250) is between 0.005 and 0.01. The corresponding Minitab printout is shown below.

```
Paired T-Test and CI: Before, After

Paired T for Before - After
                    N       Mean      StDev    SE Mean
Before              8      149.5       29.9       10.6
After               8      143.4       28.5       10.1
Difference          8       6.13       5.33       1.88

95% lower bound for mean difference: 2.55
T-Test of mean difference = 0 (vs > 0): T-Value = 3.25  P-Value = 0.007
```

11.51 p/c/m The null and alternative hypotheses:

H_0: $\mu_d = 0$ (there is no difference in worker productivity on the two days)

H_1: $\mu_d \neq 0$ (there is a difference in worker productivity on the two days)

The Minitab printout is shown below. For this two-tail test, the p-value (0.170) is not less than the 0.01 level of significance being used to reach a conclusion, so H_0 is not rejected. There is no reason to believe that worker productivity differs for Monday versus Thursday. For a true H_0, the probability is 0.170 of getting a mean difference in productivity this far from zero just by chance. The 99% confidence interval for μ_d is from -4.663 to 1.774. Since zero is within the interval, we conclude that μ_d could be zero. This is the same conclusion reached in the hypothesis test.

```
Paired T-Test and CI: Monday, Thursday

Paired T for Monday - Thursday
                N      Mean     StDev    SE Mean
Monday          9     54.67     10.26       3.42
Thursday        9     56.11     10.90       3.63
Difference      9    -1.444     2.877      0.959

99% CI for mean difference: (-4.663, 1.774)
T-Test of mean difference = 0 (vs not = 0): T-Value = -1.51   P-Value = 0.170
```

11.53 c/a/m The null and alternative hypotheses are:

H_0: $\pi_1 = \pi_2$, or $\pi_1 - \pi_2 = 0$ (the population proportions are equal)

H_1: $\pi_1 \neq \pi_2$, or $\pi_1 - \pi_2 \neq 0$ (the population proportions are not equal)

Based on the summary statistics, we will use Excel worksheet template tm2ptest.xls to carry out this test. The test is two-tail, and the p-value (0.0286) is less than the 0.05 level of significance being used to reach a conclusion, so H_0 is rejected. For a true H_0, the probability is just 0.0286 that a difference this great between the sample proportions could occur by chance.

	A	B	C	D	E
1	2-Sample Z-Test, Comparing Proportions from Independent Samples				
2	When the Hypothesized Difference is 0				
3			Calculated Values:		
4			pbar =	0.349	
5	Summary of Sample Data:		Std. Error =	0.032	
6	Proportion for Sample 1 (p1)	0.31	z =	-2.189	
7	Size of Sample 1 (n1)	400	p-value if the test is:		
8	Proportion for Sample 2 (p2)	0.38	Left-Tail	**Two-Tail**	Right-Tail
9	Size of Sample 2 (n2)	500	0.0143	**0.0286**	0.9857
10					
11	Confidence level desired	0.95			
12	Lower confidence limit	-0.132			
13	Upper confidence limit	-0.008			

11.55 p/a/m The null and alternative hypotheses are:

H_0: $\pi_1 \geq \pi_2$, or $\pi_1 - \pi_2 \geq 0$ (aspirin is not more effective than the placebo)

H_1: $\pi_1 < \pi_2$, or $\pi_1 - \pi_2 < 0$ (aspirin is more effective than the placebo)

The pooled estimate of the population proportions is:

$$\bar{p} = \frac{n_1 p_1 + n_2 p_2}{n_1 + n_2} = \frac{(3429)(0.043) + (1710)(0.046)}{3429 + 1710} = 0.044$$

The test statistic is: $z = \dfrac{p_1 - p_2}{\sqrt{\bar{p}(1-\bar{p})(\frac{1}{n_1} + \frac{1}{n_2})}} = \dfrac{0.043 - 0.046}{\sqrt{0.044(1-0.044)(\frac{1}{3429} + \frac{1}{1710})}} = -0.49$

For a left-tail test at the 0.01 significance level, the critical value is z = -2.33. The decision rule is "Reject H_0 if the calculated z < -2.33, otherwise do not reject." Since the calculated test statistic falls in the nonrejection region, do not reject H_0. At the 0.01 level, we conclude that the observed difference in the two sample proportions was due to chance. From this test, we are not able to conclude that aspirin is effective in reducing the incidence of heart attacks. Note that the numerator in our z statistic is $p_1 - p_2$, instead of $(p_1 - p_2) - 0$. This is consistent with the "case 1" scenario in which the hypothesized difference between the population proportions is always zero.

Based on the summary information in this exercise, the Test Statistics.xls workbook provided with Data Analysis Plus is used in generating the printout below. (Note the use of Case 1, applicable when the hypothesized value of the difference is zero.)

	A	B	C	D	E
1	z-Test of the Difference Between Two Proportions (Case 1)				
2					
3		Sample 1	Sample 2	z Stat	-0.494
4	Sample proportion	0.043	0.046	P(Z<=z) one-tail	0.3106
5	Sample size	3429	1710	z Critical one-tail	2.3263
6	Alpha	0.01		P(Z<=z) two-tail	0.6212
7				z Critical two-tail	2.5758

11.57 p/a/m The null and alternative hypotheses are:

H_0: $\pi_1 = \pi_2$, or $\pi_1 - \pi_2 = 0$ (the proportions are equal for the two years)

H_1: $\pi_1 \neq \pi_2$, or $\pi_1 - \pi_2 \neq 0$ (the proportions are not equal for the two years)

The pooled estimate of the population proportions is:

$$\bar{p} = \frac{n_1 p_1 + n_2 p_2}{n_1 + n_2} = \frac{(213)(0.38) + (219)(0.438)}{213 + 219} = 0.409$$

The test statistic is: $z = \dfrac{p_1 - p_2}{\sqrt{\bar{p}(1-\bar{p})(\frac{1}{n_1} + \frac{1}{n_2})}} = \dfrac{0.380 - 0.438}{\sqrt{0.409(1-0.409)(\frac{1}{213} + \frac{1}{219})}} = -1.226$

For a two-tail test at the 0.05 significance level, the critical values are z = -1.96 and z = 1.96. The decision rule is: "Reject H_0 if the calculated z < -1.96 or > 1.96, otherwise do not reject." Since the calculated test statistic falls in the nonrejection region, do not reject H_0. At the 0.05 level, it appears the difference in the two proportions could have been the result of chance variation from one year to the next. Based on the summary information in this exercise, the Test Statistics.xls workbook provided with Data Analysis Plus was used in generating the printout below. (Note the use of Case 1, applicable when the hypothesized value of the difference is zero.)

	A	B	C	D	E
1	z-Test of the Difference Between Two Proportions (Case 1)				
2					
3		Sample 1	Sample 2	z Stat	-1.226
4	Sample proportion	0.380	0.438	P(Z<=z) one-tail	0.1102
5	Sample size	213	219	z Critical one-tail	1.6449
6	Alpha	0.05		P(Z<=z) two-tail	0.2203
7				z Critical two-tail	1.9600

11.59 p/a/m The null and alternative hypotheses are:

H_0: $\pi_1 = \pi_2$, or $\pi_1 - \pi_2 = 0$ (the proportions of multiple transactions are the same)

H_1: $\pi_1 \neq \pi_2$, or $\pi_1 - \pi_2 \neq 0$ (the proportions of multiple transactions are not the same)

The pooled estimate of the population proportions is:

$$\bar{p} = \frac{n_1 p_1 + n_2 p_2}{n_1 + n_2} = \frac{(300)(0.42) + (250)(0.50)}{300 + 250} = 0.456$$

The test statistic is: $z = \dfrac{p_1 - p_2}{\sqrt{\bar{p}(1-\bar{p})(\dfrac{1}{n_1} + \dfrac{1}{n_2})}} = \dfrac{0.42 - 0.50}{\sqrt{0.456(1-0.456)(\dfrac{1}{300} + \dfrac{1}{250})}} = -1.88$

For a two-tail test at the 0.10 significance level, the critical values are z = -1.645 and z = 1.645. The decision rule is: "Reject H_0 if the calculated z < -1.645 or > 1.645, otherwise do not reject." Since the calculated test statistic falls in the rejection region, reject H_0. At the 0.10 level, we can conclude that males and females differ in terms of making multiple transactions. Using the standard normal table, p-value = 2P(z ≤ -1.88) = 2(0.5000 - 0.4699) = 0.0602.

Based on the summary information in this exercise, the Test Statistics.xls and Estimators.xls workbooks provided with Data Analysis Plus were used in generating the printouts below. In this two-tail test, the p-value (0.0607) is less than 0.10, the level of significance used in reaching a conclusion, so we reject H_0 and conclude that the population proportions are not equal. For a true H_0, there would be only a 0.0607 probability of obtaining this great a difference between the sample proportions by chance. The 90% confidence interval for ($\pi_1 - \pi_2$) is from -0.150 to -0.010. Since zero is not within the interval, we conclude that ($\pi_1 - \pi_2$) is some value other than zero. This is the same conclusion reached in the hypothesis test.

	A	B	C	D	E
1	z-Test of the Difference Between Two Proportions (Case 1)				
2					
3		Sample 1	Sample 2	z Stat	-1.876
4	Sample proportion	0.42	0.50	P(Z<=z) one-tail	0.0304
5	Sample size	300	250	z Critical one-tail	1.2816
6	Alpha	0.10		P(Z<=z) two-tail	0.0607
7				z Critical two-tail	1.6449

	A	B	C	D	E	F
1	z-Estimate of the Difference Between Two Proportions					
2						
3		Sample 1	Sample 2	Confidence Interval Estimate		
4	Sample proportion	0.42	0.50	-0.08	plus/minus	0.070
5	Sample size	300	250	Lower confidence limit		-0.150
6	Confidence level	0.9		Upper confidence limit		-0.010

11.61 p/a/m The null and alternative hypotheses are:

H_0: $\pi_1 \leq \pi_2$, or $\pi_1 - \pi_2 \leq 0$ (the illness rate is no higher for soldiers who were near the release)

H_1: $\pi_1 > \pi_2$, or $\pi_1 - \pi_2 > 0$ (the illness rate is higher for soldiers who were near the release)

The pooled estimate of the population proportions is:

$$\overline{p} = \frac{n_1 p_1 + n_2 p_2}{n_1 + n_2} = \frac{(81)(0.309) + (52,000)(0.235)}{81 + 52,000} = 0.235$$

The test statistic is: $z = \dfrac{p_1 - p_2}{\sqrt{\overline{p}(1-\overline{p})(\frac{1}{n_1} + \frac{1}{n_2})}} = \dfrac{0.309 - 0.235}{\sqrt{0.235(1-0.235)(\frac{1}{81} + \frac{1}{52,000})}} = 1.569$

For a right-tail test at the 0.10 significance level, the critical value is z = 1.28. The decision rule is: "Reject H_0 if the calculated z > 1.28, otherwise do not reject." Since the calculated test statistic falls in the rejection region, reject H_0. At the 0.10 level, the soldiers in the vicinity of the nerve gas release do have a significantly higher rate of illness.

Based on the summary information in this exercise, the Test Statistics.xls workbook provided with Data Analysis Plus was used in generating the printout below. In this right-tail test, the p-value (0.0583) is less than 0.10, the level of significance used in reaching a conclusion, so we reject H_0. For a true H_0, there is a 0.0583 probability that p_1 would be this much larger than p_2 by chance.

	A	B	C	D	E
1	z-Test of the Difference Between Two Proportions (Case 1)				
2					
3		Sample 1	Sample 2	z Stat	1.569
4	Sample proportion	0.309	0.235	P(Z<=z) one-tail	0.0583
5	Sample size	81	52000	z Critical one-tail	1.2816
6	Alpha	0.10		P(Z<=z) two-tail	0.1166
7				z Critical two-tail	1.6449

11.63 p/c/m The null and alternative hypotheses are:

H_0: $\pi_1 = \pi_2$, or $\pi_1 - \pi_2 = 0$ (the proportions having a business home page are the same)

H_1: $\pi_1 \neq \pi_2$, or $\pi_1 - \pi_2 \neq 0$ (the proportions are not the same)

Using Data Analysis Plus, we obtain the printouts below. In this two-tail test, the p-value (0.077) is not less than 0.01, the level of significance used in reaching a conclusion, so we do not reject H_0. The population proportions could be equal. For a true H_0, there would be a 0.077 probability of obtaining this great a difference between the sample proportions by chance. The 99% confidence interval for $(\pi_1 - \pi_2)$ is from -0.032 to 0.172. Since zero is within the interval, we conclude that $(\pi_1 - \pi_2)$ could be zero. This is the same conclusion reached in the hypothesis test.

	A	B	C	D
1	z-Test: Two Proportions			
2				
3			Female	Male
4	Sample Proportions		0.23	0.16
5	Observations		200	200
6	Hypothesized Difference		0	
7	z Stat		1.767	
8	P(Z<=z) one tail		0.039	
9	z Critical one-tail		2.326	
10	P(Z<=z) two-tail		0.077	
11	z Critical two-tail		2.576	

	A	B	C	D
1	z-Estimate: Two Proportions			
2				
3			Female	Male
4	Sample Proportions		0.23	0.16
5	Observations		200	200
6				
7	LCL		-0.032	
8	UCL		0.172	

11.65 p/c/m The null and alternative hypotheses are:

$H_0: \pi_1 = \pi_2$, or $\pi_1 - \pi_2 = 0$ (the population proportions for females 25-34 are the same)

$H_1: \pi_1 \neq \pi_2$, or $\pi_1 - \pi_2 \neq 0$ (the proportions are not the same)

Using Data Analysis Plus, we obtain the printouts below. In this two-tail test, the p-value (0.1866) is not less than 0.10, the level of significance used in reaching a conclusion, so we do not reject H_0.

The population proportions could be equal. For a true H_0, there is a 0.1866 probability of obtaining this great a difference between the sample proportions by chance. The 90% confidence interval for $(\pi_1 - \pi_2)$ is from -0.055 to 0.005. Since zero is within the interval, we conclude that $(\pi_1 - \pi_2)$ could be zero. This is the same conclusion reached in the hypothesis test.

	A	B	C	D
1	z-Test: Two Proportions			
2				
3			sail	horse
4	Sample Proportions		0.122	0.147
5	Observations		500	1000
6	Hypothesized Difference		0	
7	z Stat		-1.321	
8	P(Z<=z) one tail		0.0933	
9	z Critical one-tail		1.282	
10	P(Z<=z) two-tail		0.1866	
11	z Critical two-tail		1.645	

	A	B	C	D
1	z-Estimate: Two Proportions			
2				
3			sail	horse
4	Sample Proportions		0.122	0.147
5	Observations		500	1000
6				
7	LCL		-0.055	
8	UCL		0.005	

11.67 c/a/d The null and alternative hypotheses are:

$$H_0 : \sigma_1^2 = \sigma_2^2 \quad \text{or} \quad \frac{\sigma_1^2}{\sigma_2^2} = 1 \quad \text{and} \quad H_1 : \sigma_1^2 \neq \sigma_2^2 \quad \text{or} \quad \frac{\sigma_1^2}{\sigma_2^2} \neq 1$$

Since $s_1 < s_2$, the test statistic is: $F = \dfrac{s_2^2}{s_1^2} = \dfrac{133.5^2}{103.1^2} = 1.68$

For this test, $\alpha = 0.10$, $v_1 = (n_2 - 1) = 40$, and $v_2 = (n_1 - 1) = 27$. Referring to the F table with an upper-tail area of $0.10/2 = 0.05$, with $v_1 = 40$ and $v_2 = 27$, the critical value is $F = 1.84$. The decision rule is "Reject H_0 if the calculated $F > 1.84$, otherwise do not reject." Since the calculated test statistic falls in the nonrejection region, do not reject H_0. At the 0.10 level, we cannot conclude that the population variances are different. Since we cannot reject H_0 at the 0.10 level of significance, we also will not be able to reject H_0 at the 0.05 and 0.02 levels. The critical value will increase as the level of significance decreases. (Note: when $\alpha = 0.05$ the critical value is $F = 2.07$ and when $\alpha = 0.02$ the critical value is $F = 2.38$.) Using the Test Statistics.xls workbook that accompanies Data Analysis Plus, we obtain the printout below. Although it isn't really necessary when using this workbook's F-test based on summary data, we have been consistent with the procedure described in the text by designating the sample with the larger variance as sample 1. For this two-tail test, the p-value is shown as 0.161.

	A	B	C	D	E
1	F-Test of the Ratio of Two Variances				
2					
3		Sample 1	Sample 2	F Stat	1.677
4	Sample variance	17822.25	10629.61	P(F<=f) one-tail	0.080
5	Sample size	41	28	f Critical one-tail	1.603
6	Alpha	0.10		P(F<=f) two-tail	0.161
7				f Critical two-tail	0.566

11.69 p/a/d For the data in exercise 11.11, $s_1^2 = 4.41$, $n_1 = 11$, $s_2^2 = 3.24$, and $n_2 = 11$. A pocket calculator and formulas could be used for this exercise, but we will use the computer. The null and alternative hypotheses are:

$$H_0 : \sigma_1^2 = \sigma_2^2 \quad \text{or} \quad \frac{\sigma_1^2}{\sigma_2^2} = 1 \quad \text{and} \quad H_1 : \sigma_1^2 \neq \sigma_2^2 \quad \text{or} \quad \frac{\sigma_1^2}{\sigma_2^2} \neq 1$$

Since $s_1 > s_2$, the test statistic is $F = \dfrac{s_1^2}{s_2^2} = \dfrac{4.41}{3.24} = 1.361$, as in the first of the printouts below. The printouts were generated using the Test Statistics.xls workbook that accompanies Data Analysis Plus. For the two-tail test of whether the variances in exercise 11.11 could be equal, the p-value is 0.6351, as shown in cell E6 in the first printout below. This is not less than the 0.05 level of significance being used to reach a conclusion, so H_0 is not rejected. We were justified in assuming the population variances were equal.

If the standard deviation for the FoodFarm jars had been 3.0 grams, the test result would have been as shown in the second of the printouts below and the two-tail p-value would have been 0.1226. Again, the p-value would not be less than the 0.05 level of significance and we would not reject H_0. The conclusion has not changed.

	A	B	C	D	E
1	F-Test of the Ratio of Two Variances				
2					
3		Sample 1	Sample 2	F Stat	1.361
4	Sample variance	4.41	3.24	P(F<=f) one-tail	0.3176
5	Sample size	11	11	f Critical one-tail	2.9782
6	Alpha	0.05		P(F<=f) two-tail	0.6351
7				f Critical two-tail	0.2690

	A	B	C	D	E
1	F-Test of the Ratio of Two Variances				
2					
3		Sample 1	Sample 2	F Stat	2.778
4	Sample variance	9.00	3.24	P(F<=f) one-tail	0.0613
5	Sample size	11	11	f Critical one-tail	2.9782
6	Alpha	0.05		P(F<=f) two-tail	0.1226
7				f Critical two-tail	0.2690

11.71 p/a/d A pocket calculator and formulas could be used for this exercise, but we will use the computer. The null and alternative hypotheses are:

$$H_0 : \sigma_1^2 \geq \sigma_2^2 \quad \text{or} \quad \frac{\sigma_1^2}{\sigma_2^2} \geq 1 \quad \text{and} \quad H_1 : \sigma_1^2 < \sigma_2^2 \quad \text{or} \quad \frac{\sigma_1^2}{\sigma_2^2} < 1$$

Note: Because $s_2 > s_1$, if we were using a pocket calculator and the procedure described in the text, the test statistic would be calculated as $F = \dfrac{s_2^2}{s_1^2} = \dfrac{0.49}{0.16} = 3.0625$. However, when we use the F-test in the Test Statistics.xls workbook that accompanies Data Analysis Plus, there is no such need to designate the sample with the greater variance as sample 1. As a result, we will maintain the "sample 1" and "sample 2" designations as they are described in the exercise and the test statistic will be the inverse of the one shown above, or $1/3.0625 = 0.3265$. This value is automatically calculated and displayed in the printout.

For this one-tail test, the p-value is 0.004. This is less than the 0.025 level of significance being used to reach a conclusion, so H_0 is rejected. We conclude that vendor 2 (sample 2 in the printout) has a higher population variance (and, hence, standard deviation) of shipping times than its more-consistent competitor.

	A	B	C	D	E
1	F-Test of the Ratio of Two Variances				
2					
3		Sample 1	Sample 2	F Stat	0.3265
4	Sample variance	0.16	0.49	P(F<=f) one-tail	0.004
5	Sample size	24	30	f Critical one-tail	2.167
6	Alpha	0.025		P(F<=f) two-tail	0.008
7				f Critical two-tail	0.394

11.73 p/c/d The null and alternative hypotheses are:

$$H_0 : \sigma_1^2 = \sigma_2^2 \quad \text{or} \quad \frac{\sigma_1^2}{\sigma_2^2} = 1 \quad \text{and} \quad H_1 : \sigma_1^2 \neq \sigma_2^2 \quad \text{or} \quad \frac{\sigma_1^2}{\sigma_2^2} \neq 1$$

In using Excel in this two-tail test at the 0.05 level, we will specify the sample with the larger variance as sample number 1 (left column). In these data, the sample variance was higher for the subscribers to internet provider B. Although we are performing a two-tail test at the 0.05 level, specify 0.025 as the level of significance -- this is because Excel always assumes we are doing a one-tail test.

The Excel printout and a portion of its Minitab counterpart are shown below. The p-value for this two-tail test is $2(0.009) = 0.018$, which is less than the $\alpha = 0.05$ level of significance on which we are basing our conclusion. At the 0.05 level, we reject H_0 and conclude that these two internet providers do not have the same variance in the ages of their respective customers.

	E	F	G
1	F-Test Two-Sample for Variances		
2			
3		*providerB*	*providerA*
4	Mean	39.600	34.620
5	Variance	69.633	35.138
6	Observations	50	50
7	df	49	49
8	F	1.982	
9	P(F<=f) one-tail	0.009	
10	F Critical one-tail	1.762	

```
Test for Equal Variances

F-Test (normal distribution)

Test Statistic: 1.982
P-Value      : 0.018
```

CHAPTER EXERCISES

11.75 p/a/m The null and alternative hypotheses are:

$H_0: \mu_1 \leq \mu_2$, or $\mu_1 - \mu_2 \leq 0$ (Emily's suspicion is not warranted)

$H_1: \mu_1 > \mu_2$, or $\mu_1 - \mu_2 > 0$ (Emily's suspicion is warranted)

$\bar{x}_1$ and $\bar{x}_2$ are the average repair estimates given to women and men.
The pooled estimate of the common variance, and the test statistic, t, are:

$$s_p^2 = \frac{(12-1)(28)^2 + (9-1)(21)^2}{12+9-2} = 639.579 \quad \text{and} \quad t = \frac{(85-65)-0}{\sqrt{639.579(\frac{1}{12}+\frac{1}{9})}} = 1.793$$

For a right-tail area of 0.05 and d.f. = $(n_1 + n_2 - 2) = (12 + 9 - 2) = 19$, the critical value is t = 1.729. The decision rule is: "Reject H_0 if the calculated t > 1.729, otherwise do not reject." Since the calculated test statistic falls in the rejection region, reject H_0. At the 0.05 level, Emily's suspicion appears to be warranted. We can conclude that TV repair shops tend to charge women more than they do men.

Using the t table with d.f. = 19, p-value = $P(t \geq 1.793)$ is between 0.025 and 0.05. Based on the summary information in this exercise, we have used Excel worksheet template tmt2pool.xls in generating the following printout corresponding to this solution.

	A	B	C	D	E
1	2-Sample t-Test Comparing Means, Pooled Variances (using summary stats)				
2	Hypothesized Difference (mu1 - mu2)	0.000			
3			*Calculated Values:*		
4	*Summary of Sample Data:*		pooled var =	639.579	
5	Mean for Sample 1, xbar1	85.000	d.f. =	19	
6	Std. Dev. For Sample 1, s1	28.000	Std. Error =	11.152	
7	Size of Sample 1, n1	12	t =	1.793	
8	Mean for Sample 2, xbar2	65.000	*p-Value If the Test Is:*		
9	Std. Dev. For Sample 2, s2	21.000	Left-Tail	Two-Tail	**Right-Tail**
10	Size of Sample 2, n2	9	0.9556	0.0888	**0.0444**

11.77 p/a/m The null and alternative hypotheses are:

H_0: $\mu_1 = \mu_2$, or $\mu_1 - \mu_2 = 0$ (the two designs are equal at handling air pressure)

H_1: $\mu_1 \neq \mu_2$, or $\mu_1 - \mu_2 \neq 0$ (the two designs are not equal at handling air pressure)

$\bar{x}_1$ and $\bar{x}_2$ are the average air pressures handled by designs A and B.

The pooled estimate of the common variance, and the test statistic, t, are:

$$s_p^2 = \frac{(4-1)(250)^2 + (6-1)(230)^2}{4+6-2} = 56,500 \quad \text{and} \quad t = \frac{(1400-1620)-0}{\sqrt{56,500(\frac{1}{4}+\frac{1}{6})}} = -1.434$$

For a right-tail area of 0.10/2 = 0.05 and d.f. = ($n_1 + n_2 - 2$) = (4 + 6 - 2) = 8, the critical values are t = -1.860 and t = 1.860. The decision rule is: "Reject H_0 if the calculated t < -1.860 or > 1.860, otherwise do not reject." Since the calculated test statistic falls in the nonrejection region, do not reject H_0. At the 0.10 level, we cannot conclude that the two designs differ in the average amount of air pressure they can withstand. The observed difference between the two designs' means is judged to have been due to chance. Using the t table with d.f. = 8, P(t ≤ -1.434) is between 0.05 and 0.10. Therefore, p-value = 2P(t ≤ -1.434) is between 0.10 and 0.20.

Given the summary data, we can also carry out this pooled-variances t-test using the Test Statistics.xls and Estimators.xls workbooks that accompany Data Analysis Plus. Recall that it is first necessary to square the sample standard deviations (250 and 230) to obtain the sample variances (62,500 and 52,900). For this two-tail test, the p-value (0.190) is not less than the 0.10 level of significance being used to reach a conclusion, so H_0 is not rejected. For a true H_0, there is a 0.190 probability that this much of a difference between the sample means could occur by chance. The 90% confidence interval for ($\mu_1 - \mu_2$) is from -505.32 to 65.32, This interval includes 0, so the actual difference between the population means could be 0. The conclusion using the 90% confidence interval is consistent with the results of the two-tail hypothesis test at the 0.10 level of significance.

	A	B	C	D	E
1	t-Test of the Difference Between Two Means (Equal-Variances)				
2					
3		Sample 1	Sample 2	t Stat	-1.434
4	Mean	1400	1620	P(T<=t) one-tail	0.095
5	Variance	62500	52900	t Critical one-tail	1.397
6	Observations	4	6	P(T<=t) two-tail	0.190
7	Pooled Variance estimate	56500		t Critical two-tail	1.860
8	Hypothesized difference	0			
9	Alpha	0.10			

	A	B	C	D	E	F
1	t-Estimate of the Difference Between Two Means (Equal-Variances)					
2						
3		Sample 1	Sample 2	Confidence Interval Estimate		
4	Mean	1400	1620	-220.0	plus/minus	285.32
5	Variance	62500	52900	Lower confidence limit		-505.32
6	Sample size	4	6	Upper confidence limit		65.32
7	Pooled Variance	56500				
8	Confidence level	0.90				

11.79 p/a/m The null and alternative hypotheses are:

$H_0: \mu_1 \leq \mu_2$, or $\mu_1 - \mu_2 \leq 0$ (the modifications are not effective)

$H_1: \mu_1 > \mu_2$, or $\mu_1 - \mu_2 > 0$ (the modifications are effective)

$\bar{x}_1$ and $\bar{x}_2$ are the average times for the standard and new designs.

The pooled estimate of the common variance, and the test statistic, t, are:

$$s_p^2 = \frac{(10-1)(1.3)^2 + (12-1)(0.9)^2}{10+12-2} = 1.206 \quad \text{and} \quad t = \frac{(22.5-21.9)-0}{\sqrt{1.206(\frac{1}{10}+\frac{1}{12})}} = 1.276$$

For a right-tail area of 0.10 and d.f. = $(n_1 + n_2 - 2) = (10 + 12 - 2) = 20$, the critical value is t = 1.325. The decision rule is: "Reject H_0 if the calculated t > 1.325, otherwise do not reject." Since the calculated test statistic falls in the nonrejection region, do not reject H_0. At the 0.10 level, we cannot conclude that the modifications are effective in improving his racer's speed. The observed difference between the two cars' means is judged to have been due to chance. Using the t table with d.f. = 20, p-value = $P(t \geq 1.276)$ is greater than 0.10.

Given the summary data, we can also carry out this pooled-variances t-test using the Test Statistics.xls workbook that accompanies Data Analysis Plus. Recall that it is first necessary to square the sample standard deviations (1.3 and 0.9) to obtain the sample variances (1.69 and 0.81). For this right-tail test, the p-value (0.108) is not less than the 0.10 level of significance being used to reach a conclusion, so H_0 is not rejected. For a true H_0, there is a 0.108 probability that the mean for sample 2 could be this much less than that for sample 1 due to chance.

	A	B	C	D	E
1	t-Test of the Difference Between Two Means (Equal-Variances)				
2					
3		Sample 1	Sample 2	t Stat	1.276
4	Mean	22.5	21.9	P(T<=t) one-tail	0.108
5	Variance	1.69	0.81	t Critical one-tail	1.325
6	Observations	10	12	P(T<=t) two-tail	0.217
7	Pooled Variance estimate	1.206		t Critical two-tail	1.725
8	Hypothesized difference	0			
9	Alpha	0.10			

11.81 p/a/m The null and alternative hypotheses are:

$H_0: \mu_1 \leq \mu_2$, or $\mu_1 - \mu_2 \leq 0$ (vehicles are not faster on Monday than on Wednesday)

$H_1: \mu_1 > \mu_2$, or $\mu_1 - \mu_2 > 0$ (vehicles are faster on Monday than on Wednesday)

$\bar{x}_1$ and $\bar{x}_2$ are the average speeds for vehicles sampled on Monday and Wednesday.

Assuming $\sigma_1 \neq \sigma_2$, we will use the unequal-variances t-test and the computer. Hand calculations for the t statistic and d.f. can be carried out as shown below.

$$t = \frac{(\bar{x}_1 - \bar{x}_2) - (\mu_1 - \mu_2)_0}{\sqrt{\frac{s_1^2}{n_1} + \frac{s_2^2}{n_2}}} = \frac{(59.4 - 56.3) - 0}{\sqrt{\frac{3.7^2}{16} + \frac{4.4^2}{20}}} = 2.296$$

$$\text{d.f.} = \frac{[s_1^2/n_1 + s_2^2/n_2]^2}{\frac{[s_1^2/n_1]^2}{(n_1-1)} + \frac{[s_2^2/n_2]^2}{(n_2-1)}} = \frac{[3.7^2/16 + 4.4^2/20]^2}{\frac{[3.7^2/16]^2}{(16-1)} + \frac{[4.4^2/20]^2}{(20-1)}} = 33.89$$

In the d.f. = 34 row of the t distribution table, the critical value for a right-tail test at the 0.05 level of significance is t = 1.691. The calculated t (2.296) exceeds 1.691, so we reject H_0 and conclude that speeds are faster on Monday than on Wednesday. Based on the t table alone, the p-value is between 0.025 and 0.01.

Given the summary data, we can also carry out this unequal-variances t-test using the Test Statistics.xls workbook that accompanies Data Analysis Plus. Recall that it is first necessary to square the sample standard deviations (3.7 and 4.4) to obtain the sample variances (13.69 and 19.36). Based on the t table alone, the p-value is between 0.025 and 0.01. The exact p-value, shown in the printout below, is 0.0141.

	A	B	C	D	E
1	t-Test of the Difference Between Two Means (Unequal-Variances)				
2					
3		Sample 1	Sample 2		
4	Mean	59.4	56.3	t Stat	2.296
5	Variance	13.69	19.36	P(T<=t) one-tail	0.0141
6	Sample size	16	20	t Critical one-tail	1.6924
7	Degrees of freedom	33.89		P(T<=t) two-tail	0.0282
8	Hypothesized difference	0		t Critical two-tail	2.0345
9	Alpha	0.05			

11.83 p/a/m The null and alternative hypotheses are:

$H_0: \mu_1 = \mu_2$, or $\mu_1 - \mu_2 = 0$ and $H_1: \mu_1 \neq \mu_2$, or $\mu_1 - \mu_2 \neq 0$

Assuming $\sigma_1 \neq \sigma_2$, we will use the unequal-variances t-test and the computer. Hand calculations for the t statistic and d.f. can be carried out as shown below.

$$t = \frac{(\overline{x}_1 - \overline{x}_2) - (\mu_1 - \mu_2)_0}{\sqrt{\dfrac{s_1^2}{n_1} + \dfrac{s_2^2}{n_2}}} = \frac{(19.7 - 16.3) - 0}{\sqrt{\dfrac{7.3^2}{48} + \dfrac{4.1^2}{40}}} = 2.75$$

$$d.f. = \frac{[s_1^2 / n_1 + s_2^2 / n_2]^2}{\dfrac{[s_1^2 / n_1]^2}{(n_1 - 1)} + \dfrac{[s_2^2 / n_2]^2}{(n_2 - 1)}} = \frac{[7.3^2 / 48 + 4.1^2 / 40]^2}{\dfrac{[7.3^2 / 48]^2}{(48 - 1)} + \dfrac{[4.1^2 / 40]^2}{(40 - 1)}} = 76.16$$

Using the d.f. = 76 row of the t distribution table, the critical values for a two-tail test at the 0.01 level are t = -2.642 and t = +2.642. The calculated t (2.75) exceeds the upper critical value, so we reject H_0 and conclude that the population means are not the same. Based on the t table alone, the p-value is less than 0.01.

Given the summary data, we can carry out this unequal-variances t-test and generate the confidence interval using the Test Statistics.xls and Estimators.xls workbooks that accompany Data Analysis Plus. Recall that it is first necessary to square the sample standard deviations (7.3 and 4.1) to obtain the sample variances (53.29 and 16.81). The p-value (0.0075) is less than the 0.01 level of significance being used to reach a conclusion, so H_0 is rejected. For a true H_0, there is only a 0.0075 probability that this much of a difference between the sample means could occur by chance. The 99% confidence interval for ($\mu_1 - \mu_2$) is from 0.13 to 6.67 minutes. This interval does not include 0, so we conclude that ($\mu_1 - \mu_2$) is some value other than 0. The conclusion using the 99% confidence interval is consistent with the result of the two-tail hypothesis test at the 0.01 level of significance.

	A	B	C	D	E
1	t-Test of the Difference Between Two Means (Unequal-Variances)				
2					
3		Sample 1	Sample 2		
4	Mean	19.7	16.3	t Stat	2.748
5	Variance	53.29	16.81	P(T<=t) one-tail	0.0037
6	Sample size	48	40	t Critical one-tail	2.3764
7	Degrees of freedom	76.16		P(T<=t) two-tail	0.0075
8	Hypothesized difference	0		t Critical two-tail	2.6421
9	Alpha	0.01			

	A	B	C	D	E	F
1	t-Estimate of the Difference Between Two Means (Unequal-Variances)					
2						
3		Sample 1	Sample 2	Confidence Interval Estimate		
4	Mean	19.7	16.3	3.40	plus/minus	3.27
5	Variance	53.29	16.81	Lower confidence limit		0.13
6	Sample size	48	40	Upper confidence limit		6.67
7	Degrees of freedom	76.16				
8	Confidence level	0.99				

11.85 p/a/m The null and alternative hypotheses are:

H_0: $\mu_1 = \mu_2$, or $\mu_1 - \mu_2 = 0$ (the average stay is the same at the two times)

H_1: $\mu_1 \neq \mu_2$, or $\mu_1 - \mu_2 \neq 0$ (the average stay is not the same at the two times)

$\bar{x}_1$ and $\bar{x}_2$ are the average stays in the store for the morning and afternoon customer samples.

Assuming $\sigma_1 \neq \sigma_2$, we will use the unequal-variances t-test and the computer. Hand calculations for the t statistic and d.f. can be carried out as shown below.

$$t = \frac{(\bar{x}_1 - \bar{x}_2) - (\mu_1 - \mu_2)_0}{\sqrt{\dfrac{s_1^2}{n_1} + \dfrac{s_2^2}{n_2}}} = \frac{(3.0 - 3.5) - 0}{\sqrt{\dfrac{0.9^2}{35} + \dfrac{1.8^2}{43}}} = -1.593$$

$$d.f. = \frac{[s_1^2 / n_1 + s_2^2 / n_2]^2}{\dfrac{[s_1^2 / n_1]^2}{(n_1 - 1)} + \dfrac{[s_2^2 / n_2]^2}{(n_2 - 1)}} = \frac{[0.9^2 / 35 + 1.8^2 / 43]^2}{\dfrac{[0.9^2 / 35]^2}{(35 - 1)} + \dfrac{[1.8^2 / 43]^2}{(43 - 1)}} = 64.27$$

Using the d.f. = 64 row of the t distribution table, the critical values for a two-tail test at the 0.05 level are t = -1.998 and t = +1.998. The calculated t (-1.593) falls within these limits and we do not reject H_0. We conclude that the population means could be equal. Based on the t table alone, the p-value is between 0.20 and 0.10.

Given the summary data, we can carry out this unequal-variances t-test and generate the confidence interval using the Test Statistics.xls and Estimators.xls workbooks that accompany Data Analysis Plus. Recall that it is first necessary to square the sample standard deviations (0.9 and 1.8) to obtain the sample variances (0.81 and 3.24). The p-value (0.1160) is not less than the 0.05 level of significance being used to reach a conclusion, so H_0 is not rejected. For a true H_0, there is a 0.1160 probability that this much of a difference between the sample means could occur by chance. The 95% confidence interval for ($\mu_1 - \mu_2$) is from -1.13 to 0.13 minutes. This interval includes 0, so we conclude that ($\mu_1 - \mu_2$) could be 0. The conclusion using the 95% confidence interval is consistent with the result of the two-tail hypothesis test at the 0.05 level of significance.

	A	B	C	D	E
1	t-Test of the Difference Between Two Means (Unequal-Variances)				
2					
3		Sample 1	Sample 2		
4	Mean	3.0	3.5	t Stat	-1.593
5	Variance	0.81	3.24	P(T<=t) one-tail	0.0580
6	Sample size	35	43	t Critical one-tail	1.6690
7	Degrees of freedom	64.27		P(T<=t) two-tail	0.1160
8	Hypothesized difference	0		t Critical two-tail	1.9977
9	Alpha	0.05			

	A	B	C	D	E	F
1	t-Estimate of the Difference Between Two Means (Unequal-Variances)					
2						
3		Sample 1	Sample 2	Confidence Interval Estimate		
4	Mean	3.0	3.5	-0.50	plus/minus	0.63
5	Variance	0.81	3.24	Lower confidence limit		-1.13
6	Sample size	35	43	Upper confidence limit		0.13
7	Degrees of freedom	64.27				
8	Confidence level	0.95				

11.87 p/a/m Since the samples are dependent, we will use the t-test for paired observations.
The null and alternative hypotheses are:

$H_0: \mu_d \le 0$ (the additive is not effective in increasing engine rpm)

$H_1: \mu_d > 0$ (the additive is effective in increasing engine rpm)

For each generator, $d = x_1 - x_2$, with x_1 = speed of the engine with an oil additive, and x_2 = speed of the engine without an oil additive.

The test statistic is: $t = \dfrac{\overline{d}}{s_d / \sqrt{n}} = \dfrac{23}{3.5 / \sqrt{14}} = 24.588$

For a right-tail area of 0.05 and d.f. = (n - 1) = (14 - 1) = 13, the critical value is t = 1.771. The decision rule is: "Reject H_0 if the calculated t > 1.771, otherwise do not reject." Since the calculated test statistic falls in the rejection region, reject H_0. At the 0.05 level, we can conclude the additive is effective in increasing engine rpm.

11.89 p/a/m The null and alternative hypotheses are:

$H_0: \pi_1 \le \pi_2$, or $\pi_1 - \pi_2 \le 0$ (the manager's idea is not effective)

$H_1: \pi_1 > \pi_2$, or $\pi_1 - \pi_2 > 0$.   (the manager's idea is effective)

p_1 and p_2 are the proportion of each sample who took at least one towel, those provided with white and green, respectively. The pooled estimate of the population proportions is:

$\overline{p} = \dfrac{n_1 p_1 + n_2 p_2}{n_1 + n_2} = \dfrac{(120)(0.35) + (160)(0.25)}{120 + 160} = 0.293$

The test statistic is: $z = \dfrac{p_1 - p_2}{\sqrt{\overline{p}(1 - \overline{p})(\dfrac{1}{n_1} + \dfrac{1}{n_2})}} = \dfrac{0.35 - 0.25}{\sqrt{0.293(1 - 0.293)(\dfrac{1}{120} + \dfrac{1}{160})}} = 1.82$

For a right-tail test at the 0.01 significance level, the critical value is z = 2.33. The decision rule is: "Reject H_0 if the calculated z > 2.33, otherwise do not reject." Since the calculated test statistic falls in the nonrejection region, do not reject H_0. At the 0.01 level, we cannot conclude that the manager's idea is effective in reducing the rate of towel theft. The observed difference between the groups' means is judged to have been due to chance.

Given the summary information in this exercise, the Test Statistics.xls workbook provided with Data Analysis Plus was used for the printout below. In this right-tail test, the p-value (0.0344) is not less than 0.01, the level of significance used in reaching a conclusion, so we do not reject H_0. For a true H_0, there is a 0.0344 probability that p_1 could be this much larger than p_2 simply by chance.

	A	B	C	D	E
1	z-Test of the Difference Between Two Proportions (Case 1)				
2					
3		Sample 1	Sample 2	z Stat	1.820
4	Sample proportion	0.35	0.25	P(Z<=z) one-tail	0.0344
5	Sample size	120	160	z Critical one-tail	2.3263
6	Alpha	0.01		P(Z<=z) two-tail	0.0688
7				z Critical two-tail	2.5758

11.91 p/a/m The null and alternative hypotheses are:

H_0: $\pi_1 \leq \pi_2$, or $\pi_1 - \pi_2 \leq 0$ (Ms. Smith is not a more effective auditor)

H_1: $\pi_1 > \pi_2$, or $\pi_1 - \pi_2 > 0$.   (Ms. Smith is a more effective auditor)

p_1 and p_2 are the proportion of each sample that had to pay additional taxes when audited by Ms. Smith and Mr. Burke. The pooled estimate of the population proportions is:

$$\bar{p} = \frac{n_1 p_1 + n_2 p_2}{n_1 + n_2} = \frac{(200)(0.30) + (100)(0.19)}{200 + 100} = 0.263$$

The test statistic is: $z = \dfrac{p_1 - p_2}{\sqrt{\bar{p}(1-\bar{p})(\frac{1}{n_1} + \frac{1}{n_2})}} = \dfrac{0.30 - 0.19}{\sqrt{0.263(1 - 0.263)(\frac{1}{200} + \frac{1}{100})}} = 2.04$

For a right-tail test at the 0.01 significance level, the critical value is z = 2.33. The decision rule is: "Reject H_0 if the calculated z > 2.33, otherwise do not reject." Since the calculated test statistic falls in the nonrejection region, do not reject H_0. At the 0.01 level, the manager cannot conclude that Ms. Smith is a more effective auditor than Mr. Burke. The observed difference between the auditors' means is judged to have been due to chance.

Given the summary information in this exercise, the Test Statistics.xls workbook provided with Data Analysis Plus was used for the printout below. In this right-tail test, the p-value (0.0207) is not less than 0.01, the level of significance used in reaching a conclusion, so we do not reject H_0. For a true H_0, there is a 0.0207 probability that p_1 could be this much larger than p_2 simply by chance.

	A	B	C	D	E
1	z-Test of the Difference Between Two Proportions (Case 1)				
2					
3		Sample 1	Sample 2	z Stat	2.039
4	Sample proportion	0.30	0.19	P(Z<=z) one-tail	0.0207
5	Sample size	200	100	z Critical one-tail	2.3263
6	Alpha	0.01		P(Z<=z) two-tail	0.0414
7				z Critical two-tail	2.5758

11.93 p/a/m Since the samples are dependent, we will use the t-test for paired observations. The null and alternative hypotheses are:

H_0: $\mu_d \leq 0$ (the course is not effective)

H_1: $\mu_d > 0$ (the course is effective)

For each subject, $d = x_1 - x_2$, with x_1 = words per minute after the course, and x_2 = words per minute before the course.

The test statistic is: $t = \dfrac{\overline{d}}{s_d / \sqrt{n}} = \dfrac{200}{75 / \sqrt{10}} = 8.433$

For a right-tail area of 0.10 and d.f. = $(n - 1) = (10 - 1) = 9$, the critical value is t = 1.383. The decision rule is: "Reject H_0 if the calculated t > 1.383, otherwise do not reject." Since the calculated test statistic falls in the rejection region, reject H_0. At the 0.10 level, the sample information suggests that the course is effective in improving students' reading speed. Using the t table with d.f. = 9, p-value = $P(t \geq 8.433)$ is less than 0.005.

11.95 p/a/m The null and alternative hypotheses are:

H_0: $\pi_1 \leq \pi_2$, or $\pi_1 - \pi_2 \leq 0$ (the college proportion is no higher than the below-college proportion)

H_1: $\pi_1 > \pi_2$, or $\pi_1 - \pi_2 > 0$ (the college proportion is higher)

This exercise could be solved using formulas and pocket calculator, but we will use the Test Statistics.xls workbook provided with Data Analysis Plus. In this right-tail test, the p-value (0.001) is less than 0.01, the level of significance used in reaching a conclusion, so we reject H_0. For a true H_0, there is only a 0.001 probability that p_1 would be this much larger than p_2 simply by chance.

	A	B	C	D	E
1	z-Test of the Difference Between Two Proportions (Case 1)				
2					
3		Sample 1	Sample 2	z Stat	3.086
4	Sample proportion	0.80	0.60	P(Z<=z) one-tail	0.001
5	Sample size	100	100	z Critical one-tail	2.326
6	Alpha	0.01		P(Z<=z) two-tail	0.002
7				z Critical two-tail	2.576

11.97 p/a/d In exercise 11.75, $s_1^2 = 784$, $n_1 = 12$, $s_2^2 = 441$, and $n_2 = 9$. A pocket calculator and formulas could be used, but we will use the computer. The null and alternative hypotheses are:

$$H_0 : \sigma_1^2 = \sigma_2^2 \ \text{ or } \ \frac{\sigma_1^2}{\sigma_2^2} = 1 \quad \text{and} \quad H_1 : \sigma_1^2 \neq \sigma_2^2 \ \text{ or } \ \frac{\sigma_1^2}{\sigma_2^2} \neq 1$$

Since $s_1 > s_2$, the test statistic is: $F = \dfrac{s_1^2}{s_2^2} = \dfrac{784}{441} = 1.778$, as shown in the first printout below.

The printouts below were generated using the Test Statistics.xls workbook that accompanies Data Analysis Plus. Recall that it is first necessary to square the sample standard deviations to get the sample variances. For the two-tail test of whether the variances in exercise 11.75 could be equal, the p-value is 0.424, as shown in cell E6 of the first printout. This is not less than the 0.05 level of significance being used to reach a conclusion, so H_0 is not rejected. We were justified in assuming equal population variances. If the standard deviation for Emily had been $35, the test result would have been as shown in the second printout below and the two-tail p-value would have been 0.158. Again, the p-value is not less than the 0.05 level of significance and we would not reject H_0. The conclusion has not changed.

	A	B	C	D	E
1	F-Test of the Ratio of Two Variances				
2					
3		Sample 1	Sample 2	F Stat	1.778
4	Sample variance	784.00	441.00	P(F<=f) one-tail	0.212
5	Sample size	12	9	f Critical one-tail	3.313
6	Alpha	0.05		P(F<=f) two-tail	0.424
7				f Critical two-tail	0.273

	A	B	C	D	E
1	F-Test of the Ratio of Two Variances				
2					
3		Sample 1	Sample 2	F Stat	2.778
4	Sample variance	1225.00	441.00	P(F<=f) one-tail	0.079
5	Sample size	12	9	f Critical one-tail	3.313
6	Alpha	0.05		P(F<=f) two-tail	0.158
7				f Critical two-tail	0.273

11.99 p/c/m The null and alternative hypotheses are:

H_0: $\mu_1 = \mu_2$, or $\mu_1 - \mu_2 = 0$ and H_1: $\mu_1 \neq \mu_2$, or $\mu_1 - \mu_2 \neq 0$

Assuming $\sigma_1 \neq \sigma_2$, we use the unequal-variances t-test and Minitab. The printout is shown below. For this two-tail test, the p-value (0.140) is not less than the 0.05 level of significance being used to reach a conclusion, so H_0 is not rejected. We conclude that the population mean salaries do not differ for graduates hired by consulting firms versus national-level corporations. For a true H_0, there is a 0.140 probability that this much of a difference between the sample means could occur by chance. The 95% confidence interval for ($\mu_1 - \mu_2$) is from -$1.95 thousand to +$12.96 thousand. This interval includes 0, so we conclude that ($\mu_1 - \mu_2$) could be 0. The conclusion using the 95% confidence interval is consistent with the result of the two-tail hypothesis test at the 0.05 level of significance.

```
Two-Sample T-Test and CI: Consult, Corp

Two-sample T for Consult vs Corp
             N      Mean     StDev   SE Mean
Consult     15     64.31      9.16       2.4
Corp        12     58.80      9.41       2.7

Difference = mu Consult - mu Corp
Estimate for difference:  5.51
95% CI for difference: (-1.95, 12.96)
T-Test of difference = 0 (vs not =): T-Value = 1.53   P-Value = 0.140   DF = 23
```

143

11.101 p/c/m For each person, d = vertical leap with standard shoe - vertical leap with new shoe.
The null and alternative hypotheses are:

H_0: $\mu_d \geq 0$ (the new shoe is no better than the standard shoe)

H_1: $\mu_d < 0$ (the new shoe is better)

The Minitab printout is shown below. For this left-tail test, the p-value (0.135) is not less than the 0.05 level of significance being used to reach a conclusion, so H_0 is not rejected. The new shoe may be no better than the standard model. For a true H_0, the probability is 0.135 of getting this large of a negative mean change in vertical leap height just by chance.

```
Paired T-Test and CI: HtCurrent, HtNew

Paired T for HtCurrent - HtNew
                    N      Mean     StDev    SE Mean
HtCurrent          10     26.21      3.08       0.97
HtNew              10     26.63      3.71       1.17
Difference         10    -0.420     1.132      0.358

95% upper bound for mean difference: 0.236
T-Test of mean difference = 0 (vs < 0): T-Value = -1.17  P-Value = 0.135
```

11.103 p/c/m The null and alternative hypotheses are:

H_0: $\pi_1 = \pi_2$, or $\pi_1 - \pi_2 = 0$ (the population proportions for males and females are the same)

H_1: $\pi_1 \neq \pi_2$, or $\pi_1 - \pi_2 \neq 0$ (the proportions are not the same)

Using Data Analysis Plus, we obtain the printouts below. In this two-tail test, the p-value (0.414) is not less than 0.10, the level of significance used in reaching a conclusion, so we do not reject H_0. The population proportions could be equal. For a true H_0, there is a 0.414 probability of obtaining this great a difference between the sample proportions by chance. The 90% confidence interval for ($\pi_1 - \pi_2$) is from -0.1204 to 0.0404. Since zero is within the interval, we conclude that ($\pi_1 - \pi_2$) could be zero. This is the same conclusion reached in the hypothesis test.

	A	B	C	D
1	z-Test: Two Proportions			
2				
3			male	female
4	Sample Proportions		0.58	0.62
5	Observations		200	200
6	Hypothesized Difference		0	
7	z Stat		-0.817	
8	P(Z<=z) one tail		0.207	
9	z Critical one-tail		1.282	
10	P(Z<=z) two-tail		0.414	
11	z Critical two-tail		1.645	

	A	B	C	D
1	z-Estimate: Two Proportions			
2				
3			male	female
4	Sample Proportions		0.58	0.62
5	Observations		200	200
6				
7	LCL		-0.1204	
8	UCL		0.0404	

CHAPTER 12
ANALYSIS OF VARIANCE TESTS

Note: Although it is possible to solve many of the problem-type exercises with a pocket calculator and formulas, statistical software with ANOVA capabilities is highly recommended.

SECTION EXERCISES

12.1 d/p/e A designed experiment is an experiment in which treatments are randomly assigned to the participants or test units.

12.3 d/p/m For an experiment to be balanced it is necessary to have an equal number of persons or test units receiving each treatment.

12.5 d/p/e In an experiment, the independent variable is the variable that is manipulated for the purpose of determining its effect on the value of the dependent variable. The dependent variable is the variable for which a value is measured or observed. The independent variable can be either quantitative or qualitative and the dependent variable is quantitative.

12.7 d/p/m This experiment represents a designed experiment since the two treatments (methods of teaching) are randomly assigned to the students.

12.9 d/p/m The dependent variable is the number of absences. It is a quantitative variable.
The independent variable is the college in which the professor teaches. It is a qualitative variable.

12.11 d/p/m The dependent variable is the strength of the rivets. It is a quantitative variable.
The independent variable is the supplier. It is a qualitative variable.

12.13 d/p/e The purpose of the one-way analysis of variance is to examine two or more independent samples to determine if their population means could be equal.

12.15 d/p/m The F distribution is applicable to the analysis of variance because it's the sampling distribution for the ratio of the two sample variances whenever two random samples are repeatedly drawn from the same, normally distributed population.

12.17 d/p/m The error mean square MSE is an estimate of the variance σ^2 which we assume is common to all the populations.

12.19 d/p/m If the populations in an experiment really do have the same mean, the calculated F statistic will be approximately one. This is because the two sources of variation, that between treatment means and that within treatments, are approximately equal.

12.21 c/p/e $H_0: \mu_1 = \mu_2$ $H_1:$ The population means are not equal
The calculated test statistic is $F = 3.60$.
For this test, $\alpha = 0.05$, $t = 2$ treatments, and $N = 20 + 15 = 35$. The d.f. associated with the numerator of F is $v_1 = (t - 1) = (2 - 1) = 1$ and the d.f. associated with the denominator of F is $v_2 = (N - t) = (35 - 2) = 33$. Using the F table, the critical value is $F(0.05, 1, 33) = 4.17$. (Since $v_2 = 33$ is not in the table, the closest value was used.) The decision rule is: "Reject H_0 if the calculated $F > 4.17$, otherwise do not reject." Since the calculated test statistic falls in the nonrejection region, do not reject H_0.
At the 0.05 level, we conclude that the population means could be equal. Using the F tables with $v_1 = 1$ and $v_2 = 33$, p-value = $P(F \geq 3.60)$ is greater than 0.05.

12.23 p/c/m To determine if the average number of absences differ, we test:

H_0: $\mu_1 = \mu_2 = \mu_3$ and H_1: The population means are not equal

$$\bar{x}_{.1} = \frac{8+10+6+8+4+8}{6} = 7.33, \quad \bar{x}_{.2} = 6.11, \quad \bar{x}_{.3} = 8.75; \text{ grand mean, } \bar{\bar{x}} = \frac{8+10+6+...+7}{23} = 7.35$$

$$SSTR = \sum_{j=1}^{t} n_j (\bar{x}_{.j} - \bar{\bar{x}})^2 = 6(7.33-7.35)^2 + 9(6.11-7.35)^2 + 8(8.75-7.35)^2 = 29.52$$

$$SSE = \sum_{j=1}^{t} \sum_{i=1}^{n_j} (x_{ij} - \bar{x}_{.j})^2 = [(8-7.33)^2 + (10-7.33)^2 + (6-7.33)^2 + (8-7.33)^2$$

$$+(4-7.33)^2 + (8-7.33)^2] + [(5-6.11)^2 + (7-6.11)^2 + (6-6.11)^2 + (7-6.11)^2 + (7-6.11)^2$$

$$+(6-6.11)^2 + (8-6.11)^2 + (8-6.11)^2 + (1-6.11)^2] + [(9-8.75)^2 + (10-8.75)^2 + (10-8.75)^2$$

$$+(9-8.75)^2 + (7-8.75)^2 + (5-8.75)^2 + (13-8.75)^2 + (7-8.75)^2] = 99.72$$

SST = SSTR + SSE = 29.52 + 99.72 = 129.24

Treatments d.f. = t - 1 = 3 - 1 = 2

Error d.f. = N - t = 23 - 3 = 20

Total d.f. = N - 1 = 23 - 1 = 22

$$MSTR = \frac{SSTR}{t-1} = \frac{29.52}{2} = 14.76, \quad MSE = \frac{SSE}{N-t} = \frac{99.72}{20} = 4.99, \quad F = \frac{MSTR}{MSE} = \frac{14.76}{4.99} = 2.96$$

The ANOVA table is:

Variation Source	Sum of Squares	Degrees of Freedom	Mean Square	F
Treatment	29.52	2	14.76	2.96
Error	99.72	20	4.99	
Total	129.24	22		

The calculated test statistic is F = 2.96

Using the F table, the critical value is F(0.05, 2, 20) = 3.49. The decision rule is: "Reject H_0 if the calculated F > 3.49, otherwise do not reject." Since the calculated test statistic falls in the nonrejection region, do not reject H_0. At the 0.05 level, we conclude that the average number of absences for the three colleges do not differ significantly. Using the F table with $v_1 = 2$ and $v_2 = 20$, p-value = $P(F \geq 2.96)$ is greater than 0.05. The Minitab printout for this problem is given below. Some values differ slightly due to rounding.

```
One-way ANOVA: Engineer, Business, Fine Arts

Analysis of Variance
Source      DF        SS        MS        F         P
Factor       2     29.50     14.75      2.96     0.075
Error       20     99.72      4.99
Total       22    129.22
                                    Individual 95% CIs For Mean
                                    Based on Pooled StDev
Level        N      Mean     StDev   --+---------+---------+---------+----
Engineer     6     7.333     2.066           (-----------*-----------)
Business     9     6.111     2.147   (---------*---------)
Fine Art     8     8.750     2.435                   (---------*---------)
                                    --+---------+---------+---------+----
Pooled StDev =     2.233            4.8       6.4       8.0       9.6
```

146

The Excel counterpart is shown below:

	E	F	G	H	I	J	K
1	Anova: Single Factor						
2							
3	SUMMARY						
4	Groups	Count	Sum	Average	Variance		
5	Engineering	6	44	7.333	4.267		
6	Business	9	55	6.111	4.611		
7	Fine Arts	8	70	8.750	5.929		
8							
9							
10	ANOVA						
11	Source of Variation	SS	df	MS	F	P-value	F crit
12	Between Groups	29.495	2	14.748	2.958	0.075	3.493
13	Within Groups	99.722	20	4.986			
14							
15	Total	129.217	22				

12.25 p/c/m To determine if the average breaking strengths of the rivets differ for the suppliers, we test:
$H_0: \mu_1 = \mu_2 = \mu_3 = \mu_4$ and H_1: The population means are not equal.
The Minitab printout for this problem is shown below:

```
One-way ANOVA: Supplier A, Supplier B, Supplier C, Supplier D

Analysis of Variance
Source     DF        SS        MS       F       P
Factor      3      5709      1903    1.86   0.153
Error      36     36780      1022
Total      39     42488

                             Individual 95% CIs For Mean
                             Based on Pooled StDev
Level       N      Mean    StDev   ---------+---------+---------+-------
Supplier   10    489.70    34.82                 (---------*---------)
Supplier   10    472.80    31.41           (---------*----------)
Supplier   10    464.00    30.10   (---------*---------)
Supplier   10    493.00    31.33                  (----------*---------)
                                   ---------+---------+---------+-------
Pooled StDev =     31.96                  460       480       500
```

The Excel counterpart is shown below:

	G	H	I	J	K	L	M
1	Anova: Single Factor						
2							
3	SUMMARY						
4	Groups	Count	Sum	Average	Variance		
5	Supplier A	10	4897	489.70	1212.46		
6	Supplier B	10	4728	472.80	986.84		
7	Supplier C	10	4640	464.00	906.00		
8	Supplier D	10	4930	493.00	981.33		
9							
10							
11	ANOVA						
12	Source of Variation	SS	df	MS	F	P-value	F crit
13	Between Groups	5708.68	3	1902.89	1.86	0.153	4.38
14	Within Groups	36779.70	36	1021.66			
15							
16	Total	42488.38	39				

Because p-value = 0.153 is not < α = 0.01 level of significance for the test, do not reject H_0. At the 0.01 level, we conclude that the mean breaking strengths of the rivets do not differ significantly for the four suppliers.

12.27 pcm The null and alternative hypotheses are H_0: $\mu_1 = \mu_2 = \mu_3$ and H_1: The population means are not equal. The Minitab printout for this problem is shown below:

```
One-way ANOVA: Firm1, Firm2, Firm3

Analysis of Variance
Source      DF        SS        MS        F        P
Factor       2     17846      8923     3.97    0.043
Error       14     31429      2245
Total       16     49275
                                  Individual 95% CIs For Mean
                                  Based on Pooled StDev
Level        N      Mean     StDev   ---------+---------+---------+-------
Firm1        5    223.80     40.12                  (--------*--------)
Firm2        6    148.17     54.79   (--------*-------)
Firm3        6    158.50     44.68     (--------*-------)
                                    ---------+---------+---------+-------
Pooled StDev =   47.38              150       200       250
```

Because p-value = 0.043 is < α = 0.05 level of significance for the test, reject H_0. At the 0.05 level, we conclude that the average contributions made by the clerical workers differ for the three corporations.

12.29 p/c/d

a. To determine if the average lifetime differs for the three outlets, we test:

H_0: $\mu_1 = \mu_2 = \mu_3$ and H_1: The population means are not equal.

b. The Minitab printout for this problem is shown below:

```
One-way ANOVA: 3Below, 2Below, EqLine

Analysis of Variance
Source      DF        SS        MS        F        P
Factor       2     228.4     114.2     2.93    0.092
Error       12     468.0      39.0
Total       14     696.4
                                  Individual 95% CIs For Mean
                                  Based on Pooled StDev
Level        N      Mean     StDev   ----+---------+---------+---------+-
3Below       5    55.000     6.595                  (---------*---------)
2Below       5    48.800     6.535            (---------*---------)
EqLine       5    45.600     5.550     (---------*---------)
                                     ----+---------+---------+---------+-
Pooled StDev =   6.245              42.0      48.0      54.0      60.0
```

Because p-value = 0.092 is not < α = 0.01 level of significance for the test, do not reject H_0. At the 0.01 level, we conclude that the average lifetimes of the bulbs when placed in the different outlets do not differ significantly.

c. For a confidence level of 95%, the right-tail area of interest is (1 - 0.95)/2 = 0.025 with d.f. = 12. Referring to the t table, t = 2.179. Using the pooled estimate for the population standard deviation (6.245), we get the results below. Note: Recall that the pooled standard deviation provided by Minitab is really the square root of the MSE, or 6.245 is the square root of the 39.0 shown in the Minitab printout. Since Excel does not show this square root, it will be necessary for Excel users to first locate the MSE value, then take the square root.

The 95% confidence interval for μ_1 is:

$$\overline{x}_1 \pm t\frac{s_p}{\sqrt{n_1}} = 55 \pm 2.179(\frac{6.245}{\sqrt{5}}) = 55 \pm 6.086 \text{ or from } 48.914 \text{ to } 61.086$$

The 95 % confidence interval for μ_2 is:

$$\overline{x}_2 \pm t\frac{s_p}{\sqrt{n_2}} = 48.8 \pm 2.179(\frac{6.245}{\sqrt{5}}) = 48.8 \pm 6.086 \text{ or from } 42.714 \text{ to } 54.886$$

The 95 % confidence interval for μ_3 is:

$$\bar{x}_3 \pm t \frac{s_p}{\sqrt{n_3}} = 45.6 \pm 2.179(\frac{6.245}{\sqrt{5}}) = 45.6 \pm 6.086 \text{ or from } 39.514 \text{ to } 51.686$$

12.31 c/a/m Error d.f. = Total d.f. - Treatments d.f. = 29 - 2 = 27

$$MSE = \frac{SSE}{\text{Error d.f.}} = \frac{30,178.0}{2} = 1117.7 \quad \text{and} \quad F = \frac{MSTR}{MSE} = \frac{3376}{1117.7} = 3.02$$

The completed ANOVA table is:

Variation Source	Sum of Squares	Degrees of Freedom	Mean Square	F
Treatment	6752.0	2	3376.0	3.02
Error	30178.0	27	1117.7	
Total	36930.0	29		

H_0: $\mu_1 = \mu_2 = \mu_3$ and H_1: The population means are not equal
The calculated test statistic is F = 3.02.
Using the F table, the critical value is F(0.05, 2, 27) = 3.35. The decision rule is: "Reject H_0 if the calculated F > 3.35, otherwise do not reject." Since the calculated test statistic falls in the nonrejection region, do not reject H_0. At the 0.05 level, we conclude that the population means are not significantly different.

12.33 p/p/e The null and alternative hypotheses are H_0: $\mu_1 = \mu_2 = \mu_3$ and H_1: The population means are not equal. The test statistic is F = 2.45 and p-value = 0.092. Interpreting the output, we can say that if the population means are equal, the probability of obtaining a test statistic this large or larger is 0.092. If the specified level of significance is less than p-value = 0.092, do not reject H_0. We would conclude that the average job-satisfaction scores for the three departments are not significantly different.

12.35 c/a/m For a confidence level of 95%, the right-tail area of interest is (1 - 0.95)/2 = 0.025 with d.f. = 92. Referring to the t table, t = 1.986. Using the pooled estimate for the population standard deviation (3.775),
The 95% confidence interval for μ_1 is:

$$\bar{x}_1 \pm t \frac{s_p}{\sqrt{n_1}} = 17.64 \pm 1.986(\frac{3.775}{\sqrt{30}}) = 17.64 \pm 1.369 \text{ or from } 16.271 \text{ to } 19.009$$

The 95% confidence interval for μ_2 is:

$$\bar{x}_2 \pm t \frac{s_p}{\sqrt{n_2}} = 15.4 \pm 1.986(\frac{3.775}{\sqrt{25}}) = 15.4 \pm 1.499 \text{ or from } 13.901 \text{ to } 16.899$$

The 95% confidence interval for μ_3 is:

$$\bar{x}_3 \pm t \frac{s_p}{\sqrt{n_3}} = 16.875 \pm 1.986(\frac{3.775}{\sqrt{40}}) = 16.875 \pm 1.185 \text{ or from } 15.69 \text{ to } 18.06$$

12.37 d/p/m When there are two treatments, one-way ANOVA is equivalent to the pooled-variances t-test. Each tests whether two population means could be equal, and each relies on a pooled estimate for the variance that the populations are assumed to share. In addition, both tests assume the population to be approximately normally distributed and the samples to be independent. The p-value is the same for both tests and the results are the same for a specified level of significance.

12.39 p/c/m The Minitab output below gives the pooled-variances t-test for the data of exercise 12.30, with $H_0: \mu_1 = \mu_2$ and $H_1: \mu_1 \neq \mu_2$:

```
Two-Sample T-Test and CI: Alone, WithPass

Two-sample T for Alone vs WithPass

            N     Mean    StDev   SE Mean
Alone      10    63.70     9.36      3.0
WithPass   12    56.92     7.97      2.3

Difference = mu Alone - mu WithPass
Estimate for difference:  6.78
95% CI for difference: (-0.92, 14.48)
T-Test of difference = 0 (vs not =): T-Value = 1.84   P-Value = 0.081   DF = 20
Both use Pooled StDev = 8.62
```

From the printout, the test statistic is t = 1.84 and the p-value = 0.081. Interpreting this output, we can say that if the population means are equal, the probability of obtaining a test statistic this extreme or more extreme is 0.081. Since the p-value = 0.081 is not less than $\alpha = 0.025$, do not reject H_0. At the 0.025 level, we conclude that the average speed at which a car is driven is not affected by the occupancy of the vehicle. This is the same conclusion that was reached in exercise 12.30.

12.41 d/p/e The assumptions for the randomized block design are:
1. The one observation in each treatment-block combination has been randomly selected from a normally distributed population.
2. The variances are equal for the values in the respective populations.
3. There is no interaction between the blocks and treatments.

12.43 d/p/m $x_{ij} = \mu + \tau_j + \beta_i + \varepsilon_{ij}$
x_{ij} is an individual observation or measurement, it is the observation in the ith block for treatment j.
μ is the overall population mean for all of the treatments.
τ_j is the effect of treatment j.
β_i is the effect of block i.
ε_{ij} is the random error associated with the sampling process.
Since we are only controlling for the effect of the blocking variable and not attempting to examine its influence, the hypotheses are expressed only in terms of the treatments. The null and alternative hypotheses are:
$H_0: \tau_j = 0$ for treatments j = 1 through t. (Each treatment has no effect.)
$H_1: \tau_j \neq 0$ for at least one of the j = 1 through t treatments. (One or more treatments has an effect.)

12.45 d/p/e The treatments are no longer being applied randomly to the test units. We have tampered with the randomness of the treatment groups.

150

12.47 c/a/m The null and alternative hypotheses are:

H_0: $\mu_1 = \mu_2 = \mu_3 = \mu_4 = \mu_5$ and H_1: The population means are not equal.
The test statistic is F = 3.75.
For this test, $\alpha = 0.05$, t = 5 treatments, and n = 4 blocks. The d.f. associated with the numerator of F is $v_1 = (t - 1) = (5 - 1) = 4$ and the d.f. associated with the denominator of F is $v_2 = (t - 1)(n - 1) = (5 - 1)(4 - 1) = 12$. Using the F table, the critical value is F(0.05, 4, 12) = 3.26. The decision rule is: "Reject H_0 if the calculated F > 3.26, otherwise do not reject." Since the calculated test statistic falls in the rejection region, reject H_0. At the 0.05 level, we conclude that the population means are not equal. Using the F tables with $v_1 = 4$ and $v_2 = 12$, p-value = $P(F \geq 3.75)$ is between 0.025 and 0.05.

12.49 c/c/m To illustrate how the process works, we will first show the solution as it would come about from use of the formulas and a pocket calculator:

	Treatment 1	Treatment 2	
Block A	46	31	$\overline{x}_{1.} = 38.5$
Block B	37	26	$\overline{x}_{2.} = 31.5$
Block C	44	35	$\overline{x}_{3.} = 39.5$
	$\overline{x}_{.1} = 42.33$	$\overline{x}_{.2} = 30.67$	

$$\overline{\overline{x}} = \frac{46 + 37 + 44 + 31 + 26 + 35}{6} = 36.5$$

$$SSTR = n\sum_{j=1}^{t}(\overline{x}_{.j} - \overline{\overline{x}})^2 = 3[(42.33 - 36.5)^2 + (30.67 - 36.5)^2] = 203.93$$

$$SSB = t\sum_{i=1}^{n}(\overline{x}_{i.} - \overline{\overline{x}})^2 = 2[(38.5 - 36.5)^2 + (31.5 - 36.5)^2 + (39.5 - 36.5)^2] = 76$$

$$SST = \sum_{j=1}^{t}\sum_{i=1}^{n}(x_{ij} - \overline{\overline{x}})^2 = [(46 - 36.5)^2 + (37 - 36.5)^2 + (44 - 36.5)^2] + [(31 - 36.5)^2$$

$$+ (26 - 36.5)^2 + (35 - 36.5)^2] = 289.5$$

SSE = SST - SSTR - SSB = 289.5 - 203.93 - 76 = 9.57
Treatments d.f. = t - 1 = 2 - 1 = 1
Blocks   d.f. = n - 1 = 3 - 1 = 2
Error   d.f. = (t - 1)(n - 1) = (2 - 1)(3 - 1) = 2
Total   d.f. = tn - 1 = 2(3) - 1 = 5

$$MSTR = \frac{SSTR}{t - 1} = \frac{203.93}{1} = 203.93 \quad \text{and} \quad MSB = \frac{SSB}{n - 1} = \frac{76.00}{2} = 38.00$$

$$MSE = \frac{SSE}{(t - 1)(n - 1)} = \frac{9.57}{2} = 4.79$$

$$F_T = \frac{MSTR}{MSE} = \frac{203.93}{4.79} = 42.57 \quad \text{and} \quad F_B = \frac{MSB}{MSE} = \frac{38.00}{4.79} = 7.93$$

The ANOVA table is:

Variation Source	Sum of Squares	Degrees of Freedom	Mean Square	F
Treatments	203.93	1	203.93	42.57
Blocks	76.00	2	38.00	7.93
Error	9.57	2	4.79	
Total	289.50	5		

Testing for treatment effects:

H_0: $\tau_j = 0$ for treatments j = 1, 2   and H_1: $\tau_j \neq 0$ for at least one of the treatments

The calculated test statistic is $F = \dfrac{MSTR}{MSE} = \dfrac{203.93}{4.79} = 42.57$

Using the F table, the critical value is F(0.01, 1, 2) = 98.5. The decision rule is: "Reject H_0 if the calculated F > 98.5, otherwise do not reject." Since the calculated test statistic falls in the nonrejection region, do not reject H_0. As shown in the Minitab and Excel printouts for this exercise, the p-value for the test of treatment effects is 0.022.

Testing for block effects:

H_0: $\beta_i = 0$ for i = 1, 2, 3 (the levels of the blocking variable are equal in their effect)

H_1: $\beta_i \neq 0$ for at least one value of i (at least one level has an effect different from the others)

For the block effects, the calculated test statistic is $F = \dfrac{MSB}{MSE} = \dfrac{38.00}{4.79} = 7.93$

Using the F table, the critical value is F(0.01, 2, 2) = 99.0. The decision rule is: "Reject H_0 if the calculated F > 99.00, otherwise do not reject." Since the calculated test statistic falls in the nonrejection region, do not reject H_0. At the 0.01 level, we conclude that all of the block effects could be zero. As shown in the Minitab and Excel printouts below, the p-value for the test of block effects is 0.109. First, the Minitab data layout and printout:

```
Row  Block  Treat     x

 1     1      1       46
 2     1      2       31
 3     2      1       37
 4     2      2       26
 5     3      1       44
 6     3      2       35
```

Two-way ANOVA: x versus Block, Treat

```
Analysis of Variance for x
Source     DF       SS       MS       F       P
Block       2     76.00    38.00    8.14    0.109
Treat       1    204.17   204.17   43.75    0.022
Error       2      9.33     4.67
Total       5    289.50
```

Shown below is the Excel printout for this exercise. Some values may differ from those described earlier because of rounding.

	A	B	C	D	E	F	G
1		Treat1	Treat2				
2	Block1	46	31				
3	Block2	37	26				
4	Block3	44	35				
5							
6	Anova: Two-Factor Without Replication						
7							
8	SUMMARY	Count	Sum	Average	Variance		
9	Block1	2	77	38.5	112.5		
10	Block2	2	63	31.5	60.5		
11	Block3	2	79	39.5	40.5		
12							
13	Treat1	3	127	42.333	22.333		
14	Treat2	3	92	30.667	20.333		
15							
16							
17	ANOVA						
18	Source of Variation	SS	df	MS	F	P-value	F crit
19	Rows	76	2	38	8.143	0.109	99.000
20	Columns	204.167	1	204.167	43.750	0.022	98.502
21	Error	9.333	2	4.667			
22							
23	Total	289.5	5				

12.51 p/c/m For this problem, the three battery brands are the treatments and the four toys are the blocks. The Minitab printout is shown below:

```
Two-way ANOVA: Time versus Block, Treatmnt

Analysis of Variance for Time
Source      DF      SS      MS      F      P
Block        3   6.716   2.239   3.90   0.074
Treatmnt     2   7.227   3.613   6.29   0.034
Error        6   3.447   0.574
Total       11  17.389
```

Testing for treatment effects:

To determine if the battery treatment effects could all be zero, we test:

H_0: $\tau_j = 0$ for treatments j = 1, 2, 3   and H_1: $\tau_j \neq 0$ for at least one of the treatments

The calculated test statistic is: $F = \dfrac{MSTR}{MSE} = \dfrac{3.613}{0.574} = 6.29$

Using the F table, the critical value is F(0.025, 2, 6) = 7.26. The decision rule is: "Reject H_0 if the calculated F > 7.26, otherwise do not reject." Since the calculated test statistic falls in the nonrejection region, do not reject H_0. At the 0.025 level, we conclude that the battery treatment effects could all be zero. Therefore, at the 0.025 level, the batteries are equally effective. As shown in the Minitab printout, the p-value for the test of treatment effects is 0.034.

Testing for block effects:

H_0: $\beta_i = 0$ for i = 1, 2, 3, 4 (the levels of the blocking variable are equal in their effect)

H_1: $\beta_i \neq 0$ for at least one value of i (at least one level has an effect different from the others)

For the block effect, the calculated test statistic is: $F = \dfrac{MSB}{MSE} = \dfrac{2.239}{0.574} = 3.90$

Using the F table, the critical value is $F_{(0.01, 3, 6)} = 9.78$. The decision rule is: "Reject H_0 if the calculated F > 9.78, otherwise do not reject." Since the calculated test statistic falls in the nonrejection region, do not reject H_0. At the 0.01 level, we conclude that all of the block effects could be zero. As shown in the Minitab printout, the p-value for the test of block effects is 0.074.

Shown below is the Excel printout corresponding to this exercise. Once again, the p-value for the treatment (identified here as rows or batteries) effect is 0.034, and the p-value for the block (identified here as columns, or toy types) effect is 0.074.

	A	B	C	D	E	F	G
1		Rabbit	Duck	Teeth	Turtle		
2	BatteryA	6.1	7.7	7.4	5.5		
3	BatteryB	5.6	5.7	6.9	4.9		
4	BatteryC	3.4	4.3	6.4	5		
5							
6							
7	Anova: Two-Factor Without Replication						
8							
9	SUMMARY	Count	Sum	Average	Variance		
10	BatteryA	4	26.7	6.675	1.096		
11	BatteryB	4	23.1	5.775	0.689		
12	BatteryC	4	19.1	4.775	1.603		
13							
14	Rabbit	3	15.1	5.033	2.063		
15	Duck	3	17.7	5.900	2.920		
16	Teeth	3	20.7	6.900	0.250		
17	Turtle	3	15.4	5.133	0.103		
18							
19							
20	ANOVA						
21	Source of Variation	SS	df	MS	F	P-value	F crit
22	Rows	7.227	2	3.613	6.290	0.034	7.260
23	Columns	6.716	3	2.239	3.897	0.074	6.599
24	Error	3.447	6	0.574			
25							
26	Total	17.389	11				

12.53 p/c/m For this problem, the two control-button configurations are the treatments and the seven operators are the blocks. To determine if the button treatment effects could both be zero, we test: H_0: $\tau_j = 0$ for treatments $j = 1, 2$ and H_1: $\tau_j \neq 0$ for at least one of the treatments. A condensed version of the Excel printout is shown below:

	A	B	C	D	E	F	G
1		Button1	Button2				
2	Operator A	5	6				
3	Operator B	9	6				
4	Operator C	11	8				
5	Operator D	13	10				
6	Operator E	10	7				
7	Operator F	9	10				
8	Operator G	12	9				
9							
10	Anova: Two-Factor Without Replication						
11	SUMMARY	Count	Sum	Average	Variance		
12	Operator A	2	11	5.50	0.50		
13	Operator B	2	15	7.50	4.50		
14	Operator C	2	19	9.50	4.50		
15	Operator D	2	23	11.50	4.50		
16	Operator E	2	17	8.50	4.50		
17	Operator F	2	19	9.50	0.50		
18	Operator G	2	21	10.50	4.50		
19							
20	Button1	7	69	9.86	6.81		
21	Button2	7	56	8.00	3.00		
22							
23	ANOVA						
24	Source of Variation	SS	df	MS	F	P-value	F crit
25	Rows	47.43	6	7.90	4.15	0.054	4.28
26	Columns	12.07	1	12.07	6.34	0.045	5.99
27	Error	11.43	6	1.90			
28	Total	70.93	13				

The p-value for the treatment (identified here as columns or button configurations) effect is p-value = 0.045. Since p-value = 0.045 is < α = 0.05 level of significance for the test, we reject H_0. At the 0.05 level, we conclude that the treatment effects of the two layouts are not both zero, and that the two configurations are not equally effective.

12.55 c/a/m

Treatments d.f. = 14 - (2 + 8) = 4

$$MSTR = \frac{SSTR}{t-1} = \frac{35.33}{4} = 8.83, \quad MSB = \frac{SSB}{n-1} = \frac{134.4}{2} = 67.2, \quad MSE = \frac{SSE}{(t-1)(n-1)} = \frac{16.27}{8} = 2.03$$

$$F_T = \frac{MSTR}{MSE} = \frac{8.83}{2.03} = 4.35 \quad \text{and} \quad F_B = \frac{MSB}{MSE} = \frac{67.2}{2.03} = 33.1$$

The completed ANOVA table is:

Variation Source	Sum of Squares	Degrees of Freedom	Mean Square	F
Treatments	35.33	4	8.83	4.35
Blocks	134.40	2	67.20	33.10
Error	16.27	8	2.03	
Total	186.00	14		

The null and alternative hypotheses are:

H_0: $\tau_j = 0$ for treatments $j = 1, 2, 3, 4, 5$ and H_1: $\tau_j \neq 0$ for at least one of the treatments.

The calculated test statistic is F = 4.35. Using the F table, the critical value is $F(0.05, 4, 8) = 3.84$. The decision rule is: "Reject H_0 if the calculated F > 3.84, otherwise do not reject." Since the calculated test statistic falls in the rejection region, reject H_0. At the 0.05 level, we conclude that all of the treatment effects are not zero. Therefore, the treatments are not equally effective.

12.57 d/p/m The randomized block ANOVA with two treatments and the dependent-samples t-test of Chapter 11 are equivalent. The assumptions are the same for both procedures and the results will be the same at a specified level of significance.

12.59 p/c/m In applying the dependent-samples t-test to the data in exercise 12.53, the difference within each block will be $d = x_1 - x_2$. The null and alternative hypotheses will be $H_0: \mu_d = 0$ and $H_1: \mu_d \neq 0$. The Minitab printout is shown below:

```
Paired T-Test and CI: Button1, Button2

Paired T for Button1 - Button2

                  N      Mean     StDev    SE Mean
Button1           7     9.857     2.610      0.986
Button2           7     8.000     1.732      0.655
Difference        7     1.857     1.952      0.738

95% CI for mean difference: (0.052, 3.663)
T-Test of mean difference = 0 (vs not = 0): T-Value = 2.52   P-Value = 0.045
```

Interpreting this output, we can say that if the population means were equal, the probability of obtaining a test statistic this extreme or more extreme is 0.045. Since the p-value = 0.045 is less than $\alpha = 0.05$, reject H_0. At the 0.05 level, we conclude that the population means are not equal. This is the same conclusion that was reached in exercise 12.53.

12.61 d/p/m We assume the r observations in each cell have been drawn from normally distributed populations with equal variances.

12.63 d/p/m The two-way analysis of variance and the randomized block design are similar since they both consider two factors in the model. However, in the randomized block design, the blocking variable is only used for the purpose of exerting improved control over the examination of the single factor of interest. In two-way analysis of variance, we are interested in the effects of two factors and their interaction on the dependent variable.

12.65 d/p/e Replications refers to the number of persons or test units within each cell. Within each combination of levels, there will be k = 1 through r observations or replications.

12.67 d/p/m
$x_{ijk} = \mu + \alpha_i + \beta_j + (\alpha\beta)_{ij} + \varepsilon_{ijk}$
x_{ijk} is the kth observation within the ith level for Factor A and the jth level for Factor B.
μ is the overall population mean.
α_i is the effect of the ith level of Factor A.
β_j is is the effect of the jth level of Factor B.
$(\alpha\beta)_{ij}$ is the effect of the interaction between the ith level of Factor A and the jth level of Factor B.
ε_{ijk} is the random error associated with the sampling process.

There are three sets of null and alternative hypotheses to be tested.
1. Testing for main effects, Factor A
   H_0: $\alpha_i = 0$ for each level of Factor A, i = 1 through a. (No level of Factor A has an effect.)
   H_1: $\alpha_i \neq 0$ for at least one value of i. (At least one level of Factor A has an effect.)
2. Testing for main effects, Factor B
   H_0: $\beta_j = 0$ for each level of Factor B, j = 1 through b. (No level of Factor B has an effect.)
   H_1: $\beta_j \neq 0$ for at least one value of j. (At least one level of Factor B has an effect.)
3. H_0: $(\alpha\beta)_{ij} = 0$ for each combination of i and j. (There are no interaction effects.)
   H_1: $(\alpha\beta)_{ij} \neq 0$ for at least one combination of i and j. (There is at least one interaction effect.)

12.69 c/a/m

Testing for Main Effects, Factor A.
H_0: $\alpha_i = 0$ for each value of i, i = 1 through 4
H_1: $\alpha_i \neq 0$ for at least one value of i

The calculated test statistic is $F = \dfrac{MSA}{MSE} = 3.54$

For this test, $\alpha = 0.025$, a = 4, b = 3, and r = 3. The d.f. associated with the numerator of F is $v_1 = a - 1 = 4 - 1 = 3$ and the d.f. associated with the denominator of F is $v_2 = ab(r - 1) = 4(3)(3 - 1) = 24$. Using the F table, the critical value is F(0.025, 3, 24) = 3.72. The decision rule is: "Reject H_0 if the calculated F > 3.72, otherwise do not reject." Since the calculated test statistic falls in the nonrejection region, do not reject H_0. At the 0.025 level, we conclude that no level of Factor A has an effect.

Testing for Main Effects, Factor B.
H_0: $\beta_j = 0$ for each value of j, j = 1 through 3
H_1: $\beta_j \neq 0$ for at least one value of j

The test statistic is $F = \dfrac{MSB}{MSE} = 5.55$

For this test, $\alpha = 0.025$, a = 4, b = 3, and r = 3. The d.f. associated with the numerator of F is $v_1 = b - 1 = 3 - 1 = 2$ and the d.f. associated with the denominator of F is $v_2 = ab(r - 1) = 24$. Using the F table, the critical value is F(0.025, 2, 24) = 4.32. The decision rule is: "Reject H_0 if the calculated F > 4.32, otherwise do not reject." Since the calculated test statistic falls in the rejection region, reject H_0. At the 0.025 level, we conclude that at least one level of Factor B has an effect.

Testing for Interaction Effects
H_0: $(\alpha\beta)_{ij} = 0$ for each combination of i and j
H_1: $(\alpha\beta)_{ij} \neq 0$ for at least one combination of i and j

The test statistic is $F = \dfrac{MSAB}{MSE} = 12.40$

For this test, $\alpha = 0.025$, a = 4, b = 3, and r = 3. The d.f. associated with the numerator of F is $v_1 = (a - 1)(b - 1) = (4 - 1)(3 - 1) = 6$ and the d.f. associated with the denominator of F is $v_2 = ab(r - 1) = 24$. Using the F table, the critical value is F(0.025, 6, 24) = 2.99. The decision rule is: "Reject H_0 if the calculated F > 2.99, otherwise do not reject." Since the calculated test statistic falls in the rejection region, reject H_0. At the 0.025 level, we conclude that there is some relationship between the levels of Factor A and Factor B.

12.71 c/c/d

		Factor B			
		1	2	3	
Factor A	1	13, 10	19, 15	19, 17	$\overline{x}_{1..} = 15.5$
	2	15, 15	22, 19	16, 17	$\overline{x}_{2..} = 17.33$
		$\overline{x}_{.1.} = 13.25$	$\overline{x}_{.2.} = 18.75$	$\overline{x}_{.3.} = 17.25$	

The grand mean, $\overline{\overline{x}} = \dfrac{13+10+15+\ldots+16+17}{12} = 16.42$

$SSA = rb \sum_{i=1}^{a} (\overline{x}_{i..} - \overline{\overline{x}})^2 = 2(3)[(15.5-16.42)^2 + (17.33-16.42)^2] = 10.05$

$SSB = ra \sum_{j=1}^{b} (\overline{x}_{.j.} - \overline{\overline{x}})^2 = 2(2)[(13.25-16.42)^2 + (18.75-16.42)^2 + (17.25-16.42)^2] = 64.67$

$SSE = \sum_{i=1}^{a} \sum_{j=1}^{b} (x_{ijk} - \overline{x}_{ij.})^2 = [(13-11.5)^2 + (10-11.5)^2 + (15-15)^2 + (15-15)^2]$

$+[(19-17)^2 + (15-17)^2 + (22-20.5)^2 + (19-20.5)^2]$

$+[(19-18)^2 + (17-18)^2 + (16-16.5)^2 + (17-16.5)^2] = 19.5$

Note: $\overline{x}_{11.} = \dfrac{13+10}{2} = 11.5$

$SST = \sum_{i=1}^{a} \sum_{j=1}^{b} \sum_{k=1}^{r} (x_{ijk} - \overline{\overline{x}})^2 = [(13-16.42)^2 + (10-16.42)^2 + (15-16.42)^2 + (15-16.42)^2]$

$+[(19-16.42)^2 + (15-16.42)^2 + (22-16.42)^2 + (19-16.42)^2]$

$+[(19-16.42)^2 + (17-16.42)^2 + (16-16.42)^2 + (17-16.42)^2]$

$= 110.92$

$SSAB = SST - SSA - SSB - SSE = 110.92 - 10.05 - 64.67 - 19.5 = 16.70$

A   d.f. = a - 1 = 2 - 1 = 1
B   d.f. = b - 1 = 3 - 1 = 2
AB   d.f. = (a - 1)(b - 1) = 1(2) = 2
Error d.f. = ab(r - 1) = 2(3)(2 - 1) = 6
Total d.f. = abr - 1 = 2(3)(2) - 1 = 11

$MSA = \dfrac{SSA}{a-1} = \dfrac{10.05}{1} = 10.05$, $MSB = \dfrac{SSB}{b-1} = \dfrac{64.67}{2} = 32.34$, $MSAB = \dfrac{SSAB}{(a-1)(b-1)} = \dfrac{16.7}{2} = 8.35$

$MSE = \dfrac{SSE}{ab(r-1)} = \dfrac{19.5}{6} = 3.25$

$F_A = \dfrac{MSA}{MSE} = \dfrac{10.05}{3.25} = 3.09$, $F_B = \dfrac{MSB}{MSE} = \dfrac{32.33}{3.25} = 9.95$, $F_{AB} = \dfrac{MSAB}{MSE} = \dfrac{8.35}{3.25} = 2.57$

The ANOVA table is:

Variation Source	Sum of Squares	Degrees of Freedom	Mean Square	F
Factor A	10.05	1	10.05	3.09
Factor B	64.67	2	32.34	9.95
Interaction, AB	16.70	2	8.35	2.57
Error	19.50	6	3.25	
Total	110.92	11		

Testing for Main Effects, Factor A.

H_0: $\alpha_i = 0$ for each value of i, i = 1 through 2

H_1: $\alpha_i \neq 0$ for at least one value of i

The calculated test statistic is $F = \dfrac{MSA}{MSE} = 3.09$

Using the F table, the critical value is F(0.05, 1, 6) = 5.99. The decision rule is: "Reject H_0 if the calculated F > 5.99, otherwise do not reject." Since the calculated test statistic falls in the nonrejection region, do not reject H_0. At the 0.05 level, we conclude that no level of Factor A has an effect.

Testing for Main Effects, Factor B.

H_0: $\beta_j = 0$ for each value of j, j = 1 through 3

H_1: $\beta_j \neq 0$ for at least one value of j

The test statistic is $F = \dfrac{MSB}{MSE} = 9.95$

Using the F table, the critical value is F(0.05, 2, 6) = 5.14. The decision rule is: "Reject H_0 if the calculated F > 5.14, otherwise do not reject." Since the calculated test statistic falls in the rejection region, reject H_0. At the 0.05 level, we conclude that at least one level of Factor B has an effect.

Testing for Interaction Effects

H_0: $(\alpha\beta)_{ij} = 0$ for each combination of i and j

H_1: $(\alpha\beta)_{ij} \neq 0$ for at least one combination of i and j

The test statistic is $F = \dfrac{MSAB}{MSE} = 2.57$

Using the F table, the critical value is F(0.05, 2, 6) = 5.14. The decision rule is: "Reject H_0 if the calculated F > 5.14, otherwise do not reject." Since the calculated test statistic falls in the nonrejection region, do not reject H_0. At the 0.05 level, we conclude that there is no relationship between the levels of Factor A and Factor B.

The Minitab printout is shown below. Some values differ slightly from those shown previously, due to rounding. Note the p-values shown in the Minitab printout: Factor A test, p-value = 0.129; Factor B test, p-value = 0.012, and Interaction AB test, p-value = 0.157.

Two-way ANOVA: value versus FactorA, FactorB

Analysis of Variance for value

Source	DF	SS	MS	F	P
FactorA	1	10.08	10.08	3.10	0.129
FactorB	2	64.67	32.33	9.95	0.012
Interaction	2	16.67	8.33	2.56	0.157
Error	6	19.50	3.25		
Total	11	110.92			

```
                        Individual 95% CI
FactorA    Mean   ---------+---------+---------+---------+--
1          15.50  (-----------*-----------)
2          17.33              (-----------*-----------)
                   ---------+---------+---------+---------+--
                      15.00     16.50     18.00     19.50
```

```
                        Individual 95% CI
FactorB    Mean   ------+---------+---------+---------+-----
1          13.25  (--------*--------)
2          18.75                    (--------*--------)
3          17.25               (--------*--------)
                   ------+---------+---------+---------+-----
                      12.50     15.00     17.50     20.00
```

The row and column means, along with the cell means, are shown below:

Tabulated Statistics: FactorA, FactorB

Rows: FactorA   Columns: FactorB

	1	2	3	All
1	11.500	17.000	18.000	15.500
2	15.000	20.500	16.500	17.333
All	13.250	18.750	17.250	16.417

Cell Contents --
        value:Mean

160

The corresponding Excel printout is shown below. The p-values are: Factor A test, p-value = 0.129; Factor B test, p-value = 0.012, and Interaction AB test, p-value = 0.157.

	E	F	G	H	I	J	K
1			. B1	B2	B3		
2		A1	13	19	19		
3			10	15	17		
4		A2	15	22	16		
5			15	19	17		
6							
7	Anova: Two-Factor With Replication						
8							
9	SUMMARY	B1	B2	B3	Total		
10	A1						
11	Count	2	2	2	6		
12	Sum	23	34	36	93		
13	Average	11.50	17.00	18.00	15.50		
14	Variance	4.50	8.00	2.00	12.70		
15							
16	A2						
17	Count	2	2	2	6		
18	Sum	30	41	33	104		
19	Average	15.00	20.50	16.50	17.33		
20	Variance	0.00	4.50	0.50	7.47		
21							
22	Total						
23	Count	4	4	4			
24	Sum	53	75	69			
25	Average	13.25	18.75	17.25			
26	Variance	5.58	8.25	1.58			
27							
28	ANOVA						
29	Source of Variation	SS	df	MS	F	P-value	F crit
30	Sample	10.083	1	10.083	3.103	0.129	5.987
31	Columns	64.667	2	32.333	9.949	0.012	5.143
32	Interaction	16.667	2	8.333	2.564	0.157	5.143
33	Within	19.500	6	3.25			
34	Total	110.917	11				

12.73 c/c/d The Minitab printout for this problem is shown below. The p-values are: Factor A test, p-value = 0.000; Factor B test, p-value = 0.012, and Interaction AB test, p-value = 0.031.

```
Two-way ANOVA: value versus FactorA, FactorB

Analysis of Variance for value
Source        DF       SS       MS       F       P
FactorA        2   227.06   113.53   13.58   0.000
FactorB        3   113.44    37.81    4.52   0.012
Interaction    6   142.06    23.68    2.83   0.031
Error         24   200.67     8.36
Total         35   683.22

                         Individual 95% CI
FactorA       Mean    ----+---------+---------+---------+-------
1            64.17                               (------*------)
2            59.75             (------*------)
3            58.25        (------*------)
                      ----+---------+---------+---------+-------
                      57.50     60.00     62.50     65.00

                         Individual 95% CI
FactorB       Mean    ---+---------+---------+---------+--------
1            61.33               (-------*------)
2            58.78        (-------*------)
3            59.44          (-------*------)
4            63.33                     (-------*------)
                      ---+---------+---------+---------+--------
                      57.50     60.00     62.50     65.00
```

The row and column means, along with the cell means, are shown below:

Tabulated Statistics: FactorA, FactorB

```
Rows: FactorA      Columns: FactorB

            1        2        3        4       All

1        61.000   63.333   63.667   68.667   64.167
2        60.667   57.667   60.333   60.333   59.750
3        62.333   55.333   54.333   61.000   58.250
All      61.333   58.778   59.444   63.333   60.722

  Cell Contents --
         value:Mean
```

Shown below are the data, along with a portion of the Excel printout for this problem. We will refer to this partial printout and its p-values in reaching conclusions for the three hypothesis tests.

	G	H	I	J	K	L	M
1		B1	B2	B3	B4		
2	A1	58	63	64	66		
3		64	66	68	68		
4		61	61	59	72		
5	A2	59	56	58	60		
6		64	61	60	59		
7		59	56	63	62		
8	A3	64	57	55	61		
9		62	58	56	58		
10		61	51	52	64		
11							
12	ANOVA						
13	*Source of Variation*	*SS*	*df*	*MS*	*F*	*P-value*	*F crit*
14	Sample	227.056	2	113.528	13.578	0.000	3.403
15	Columns	113.444	3	37.815	4.523	0.012	3.009
16	Interaction	142.056	6	23.676	2.832	0.031	2.508
17	Within	200.667	24	8.361			
18	Total	683.222	35				

Testing for Main Effects, Factor A.

H_0: $\alpha_i = 0$ for each value of i, i = 1 through 3

H_1: $\alpha_i \neq 0$ for at least one value of i

The calculated test statistic is F = 13.578. The p-value for this test, to three decimal places, is 0.000, which is $< \alpha = 0.05$. Reject H_0. At the 0.05 level, at least one level of Factor A has an effect.

Testing for Main Effects, Factor B.

H_0: $\beta_j = 0$ for each value of j, j = 1 through 4

H_1: $\beta_j \neq 0$ for at least one value of j

The calculated test statistic is F = 4.523. The p-value is 0.012, which is $< \alpha = 0.05$. Reject H_0. At the 0.05 level, at least one level of Factor B has an effect.

Testing for Interaction Effects

H_0: $(\alpha\beta)_{ij} = 0$ for each combination of i and j

H_1: $(\alpha\beta)_{ij} \neq 0$ for at least one combination of i and j

The calculated test statistic is F = 2.832. The p-value for this test is 0.031, which is $< \alpha = 0.05$. Reject H_0. At the 0.05 level, there appears to be some relationship between the levels of Factor A and Factor B.

12.75 p/c/d Below is a partial Excel printout that includes the data for this problem:

	E	F	G	H	I	J	K
1		Classical	Rock				
2	Method1	29	49				
3		32	45				
4	Method2	35	41				
5		32	40				
6							
7	ANOVA						
8	Source of Variation	SS	df	MS	F	P-value	F crit
9	Sample	6.125	1	6.125	1.400	0.302	7.709
10	Columns	276.125	1	276.125	63.114	0.001	7.709
11	Interaction	45.125	1	45.125	10.314	0.033	7.709
12	Within	17.500	4	4.375			
13	Total	344.875	7				

Testing for Main Effects, Factor A.

H_0: $\alpha_i = 0$ for each value of i, i = 1 through 2

H_1: $\alpha_i \neq 0$ for at least one value of i

The calculated test statistic is F = 1.400. The p-value = 0.303 is not < α = 0.05. Do not reject H_0.

At the 0.05 level, none of the assembly methods has any effect on the number of boards assembled.

Testing for Main Effects, Factor B.

H_0: $\beta_j = 0$ for each value of j, j = 1 through 2

H_1: $\beta_j \neq 0$ for at least one value of j

The calculated test statistic is F = 63.114. The p-value is 0.001, which is < α = 0.05. Reject H_0.

At the 0.05 level, at least one of the types of background music has an effect on the number of circuit boards assembled.

Testing for Interaction Effects

H_0: $(\alpha\beta)_{ij} = 0$ for each combination of i and j

H_1: $(\alpha\beta)_{ij} \neq 0$ for at least one combination of i and j

The calculated test statistic is F = 10.314. The p-value for this test is 0.033, which is < α = 0.05.
Reject H_0. At the 0.05 level, there appears to be some relationship between the assembly method used and the type of background music played.

The corresponding Minitab printout is shown below. It includes the confidence intervals that we are going to verify by use of our formulas. We will also use Minitab to generate the cell means:

```
Two-way ANOVA: boards versus method, music

Analysis of Variance for boards
Source        DF        SS        MS        F        P
method         1      6.13      6.13     1.40    0.302
music          1    276.13    276.13    63.11    0.001
Interaction    1     45.13     45.13    10.31    0.033
Error          4     17.50      4.38
Total          7    344.88

                       Individual 95% CI
method      Mean    ---------+---------+---------+---------+-
1           38.8               (--------------*-------------)
2           37.0     (--------------*---------------)
                    ---------+---------+---------+---------+-
                          36.0      38.0      40.0      42.0

                       Individual 95% CI
music       Mean    --+---------+---------+---------+---------
1           32.0    (-----*-----)
2           43.8                            (-----*----)
                    --+---------+---------+---------+---------
                     30.0      35.0      40.0      45.0
```

163

```
Tabulated Statistics: method, music

 Rows: method     Columns: music

            1       2      All

 1     30.500   47.000   38.750
 2     33.500   40.500   37.000
 All   32.000   43.750   37.875

   Cell Contents --
           boards:Mean
```

For a confidence level of 95%, the right-tail area of interest is $(1 - 0.95)/2 = 0.025$ with d.f. = 4. Referring to the t table, t = 2.776. Using the square root of the MSE for the population standard deviation (and the Excel printout's MSE = 4.375), this will be the square root of 4.375, or 2.092. The 95% confidence interval for μ_{1A}, the population mean associated with level 1 of Factor A (the assembly method -- assembly method 1) is:

$$\overline{x}_{1A} \pm t \frac{s_p}{\sqrt{n_{1A}}} = 38.75 \pm 2.776(\frac{2.092}{\sqrt{4}}) = 38.75 \pm 2.904 \text{ or from } 35.846 \text{ to } 41.654$$

The 95% confidence interval for μ_{2A}, the population mean associated with level 2 of Factor A (the assembly method -- assembly method 2) is:

$$\overline{x}_{2A} \pm t \frac{s_p}{\sqrt{n_{2A}}} = 37 \pm 2.776(\frac{2.092}{\sqrt{4}}) = 37 \pm 2.904 \text{ or from } 34.096 \text{ to } 39.904$$

The 95% confidence interval for μ_{1B}, the population mean associated with level 1 of Factor B (the music played -- classical music) is:

$$\overline{x}_{1B} \pm t \frac{s_p}{\sqrt{n_{1B}}} = 32 \pm 2.776(\frac{2.092}{\sqrt{4}}) = 32 \pm 2.904 \text{ or from } 29.096 \text{ to } 34.904$$

The 95% confidence interval for μ_{2B}, the population mean associated with level 2 of Factor B (the music played -- rock music) is:

$$\overline{x}_{2B} \pm t \frac{s_p}{\sqrt{n_{2B}}} = 43.75 \pm 2.776(\frac{2.092}{\sqrt{4}}) = 43.75 \pm 2.904 \text{ or from } 40.846 \text{ to } 46.654$$

12.77 p/c/d Shown below is the underlying data and a partial Excel printout for this problem situation:

	F	G	H	I	J	K	L
1		Sloppy	Casual	Dressy			
2	Bag	41	24	14			
3		39	29	19			
4	No Bag	52	16	27			
5		49	16	21			
6							
7	ANOVA						
8	Source of Variation	SS	df	MS	F	P-value	F crit
9	Sample	18.75	1	18.75	2.27	0.182	8.81
10	Columns	1602.67	2	801.33	97.13	0.000	7.26
11	Interaction	258.00	2	129.00	15.64	0.004	7.26
12	Within	49.50	6	8.25			
13	Total	1928.92	11				

Testing for Main Effects, Factor A

H_0: $\alpha_i = 0$ for each value of i, i = 1 through 2

H_1: $\alpha_i \neq 0$ for at least one value of i

The calculated test statistic is F = 2.27. The p-value is 0.182, which is not $< \alpha = 0.025$. Do not reject H_0. At the 0.025 level, we conclude that carrying or not carrying a shopping bag has no effect on the number of seconds it takes to get waited on.

Testing for Main Effects, Factor B

H_0: $\beta_j = 0$ for each value of j, j = 1 through 3

H_1: $\beta_j \neq 0$ for at least one value of j

The calculated test statistic is F = 97.13. The p-value is displayed as 0.000, and this is $< \alpha = 0.025$. Reject H_0. At the 0.025 level, we conclude that at least one of the dress modes has an effect on the number of seconds it takes to get waited on.

Testing for Interaction Effects

H_0: $(\alpha\beta)_{ij} = 0$ for each combination of i and j

H_1: $(\alpha\beta)_{ij} \neq 0$ for at least one combination of i and j

The calculated test statistic is F = 15.64. The p-value for this test is 0.004, which is $< \alpha = 0.025$. Reject H_0. At the 0.025 level, there appears to be some relationship between whether a shopping bag is carried and mode of dress.

The corresponding Minitab printout is shown below. It includes the confidence intervals that we are going to verify by use of our formulas. We will also use Minitab to generate the cell means:

```
Two-way ANOVA: time versus bag, dress

Analysis of Variance for time
Source        DF       SS        MS        F        P
bag            1     18.75     18.75     2.27    0.182
dress          2   1602.67    801.33    97.13    0.000
Interaction    2    258.00    129.00    15.64    0.004
Error          6     49.50      8.25
Total         11   1928.92

                         Individual 95% CI
bag          Mean    -------+---------+---------+---------+----
1            27.7    (-------------*--------------)
2            30.2                 (--------------*-------------)
                     -------+---------+---------+---------+----
                         26.0      28.0      30.0      32.0

                         Individual 95% CI
dress        Mean    ----------+---------+---------+---------+-
1            45.3                                   (----*---)
2            21.3    (----*---)
3            20.3    (---*----)
                     ----------+---------+---------+---------+-
                            24.0      32.0      40.0      48.0
```

Using Minitab, the row, column, and cell means are as shown below:

```
Tabulated Statistics: bag, dress

 Rows: bag    Columns: dress

          1        2        3      All

 1    40.000   26.500   16.500   27.667
 2    50.500   16.000   24.000   30.167
 All  45.250   21.250   20.250   28.917

 Cell Contents --
          time:Mean
```

For a confidence level of 95%, the right-tail area of interest is (1 - 0.95)/2 = 0.025 with d.f. = 6. Referring to the t table, t = 2.447. Using the square root of the MSE for the population standard deviation, our estimate of the population standard deviation is the square root of 8.25, or 2.872 minutes.

The 95% confidence interval for μ_{1A}, the population mean associated with level 1 of Factor A (shopping bag -- a bag is carried) is:

$$\overline{x}_{1A} \pm t\frac{s_p}{\sqrt{n_{1A}}} = 27.667 \pm 2.447(\frac{2.872}{\sqrt{6}}) = 27.667 \pm 2.869 \text{ or from } 24.798 \text{ to } 30.536$$

The 95% confidence interval for μ_{2A}, the population mean associated with level 2 of Factor A (shopping bag -- a bag is not carried) is:

$$\overline{x}_{2A} \pm t\frac{s_p}{\sqrt{n_{2A}}} = 30.167 \pm 2.447(\frac{2.872}{\sqrt{6}}) = 30.167 \pm 2.869 \text{ or from } 27.298 \text{ to } 33.036$$

The 95% confidence interval for μ_{1B}, the population mean associated with level 1 of Factor B (mode of dress -- sloppy) is:

$$\overline{x}_{1B} \pm t\frac{s_p}{\sqrt{n_{1B}}} = 45.25 \pm 2.447(\frac{2.872}{\sqrt{4}}) = 45.25 \pm 3.514 \text{ or from } 41.736 \text{ to } 48.764$$

The 95% confidence interval for μ_{2B}, the population mean associated with level 2 of Factor B (mode of dress -- casual) is:

$$\overline{x}_{2B} \pm t\frac{s_p}{\sqrt{n_{2B}}} = 21.25 \pm 2.447(\frac{2.872}{\sqrt{4}}) = 21.25 \pm 3.514 \text{ or from } 17.736 \text{ to } 24.764$$

The 95% confidence interval for μ_{3B}, the population mean associated with level 3 of Factor B (mode of dress -- dressy) is:

$$\overline{x}_{3B} \pm t\frac{s_p}{\sqrt{n_{3B}}} = 20.25 \pm 2.447(\frac{2.872}{\sqrt{4}}) = 20.25 \pm 3.514 \text{ or from } 16.736 \text{ to } 23.764$$

12.79 p/a/d
Testing for Main Effects, Factor A

H_0: $\alpha_i = 0$ for each value of i, i = 1 through 3

H_1: $\alpha_i \neq 0$ for at least one value of i

Because p-value = 0.043 is less than the 0.05 level of significance being used to reach a conclusion, we reject H_0. At the 0.05 level, we conclude that at least one level of keyboard configuration has an effect on the number of minutes required to type a document.

Testing for Main Effects, Factor B

H_0: $\beta_j = 0$ for each value of j, j = 1 through 2

H_1: $\beta_j \neq 0$ for at least one value of j

Because p-value = 0.104 is not less than the 0.05 level of significance being used to reach a conclusion, we do not reject H_0. At the 0.05 level, we conclude that none of the word processing packages has any effect on the number of minutes required to type a document.

Testing for Interaction Effects

H_0: $(\alpha\beta)_{ij} = 0$ for each combination of i and j

H_1: $(\alpha\beta)_{ij} \neq 0$ for at least one combination of i and j

Because p-value = 0.021 is less than the 0.05 level of significance being used to reach a conclusion, we reject H_0. At the 0.05 level, we conclude that there is some relationship between the keyboard configuration and the word processing package.

CHAPTER EXERCISES

12.81 d/p/m When we are interested in the effects of one factor on the dependent variable and the treatments are randomly assigned to all of the test units in the experiment, we would use one-way analysis of variance. If all of the test units are similar, this would be the analysis to use. If we are interested in the effects of one factor on the dependent variable but the test units can be arranged into similar groups, or "blocks" before the treatments are assigned, we would use the randomized block design. This allows us to reduce the amount of "error" variation. If we are interested in the effects of two different factors on the dependent variable, then two-way analysis of variance would be used. Then we can look at the effects of both factors simultaneously and also test for any interaction between them.

12.83 d/p/m The randomized block design has not been used in this analysis. In the randomized block design, the confidence interval for a treatment mean would not be meaningful. The treatments are applied randomly only within a block.

12.85 d/c/m The independent variable is faceplate design and the dependent variable is time to complete the assigned task.

12.87 d/c/m No. The randomized block ANOVA procedure would now be appropriate. We are not really interested in the effect of the age variable, we are only interested in using it as a blocking variable. This study is similar to the headlamp-design example in the randomized block section of the chapter.

12.89 p/c/m The Minitab and Excel printouts are shown below. The Excel printout includes the data.

```
One-way ANOVA: present1, present2, present3

Analysis of Variance
Source     DF        SS        MS        F        P
Factor      2     46.61     23.31     7.00    0.011
Error      11     36.60      3.33
Total      13     83.21
                                     Individual 95% CIs For Mean
                                     Based on Pooled StDev
Level       N      Mean     StDev   -----+---------+---------+---------+-
present1    4     8.500     1.915    (-------*-------)
present2    5    12.200     1.643                     (------*------)
present3    5    12.800     1.924                        (------*------)
                                     -----+---------+---------+---------+-
Pooled StDev =    1.824              7.5      10.0      12.5      15.0
```

	E	F	G	H	I	J	K
1			present1	present2	present3		
2			7	13	14		
3			9	10	15		
4			11	11	12		
5	Anova: Single Factor		7	13	10		
6				14	13		
7	SUMMARY						
8	Groups	Count	Sum	Average	Variance		
9	present1	4	34	8.50	3.67		
10	present2	5	61	12.20	2.70		
11	present3	5	64	12.80	3.70		
12							
13	ANOVA						
14	Source of Variation	SS	df	MS	F	P-value	F crit
15	Between Groups	46.614	2	23.307	7.005	0.011	3.982
16	Within Groups	36.600	11	3.327			
17	Total	83.214	13				

The null and alternative hypotheses are: H_0: $\mu_1 = \mu_2 = \mu_3$ and H_1: the population means are not equal. The calculated test statistic is F = 7.005. The p-value for this test is 0.011, which is $< \alpha = 0.05$. Reject H_0. At the 0.05 level, we conclude that the three sales presentations are not equally effective.

12.91 p/c/m For this problem, the four assessors are the treatments and the five homes are the blocks. The Minitab printout and a portion of the Excel printout are shown below:

```
Two-way ANOVA: amount versus home, assessor

Analysis of Variance for amount
Source      DF        SS        MS        F        P
home         4    6638.3    1659.6    38.20    0.000
assessor     3     461.2     153.7     3.54    0.048
Error       12     521.3      43.4
Total       19    7620.8
```

	G	H	I	J	K	L	M
1			person1	person2	person3	person4	
2		home1	40	48	55	53	
3		home2	49	46	52	50	
4		home3	35	47	51	48	
5		home4	60	54	70	72	
6		home5	100	81	109	88	
7	ANOVA						
8	Source of Variation	SS	df	MS	F	P-value	F crit
9	Rows	6638.30	4	1659.575	38.202	0.000	4.121
10	Columns	461.20	3	153.733	3.539	0.048	4.474
11	Error	521.30	12	43.442			
12	Total	7620.80	19				

Testing for treatment effects:
To determine if the treatment (assessor) effects could all be zero, we test:
H_0: $\tau_j = 0$ for treatments j = 1, 2, 3, 4 and H_1: $\tau_j \neq 0$ for at least one of the treatments

Referring to the Excel printout, the calculated test statistic is: $F = \dfrac{MSTR}{MSE} = \dfrac{153.733}{43.442} = 3.539$

and the p-value for the test of treatment (assessor) effects is 0.048. The p-value = 0.048 is not less than the $\alpha = 0.025$ level of significance being used in the test, so we do not reject H_0. At the 0.025 level, we conclude that the assessments provided by the four assessors are equal. As might be expected, the p-value for the test of the block (homes) effect is much stronger. To three decimal places, the p-value = 0.000.

12.93 p/c/m We wish to determine if the three brands of bathroom scales could have the same population mean for this test object, so the null and alternative hypotheses are H_0: $\mu_1 = \mu_2 = \mu_3$ and H_1: the population means are not equal. Applying one-way analysis of variance, we obtain the Minitab and Excel printouts below:

```
One-way ANOVA: BrandA, BrandB, BrandC

Analysis of Variance
Source      DF        SS        MS        F        P
Factor       2    126.72     63.36     8.52    0.005
Error       12     89.28      7.44
Total       14    216.00
                            Individual 95% CIs For Mean
                            Based on Pooled StDev
Level        N      Mean     StDev   --------+---------+---------+--------
BrandA       5    202.40      3.21                         (------*------)
BrandB       4    199.75      2.99                 (--------*-------)
BrandC       6    195.67      2.07   (------*------)
                                     --------+---------+---------+--------
Pooled StDev =     2.73               196.0     199.5     203.0
```

	E	F	G	H	I	J	K
12	Anova: Single Factor						
13	SUMMARY						
14	*Groups*	*Count*	*Sum*	*Average*	*Variance*		
15	BrandA	5	1012	202.40	10.30		
16	BrandB	4	799	199.75	8.92		
17	BrandC	6	1174	195.67	4.27		
18							
19	ANOVA						
20	*Source of Variation*	*SS*	*df*	*MS*	*F*	*P-value*	*F crit*
21	Between Groups	126.717	2	63.358	8.516	0.005	5.096
22	Within Groups	89.283	12	7.440			
23	Total	216.000	14				

Since p-value = 0.005 is < α = 0.025, the level of significance for the test, we reject H_0. At the 0.025 level, we conclude that the three brands of bathroom scales do not have the same population mean for this test object.

12.95 p/c/m For this problem, the three portfolios are the treatments and the four advisors are the blocks. The Minitab printout is shown below:

```
Two-way ANOVA: estrate versus advisor, portfol

Analysis of Variance for estrate
Source      DF        SS        MS       F        P
advisor      3     48.33     16.11    5.98    0.031
portfol      2     11.17      5.58    2.07    0.207
Error        6     16.17      2.69
Total       11     75.67
```

Since p-value = 0.207 is not less than α = 0.05, the level of significance for the test, we do not reject H_0. At the 0.05 level, we conclude that the three portfolios could be equal in terms of their estimated annual return.

12.97 p/c/m For this problem, the three publicity methods are the treatments and the three community sizes are the blocks. Minitab and Excel printouts are shown below:

```
Two-way ANOVA: Values versus CommSize, PubMeth

Analysis of Variance for Values
Source      DF        SS        MS       F        P
CommSize     2    178.67     89.33    33.50    0.003
PubMeth      2     68.67     34.33    12.87    0.018
Error        4     10.67      2.67
Total        8    258.00
```

	E	F	G	H	I	J	K
1			Method1	Method2	Method3		
2		SmCom	14	13	8		
3		MedCom	17	15	13		
4		LrgCom	27	23	17		
5							
6	Anova: Two-Factor Without Replication						
7	SUMMARY	Count	Sum	Average	Variance		
8	SmCom	3	35	11.67	10.33		
9	MedCom	3	45	15.00	4.00		
10	LrgCom	3	67	22.33	25.33		
11							
12	Method1	3	58	19.33	46.33		
13	Method2	3	51	17.00	28.00		
14	Method3	3	38	12.67	20.33		
15							
16	ANOVA						
17	Source of Variation	SS	df	MS	F	P-value	F crit
18	Rows	178.667	2	89.333	33.500	0.003	6.944
19	Columns	68.667	2	34.333	12.875	0.018	6.944
20	Error	10.667	4	2.667			
21	Total	258.000	8				

Testing for treatment effects:

To determine if the treatment (publicity method) effects could all be zero, we test:

H_0: $\tau_j = 0$ for treatments j = 1, 2, 3   and H_1: $\tau_j \neq 0$ for at least one of the treatments

Referring to the Excel printout, the calculated test statistic is: $F = \dfrac{MSTR}{MSE} = \dfrac{34.333}{2.667} = 12.875$ and the

p-value for the treatment (publicity method) effects is 0.018. Since p-value = 0.018 is $< \alpha = 0.05$, the level of significance being used in the test, we reject H_0. At the 0.05 level, we conclude that the three publicity methods are not equally effective. The p-value for the test of the block (community size) effects is even lower: 0.003.

12.99 p/c/d There are two factors: A (type style) and B (type darkness). We will apply two-way analysis of variance with two observations per cell. The Minitab printout is shown below. Factor means and confidence intervals for factor levels are included.

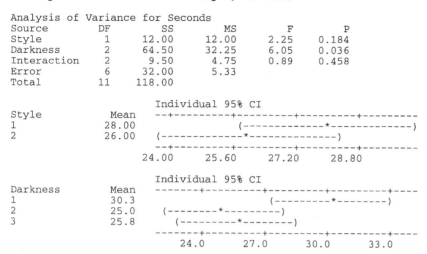

```
Two-way ANOVA: Seconds versus Style, Darkness

Analysis of Variance for Seconds
Source          DF      SS       MS       F       P
Style            1    12.00    12.00    2.25    0.184
Darkness         2    64.50    32.25    6.05    0.036
Interaction      2     9.50     4.75    0.89    0.458
Error            6    32.00     5.33
Total           11   118.00

                        Individual 95% CI
Style       Mean    --+---------+---------+---------+---------
1          28.00                      (-------------*-------------)
2          26.00     (-------------*--------------)
                    --+---------+---------+---------+---------
                    24.00     25.60     27.20     28.80

                        Individual 95% CI
Darkness    Mean    -------+---------+---------+---------+----
1          30.3                      (---------*--------)
2          25.0      (--------*---------)
3          25.8      (---------*--------)
                    -------+---------+---------+---------+----
                        24.0      27.0      30.0      33.0
```

Using Minitab, we can also generate the cell means along with the factor level means:

```
Tabulated Statistics: Style, Darkness

Rows: Style     Columns: Darkness

          1        2        3       All

1     30.500   25.500   28.000   28.000
2     30.000   24.500   23.500   26.000
All   30.250   25.000   25.750   27.000

Cell Contents --
        Seconds:Mean
```

The data setup and a portion of the Excel printout are shown below:

	A	B	C	D	E	F	G
1		TypLight	TypeMed	TypeDrk			
2	TypSty1	29	23	26			
3		32	28	30			
4	TypSty2	29	26	23			
5		31	23	24			
6							
7	Anova: Two-Factor With Replication						
8	ANOVA						
9	Source of Variation	SS	df	MS	F	P-value	F crit
10	Sample	12.000	1	12.000	2.250	0.184	5.987
11	Columns	64.500	2	32.250	6.047	0.036	5.143
12	Interaction	9.500	2	4.750	0.891	0.458	5.143
13	Within	32.000	6	5.333			
14	Total	118.000	11				

Testing for Main Effects, Factor A (type style):

H_0: $\alpha_i = 0$ for each value of i, i = 1 through 2

H_1: $\alpha_i \neq 0$ for at least one value of i

The calculated test statistic is F = 2.250. The p-value is 0.184, which is not $< \alpha = 0.05$.

Do not reject H_0. At the 0.05 level, none of the type styles has any effect on readability.

Testing for Main Effects, Factor B (type darkness)

  H_0: $\beta_j = 0$ for each value of j, j = 1 through 3

  H_1: $\beta_j \neq 0$ for at least one value of j

  The calculated test statistic is F = 6.047. The p-value is 0.036, which is $< \alpha = 0.05$. Reject H_0.

  At the 0.05 level, at least one of the levels of type darkness has an effect on readability.

Testing for Interaction Effects

  H_0: $(\alpha\beta)_{ij} = 0$ for each combination of i and j

  H_1: $(\alpha\beta)_{ij} \neq 0$ for at least one combination of i and j

  The calculated test statistic is F = 0.891. The p-value for this test is 0.458, which is not $< \alpha = 0.05$.

  Do not reject H_0. At the 0.05 level, there is no relationship between type style and type darkness.

For a confidence level of 95%, the right-tail area of interest is (1 - 0.95)/2 = 0.025 with d.f. = 6. Referring to the t table, t = 2.447. MSE = 5.333, and its square root (2.309) is the estimate for the population standard deviation. The 95% confidence interval for each factor level mean (shown in the Minitab printout) can be computed from the Excel printout as shown below:

The 95% confidence interval for μ_{1A}, the population mean associated with level 1 of Factor A (type style) is:

$$\overline{x}_{1A} \pm t\frac{s_p}{\sqrt{n_{1A}}} = 28 \pm 2.447(\frac{2.309}{\sqrt{6}}) = 28 \pm 2.307 \text{ or from } 25.693 \text{ to } 30.307$$

The 95% confidence interval for μ_{2A}, the population mean associated with level 2 of Factor A (type style) is:

$$\overline{x}_{2A} \pm t\frac{s_p}{\sqrt{n_{2A}}} = 26 \pm 2.447(\frac{2.309}{\sqrt{6}}) = 26 \pm 2.307 \text{ or from } 23.693 \text{ to } 28.307$$

The 95% confidence interval for μ_{1B}, the population mean associated with level 1 of Factor B (type darkness) is:

$$\overline{x}_{1B} \pm t\frac{s_p}{\sqrt{n_{1B}}} = 30.25 \pm 2.447(\frac{2.309}{\sqrt{4}}) = 30.25 \pm 2.825 \text{ or from } 27.425 \text{ to } 33.075$$

The 95% confidence interval for μ_{2B}, the population mean associated with level 2 of Factor B (type darkness) is:

$$\overline{x}_{2B} \pm t\frac{s_p}{\sqrt{n_{2B}}} = 25 \pm 2.447(\frac{2.309}{\sqrt{4}}) = 25 \pm 2.825 \text{ or from } 22.175 \text{ to } 27.825$$

The 95% confidence interval for μ_{3B}, the population mean associated with level 3 of Factor B (type darkness) is:

$$\overline{x}_{3B} \pm t\frac{s_p}{\sqrt{n_{3B}}} = 25.75 \pm 2.447(\frac{2.309}{\sqrt{4}}) = 25.75 \pm 2.825 \text{ or from } 22.925 \text{ to } 28.575$$

12.101 p/c/d There are two factors: A (position of the display) and B (type of display). Applying two-way analysis of variance with three observations per cell, Minitab and Excel printouts are shown below:

```
Two-way ANOVA: Tickets versus Position, Display

Analysis of Variance for Tickets
Source        DF        SS        MS        F        P
Position       1      88.9      88.9     6.67    0.024
Display        2      71.4      35.7     2.68    0.109
Interaction    2     510.1     255.1    19.13    0.000
Error         12     160.0      13.3
Total         17     830.4


                          Individual 95% CI
Position      Mean    ---------+---------+---------+---------+-
1             45.3     (---------*----------)
2             49.8                         (---------*----------)
                      ---------+---------+---------+---------+-
                          45.0      47.5      50.0      52.5


                          Individual 95% CI
Display       Mean    ---------+---------+---------+---------+-
1             45.3     (----------*----------)
2             47.2        (----------*----------)
3             50.2              (----------*----------)
                      ---------+---------+---------+---------+-
                          45.0      48.0      51.0      54.0
```

The Excel data setup and printout:

	A	B	C	D
1		Display1	Display2	Display3
2	Position1	43	39	57
3		39	38	60
4		40	43	49
5	Position2	53	58	47
6		46	55	42
7		51	50	46

	E	F	G	H	I	J	K
1	Anova: Two-Factor With Replication						
2							
3	SUMMARY	Display1	Display2	Display3	Total		
4	Position1						
5	Count	3	3	3	9		
6	Sum	122	120	166	408		
7	Average	40.667	40.000	55.333	45.333		
8	Variance	4.333	7.000	32.333	67.250		
9							
10	Position2						
11	Count	3	3	3	9		
12	Sum	150	163	135	448		
13	Average	50.000	54.333	45.000	49.778		
14	Variance	13.000	16.333	7.000	25.444		
15							
16	Total						
17	Count	6	6	6			
18	Sum	272	283	301			
19	Average	45.333	47.167	50.167			
20	Variance	33.067	70.967	47.767			
21							
22	ANOVA						
23	Source of Variation	SS	df	MS	F	P-value	F crit
24	Sample	88.9	1	88.9	6.667	0.024	4.747
25	Columns	71.4	2	35.7	2.679	0.109	3.885
26	Interaction	510.1	2	255.1	19.129	0.000	3.885
27	Within	160.0	12	13.3			
28	Total	830.4	17				

Testing for Main Effects, Factor A (position of display)

H_0: $\alpha_i = 0$ for each value of i, i = 1 through 2

H_1: $\alpha_i \neq 0$ for at least one value of i

The calculated test statistic is F = 6.667. The p-value is 0.024, which is $< \alpha = 0.025$. Reject H_0.
At the 0.025 level, we conclude that the two display positions are not equally effective.

Testing for Main Effects, Factor B (type of display)

H_0: $\beta_j = 0$ for each value of j, j = 1 through 3

H_1: $\beta_j \neq 0$ for at least one value of j

The calculated test statistic is F = 2.679. The p-value is 0.109, which is not $< \alpha = 0.025$.
Do not reject H_0. At the 0.025 level, we conclude that the three display positions are equally effective.

Testing for Interaction Effects

H_0: $(\alpha\beta)_{ij} = 0$ for each combination of i and j

H_1: $(\alpha\beta)_{ij} \neq 0$ for at least one combination of i and j

The calculated test statistic is F = 19.129. The p-value for this test is 0.000, which is $< \alpha = 0.025$.
Reject H_0. At the 0.025 level, we conclude that there is some relationship between the position of the display and the type of display.

For a confidence level of 95%, the right-tail area of interest is (1 - 0.95)/2 = 0.025 with d.f. = 12.
Referring to the t table, t = 2.179. MSE = 13.3, and its square root (3.647) is the estimate for the population standard deviation. The 95% confidence interval for each factor level mean (shown in the Minitab printout) can be computed from the Excel printout as shown below:

The 95% confidence interval for μ_{1A}, the population mean associated with level 1 of Factor A (display position) is:

$$\overline{x}_{1A} \pm t\frac{s_p}{\sqrt{n_{1A}}} = 45.333 \pm 2.179(\frac{3.647}{\sqrt{9}}) = 45.333 \pm 2.649 \text{ or from } 42.684 \text{ to } 47.982$$

The 95% confidence interval for μ_{2A}, the population mean associated with level 2 of Factor A (display position) is:

$$\overline{x}_{2A} \pm t\frac{s_p}{\sqrt{n_{2A}}} = 49.778 \pm 2.179(\frac{3.647}{\sqrt{9}}) = 49.778 \pm 2.649 \text{ or from } 47.129 \text{ to } 52.427$$

The 95% confidence interval for μ_{1B}, the population mean associated with level 1 of Factor B (type of display) is:

$$\overline{x}_{1B} \pm t\frac{s_p}{\sqrt{n_{1B}}} = 45.333 \pm 2.179(\frac{3.647}{\sqrt{6}}) = 45.333 \pm 3.244 \text{ or from } 42.089 \text{ to } 48.577$$

The 95% confidence interval for μ_{2B}, the population mean associated with level 2 of Factor B (type of display) is:

$$\overline{x}_{2B} \pm t\frac{s_p}{\sqrt{n_{2B}}} = 47.167 \pm 2.179(\frac{3.647}{\sqrt{6}}) = 47.167 \pm 3.244 \text{ or from } 43.923 \text{ to } 50.411$$

The 95% confidence interval for μ_{3B}, the population mean associated with level 3 of Factor B (type of display) is:

$$\overline{x}_{3B} \pm t\frac{s_p}{\sqrt{n_{3B}}} = 50.167 \pm 2.179(\frac{3.647}{\sqrt{6}}) = 50.167 \pm 3.244 \text{ or from } 46.923 \text{ to } 53.411$$

CHAPTER 13
CHI-SQUARE APPLICATIONS

SECTION EXERCISES

13.1 d/p/e Chi-square analysis can be used for goodness of fit tests. The null hypothesis in this case is that the sample is from the specified population, and the alternative hypothesis is that it is not from the specified population. In addition, it can be used for testing the independence of two variables. The null hypothesis here is that the variables are independent, and the alternative hypothesis is that they are not independent. chi-square analysis can also be used to compare population proportions from two or more independent samples. The null hypothesis for this test is that all the population proportions are equal, and the alternative hypothesis is that at least one proportion differs. Finally, chi-square analysis can be used to test a population variance. This test can be a one- or two-tail test where the null hypothesis is that the variance is equal to, no more than, or no less than a specified value.

13.3 d/p/e The chi-square distribution is continuous and approaches the normal distribution as d.f. increases.

13.5 d/p/m The chi-square statistic can never be negative. Recall that any number squared is positive.

13.7 c/p/e This problem requires the chi-square distribution table found in the appendix. The appropriate values for A with 8 degrees of freedom are:
a. A = 3.490   d. A = 15.507
b. A = 13.362   e. A = 17.535
c. A = 2.733   f. A = 2.180

13.9 c/p/m This problem requires the chi-square distribution table found in the appendix. The tail area must be divided in half to account for each tail. The appropriate values for A and B with 15 degrees of freedom are:
a. A = 8.547 B = 22.307   c. A = 6.262 B = 27.488
b. A = 7.261 B = 24.996   d. A = 5.229 B = 30.578

13.11 d/p/m The calculated chi-square will be very small if the expected frequencies differ very little from the observed frequencies. A small test statistic results in a "fail to reject the null hypothesis" decision, and this is what we expect if the differences between the observed and expected frequencies are small. The formula for the test statistic also shows why this calculated chi-square will be small -- each difference between the observed and expected frequencies squared and divided by the expected frequencies will be small, and the sum of several small numbers is a small number.

13.13 c/p/m The degrees of freedom are k - 1 - m. For this problem, k = 5, and m = 0, so the appropriate value is d.f. = 5 - 1 - 0 = 4.

13.15 c/p/m
a. The degrees of freedom are 8 - 1 - 1 = 6. (m = 1 in this case since we have to estimate the standard deviation to compute our expected frequencies.)
b. The critical value of chi-square with 6 degrees of freedom and alpha = 0.05 is 12.592.
c. Since the calculated value is less than the critical value, we fail to reject the null hypothesis and conclude, at the 0.05 level, that there is no evidence to suggest that the sample was not drawn from a normal population.

13.17 p/a/m The null and alternative hypotheses are:

H_0: The gifts are equally preferred.

H_1: The gifts are not equally preferred.

The degrees of freedom are 3 - 1 - 0 = 2, and the critical value of chi-square at the 0.05 level is 5.991. The calculations are shown below.

group	observed	expected	calculated*
A	11	20	4.050
B	21	20	0.050
C	28	20	3.200
			7.300

*Note that the "calculated" column is $\dfrac{[\text{observed} - \text{expected}]^2}{\text{expected}}$ and the sum at the bottom of the

"calculated" column is the calculated chi-square for this problem. Since the calculated value (7.300) is greater than the critical value (5.991), we reject the null hypothesis. There is evidence, at the 0.05 level, to suggest that the gifts are not equally preferred.

13.19 p/a/m The null and alternative hypotheses are:

H_0: Absences during any day are equally likely.

H_1: Absences during any day are not equally likely.

The degrees of freedom for this problem are 5 - 1 - 0 = 4, and the critical value of chi-square at the 0.01 level is 13.277. Using Excel worksheet template tmchifit.xls, on the disk supplied with the text, the C column shows the individual components and cell E4 shows their sum, the calculated chi-square. The calculated value (13.800) exceeds the critical value at the 0.01 level (13.277), and we reject H_0. There is evidence, at the 0.01 level, to suggest that the absences during any day are not equally likely. Alternatively, we are able to reject H_0 because p-value = 0.0080 is $< \alpha = 0.01$.

	A	B	C	D	E
1	Chi-Square Goodness-of-Fit Test			no. of cells, k =	5
2	Cell Frequencies:			no. of parameters estimated, m =	0
3	Observed (Oj):	Expected (Ej):	(Oj-Ej)^2/Ej:	df = k - 1 - m =	4
4	42	30	4.800	calculated chi-square =	13.800
5	18	30	4.800	p-value =	0.0080
6	24	30	1.200		
7	27	30	0.300		
8	39	30	2.700		

13.21 p/a/m The null and alternative hypotheses are:

H_0: The data are from a Poisson distribution with $\lambda = 1.5$

H_1: The data are not from a Poisson distribution with $\lambda = 1.5$

The degrees of freedom are 4 - 1 - 0 = 3, and the critical value of chi-square at the 0.05 level is 7.815. The expected counts are calculated with the probabilities given in the Poisson table in the appendix. Using Excel worksheet template tmchifit.xls, on the disk supplied with the text, the C column shows the individual components and cell E4 shows their sum, the calculated chi-square. The calculated value does not exceed the critical value at the 0.05 level, and we are unable to reject H_0. There is no evidence, at the 0.05 level, to suggest that the ownership is not Poisson-distributed with a mean of 1.5. Alternatively, using Excel worksheet template tmchifit.xls and the p-value approach, we are unable to reject H_0 because p-value = 0.8209 is not $< \alpha = 0.05$.

	A	B	C	D	E
1	Chi-Square Goodness-of-Fit Test			no. of cells, k =	4
2	Cell Frequencies:			no. of parameters estimated, m =	0
3	Observed (Oj):	Expected (Ej):	(Oj-Ej)^2/Ej:	df = k - 1 - m =	3
4	20	22.31	0.239	calculated chi-square =	0.919
5	35	33.47	0.070	p-value =	0.8209
6	23	25.10	0.176		
7	22	19.12	0.434		

13.23 p/a/d The null and alternative hypotheses are:

H_0: The number of defectives in each batch is binomial distributed with $\pi = 0.1$

H_1: The number of defectives in each batch is not binomial distributed with $\pi = 0.1$

For this problem, d.f. = 5 - 1 - 0 = 4, and the critical value of chi-square at the 0.05 level is 9.488.

The two lowest categories have been combined into "≤ 1", and the expected counts are computed using the probabilities in the binomial table in the appendix.

Using Excel worksheet template tmchifit.xls, on the disk supplied with the text, the C column shows the individual components and cell E4 shows their sum, the calculated chi-square. The calculated value does not exceed the critical value at the 0.05 level, and we are unable to reject H_0. There is no evidence, at the 0.05 level, to suggest that the number of defectives is not binomial distributed with $\pi = 0.1$. Alternatively, using the p-value approach, we are unable to reject H_0 because p-value = 0.1847 is not < α = 0.05.

	A	B	C	D	E
1	**Chi-Square Goodness-of-Fit Test**			no. of cells, k =	5
2	Cell Frequencies:			no. of parameters estimated, m =	0
3	Observed (Oj):	Expected (Ej):	(Oj-Ej)^2/Ej:	df = k - 1 - m =	4
4	20	13.560	3.059	calculated chi-square =	6.199
5	14	13.295	0.037	p-value =	0.1847
6	10	11.325	0.155		
7	4	6.920	1.232		
8	2	4.900	1.716		

13.25 p/c/m The null and alternative hypotheses are:

H_0: The mileages are normally distributed and H_1: The mileages are not normally distributed.

For this test we will use Data Analysis Plus and its chi-squared test for normality. Note that the d.f. calculation is d.f. = k - 1 - m, or d.f. = 5 - 1 - 2 = 2. The "m" term is 2 because we have used the sample mean and standard deviation to estimate their (unknown) population counterparts.

The calculated chi-square (4.491) does not exceed the critical value (9.210 for d.f. = 2 and the 0.01 level of significance) and we do not reject H_0. Alternatively, using the p-value approach, we are not able to reject H_0 because p-value = 0.106 is not less than α = 0.01. Our conclusion is that the population could be normally distributed.

	A	B	C	D
1	**Chi-Squared Test of Normality**			
2				
3		Miles		
4	Mean	14636.95		
5	Standard deviation	2483.45		
6	Observations	200		
7				
8	Intervals	Probability	Expected	Observed
9	(z <= -1.5)	0.0668	13.361	10
10	(-1.5 < z <= -0.5)	0.2417	48.346	56
11	(-0.5 < z <= 0.5)	0.3829	76.585	77
12	(0.5 < z <= 1.5)	0.2417	48.346	40
13	(z > 1.5)	0.0668	13.361	17
14				
15				
16	chi-squared Stat	4.491		
17	df	2		
18	p-value	0.106		
19	chi-squared Critical	9.210		

13.27 p/c/m For x = the number of defects per panel, the null and alternative hypotheses are:
H_0: x is normally distributed and H_1: x is not normally distributed.
For this test we will use Data Analysis Plus and its chi-squared test for normality. Note that the d.f. calculation is d.f. = k - 1 - m, or d.f. = 6 - 1 - 2 = 3. The "m" term is 2 because we have used the sample mean and standard deviation to estimate their (unknown) population counterparts.
The calculated chi-square (31.585) exceeds the critical value (7.815 for d.f. = 3 and the 0.05 level of significance) and we reject H_0. Alternatively, using the p-value approach, we are able to reject H_0 because p-value = 0.0000 (to four decimal places) is less than $\alpha = 0.05$. Our conclusion is that the population could not be normally distributed.

	A	B	C	D
1	**Chi-Squared Test of Normality**			
2				
3		*defects*		
4	Mean	4.983		
5	Standard deviation	1.7435		
6	Observations	300		
7				
8	Intervals	Probability	Expected	Observed
9	(z <= -2)	0.0228	6.8250	2
10	(-2 < z <= -1)	0.1359	40.7715	59
11	(-1 < z <= 0)	0.3413	102.4035	63
12	(0 < z <= 1)	0.3413	102.4035	117
13	(1 < z <= 2)	0.1359	40.7715	50
14	(z > 2)	0.0228	6.8250	9
15				
16	chi-squared Stat	31.585		
17	df	3		
18	p-value	0.0000		
19	chi-squared Critical	7.815		

13.29 d/p/e The chi-square distribution is continuous, while the counts on which the test statistic is based are discrete. The approximation will be unsatisfactory whenever one or more of the expected frequencies is small (less than five).

13.31 c/p/e The degrees of freedom are calculated as in exercise 13.30. The appropriate d.f. values are:
a. 2*3 = 6   b. 1*2 = 2   c. 3*4 = 12   d. 4*2 = 8   e. 2*6 = 12   f. 2*2 = 4

13.33 c/p/e The critical values of chi-square are:
a. 12.833   b. 15.507   c. 16.812   d. 7.779

13.35 c/p/m The null and alternative hypotheses are:
H_0: Age category is independent of type of greeting received
H_1: Age category is not independent of type of greeting received
With d.f. = (3 - 1)(3 - 1) = 4, the critical value of chi-square at the 0.025 level is 11.143. We can do this type of problem with a calculator or with Data Analysis Plus, but we will use Minitab.
The printout below shows the actual frequencies, the expected frequencies, the cell contributions to the calculated chi-square, and the p-value:

```
Expected counts are printed below observed counts
            C1        C2        C3      Total
    1       16        12         5        33
            9.90     15.18      7.92

    2        8        20         6        34
           10.20     15.64      8.16

    3        6        14        13        33
            9.90     15.18      7.92

Total      30        46        24       100

Chi-Sq =   3.759 +   0.666 +   1.077 +
           0.475 +   1.215 +   0.572 +
           1.536 +   0.092 +   3.258 = 12.650
DF = 4, P-Value = 0.013
```

The calculated chi-square (12.650) exceeds the critical value (11.143) at the 0.025 level, so we reject H_0. Using the table in the appendix, we can determine that the p-value for this problem is between 0.025 and 0.01. Alternatively, we reject H_0 because p-value = 0.013 is less than $\alpha = 0.025$ level of significance for the test. At this level, age category is not independent of the type of greeting received.

13.37 c/p/m The null and alternative hypotheses are:

H_0: Import/Domestic ownership is independent of magazine preference

H_1: Import/Domestic ownership is not independent of magazine preference

With d.f. = (3 - 1)(2 - 1) = 2, the critical value of chi-square at the 0.05 level is 5.991. We will use Excel worksheet template tmchivar.xls, on the disk supplied with the text. The printout below shows the actual frequencies, the expected frequencies, the calculated chi-square, and the p-value. The calculated chi-square exceeds the critical value, and we reject H_0. Using the table in the appendix, we can determine that the p-value for this problem is between 0.05 and 0.025.

Alternatively, because p-value = 0.043 is $< \alpha = 0.05$ level of significance for the test, we reject H_0.

At this level, Import/Domestic ownership is not independent of magazine preference.

	A	B	C	D	E	F
1	Chi-Square Test					
2	for Independence:					
3	Observed Freqs.:					
4		Import	Domestic			
5	C and D	54	19	73		
6	M Trend	25	22	47		
7	R & T	32	23	55		
8		111	64	175		
9	Expected Freqs.:					
10		Import	Domestic			
11	C and D	46.30	26.70	73.00		
12	M Trend	29.81	17.19	47.00		
13	R & T	34.89	20.11	55.00		
14		111.00	64.00	175.00		
15					no. rows	3
16					no. cols.	2
17					d.f.	2
18					calc. chi-square	6.275
19					p-value	0.043

Using Data Analysis Plus, we obtain the comparable results shown below:

	A	B	C	D
1	Contingency Table			
2				
3		Column 1	Column 2	TOTAL
4	Row 1	54	19	73
5	Row 2	25	22	47
6	Row 3	32	23	55
7	TOTAL	111	64	175
8				
9	chi-squared Stat			6.275
10	df			2
11	p-value			0.043
12	chi-squared Critical			5.992

13.39 p/a/m The null and alternative hypotheses are:

H_0: Bag preference and level of education are independent

H_1: Bag preference and level of education are not independent

Categories have been combined so each expected frequency will be ≥ 5. With d.f. = $(3 - 1)(3 - 1) = 4$, the critical value of chi-square at the 0.01 level is 13.277. We will use Excel worksheet template tmchivar.xls, on the disk supplied with the text. The printout below shows the actual frequencies, the expected frequencies, the calculated chi-square, and the p-value. The calculated chi-square exceeds the critical value, and we reject H_0. Using the table in the appendix, we can determine that the p-value for this problem is less than 0.01.

Alternatively, because p-value = 0.004 is < α = 0.01 level of significance for the test, we reject H_0. At this level, bag preference and education level are not independent.

	A	B	C	D	E	F	G
1	Chi-Square Test						
2	for Independence:						
3	Observed Freqs.:						
4		HS	Some C	C or Grad			
5	Paper	14	13	36	63		
6	Plastic	17	19	22	58		
7	No Pref	8	28	18	54		
8		39	60	76	175		
9	Expected Freqs.:						
10		HS	Some C	C or Grad			
11	Paper	14.04	21.60	27.36	63.00		
12	Plastic	12.93	19.89	25.19	58.00		
13	No Pref	12.03	18.51	23.45	54.00		
14		39.00	60.00	76.00	175.00		
15						no. rows	3
16						no. cols.	3
17						d.f.	4
18						calc. chi-square	15.360
19						p-value	0.004

Using Data Analysis Plus, and combining the third and fourth education-level categories, we obtain the comparable results shown below:

	A	B	C	D	E
1	**Contingency Table**				
2					
3		*Column 1*	*Column 2*	*Column 3*	TOTAL
4	*Row 1*	14	13	36	63
5	*Row 2*	17	19	22	58
6	*Row 3*	8	28	18	54
7	TOTAL	39	60	76	175
8					
9	chi-squared Stat			15.360	
10	df			4	
11	p-value			0.004	
12	chi-squared Critical			13.277	

13.41 p/a/m The null and alternative hypotheses are:

H_0: The quarter of the game is independent of the free throw result

H_1: The quarter of the game is not independent of the free throw result

With d.f. = (3 - 1)(4 - 1) = 6, the critical value of chi-square at the 0.025 level is 14.449. We will use Excel worksheet template tmchivar.xls, on the disk supplied with the text. The printout below includes the actual frequencies, the expected frequencies, the calculated chi-square, and the p-value.

The calculated chi-square does not exceed the critical value, and we do not reject H_0.

Alternatively, because p-value = 0.067 is not < α = 0.025 level of significance for the test, we do not reject H_0. At this level, the quarter of the game is independent of the free throw result.

	A	B	C	D	E	F	G	H
1	**Chi-Square Test**							
2	**for Independence:**							
3	*Observed Freqs.:*							
4		Q1	Q2	Q3	Q4			
5	0 points	6	12	10	17	45		
6	1 point	21	17	19	25	82		
7	2 points	26	24	11	15	76		
8		53	53	40	57	203		
9	*Expected Freqs.:*							
10		Q1	Q2	Q3	Q4			
11	0 points	11.75	11.75	8.87	12.64	45.00		
12	1 point	21.41	21.41	16.16	23.02	82.00		
13	2 points	19.84	19.84	14.98	21.34	76.00		
14		53.00	53.00	40.00	57.00	203.00		
15							no. rows	3
16							no. cols.	4
17							d.f.	6
18							calc. chi-square	11.777
19							p-value	0.067

13.43 p/c/m The null and alternative hypotheses are:

H_0: Type of return is independent of the type of IRS examiner assigned

H_1: Type of return is not independent of the type of IRS examiner assigned

The Minitab printout below includes the actual frequencies, the calculated chi-square, and the p-value. Because p-value = 0.037 is less than α = 0.05 level of significance for the test, we reject H_0. At this level, type of return is not independent of type of IRS examiner assigned.

```
Tabulated Statistics: Income, Examiner

Rows: Income     Columns: Examiner

          1        2        3      All

1        35      112       32      179
2        33       74       29      136
3        76      193       35      304
4        58      107       36      201
All     202      486      132      820

Chi-Square = 13.411, DF = 6, P-Value = 0.037

    Cell Contents -- Count
```

13.45 c/a/m The null and alternative hypotheses are:

H_0: The population proportions are equal

H_1: At least one population proportion differs

This test can be performed using Minitab. First, however, we must compute the observed counts for each cell. These are simply the proportion * n and (1 - the proportion) * n. The degrees of freedom for this test are (2 - 1)(4 - 1) = 3, and the critical value of chi-square at the 0.025 level is 9.348. Since the calculated value is less than the critical value, we do not reject H_0. At this level, there is no evidence to suggest that the population proportions are not equal. Alternatively, we fail to reject H_0 because p-value = 0.420 is not $< \alpha$ = 0.025 level of significance for the test.

```
Chi-Square Test: C1, C2, C3, C4

Expected counts are printed below observed counts

              C1       C2       C3       C4    Total
    1         45       16       35       13      109
           38.93    20.76    36.33    12.98
    2        105       64      105       37      311
          111.07    59.24   103.67    37.02
Total       150       80      140       50      420

Chi-Sq =  0.947 +  1.092 +  0.049 +  0.000 +
          0.332 +  0.383 +  0.017 +  0.000 =  2.820    DF = 3, P-Value = 0.420
```

13.47 p/a/m The null and alternative hypotheses are:

H_0: The population proportions are equal

H_1: At least one population proportion differs

We will use Minitab for this test. The degrees of freedom for this test are $(2 - 1)(3 - 1) = 2$ and the critical value of chi-square at the 0.01 level is listed in our chi-square table as 9.210. Since the calculated value does not exceed the critical value, we do not reject H_0. At this level, the population proportions could be equal. Alternatively, we do not reject H_0 because p-value = 0.094 is not $< \alpha = 0.01$ level of significance for the test.

```
Chi-Square Test: C1, C2, C3

Expected counts are printed below observed counts

            C1        C2        C3     Total
   1        78        46        38       162
         69.43     46.29     46.29

   2        72        54        62       188
         80.57     53.71     53.71

Total      150       100       100       350

Chi-Sq =  1.058 +  0.002 +  1.483 +
          0.912 +  0.002 +  1.278 = 4.735
DF = 2, P-Value = 0.094
```

13.49 p/a/m The null and alternative hypotheses are:

H_0: The population proportions are equal for women from all of the age groups

H_1: At least one population proportion differs

We will use Excel worksheet template tmchipro.xls, on the disk supplied with the text. The degrees of freedom for this test are $(2 - 1)(4 - 1) = 3$ and the critical value of chi-square at the 0.05 level is 7.815. Since the calculated value exceeds the critical value, we reject H_0. At this level, at least one population proportion differs from the others. Alternatively, we reject H_0 because p-value = 0.000 is $< \alpha = 0.05$ level of significance for the test.

	A	B	C	D	E	F	G	H
1	Chi-Square Test							
2	for Comparing Proportions							
3	Observed Freqs.:							
4		21-34	35-44	45-54	55+			
5	pref bad	83	50	53	70	256		
6	pref good	17	50	47	30	144		
7		100	100	100	100	400		
8	Expected Freqs.:							
9		21-34	35-44	45-54	55+			
10	pref bad	64.00	64.00	64.00	64.00	256.00		
11	pref good	36.00	36.00	36.00	36.00	144.00		
12		100.00	100.00	100.00	100.00	400.00		
13							no. rows	2
14							no. cols.	4
15							d.f.	3
16							calc. chi-square	30.990
17							p-value	0.000

13.51 p/a/m The null and alternative hypotheses are:

H_0: The population proportions heating with electricity are equal for these years

H_1: At least one population proportion differs

We will use Data Analysis Plus for this test. The degrees of freedom for this test are $(2 - 1)(3 - 1) = 2$ and the critical value of chi-square at the 0.05 level is listed in our chi-square table as 5.991. Since the calculated value exceeds the critical value, we reject H_0. At this level, at least one population proportion differs from the others. Alternatively, we reject H_0 because p-value = 0.000 is $< \alpha = 0.05$ level of significance for the test.

	A	B	C	D	E
1	Contingency Table				
2					
3		Column 1	Column 2	Column 3	TOTAL
4	Row 1	183	265	307	755
5	Row 2	817	735	693	2245
6	TOTAL	1000	1000	1000	3000
7					
8	chi-squared Stat			42.238	
9	df			2	
10	p-value			0.0000	
11	chi-squared Critical			5.9915	

13.53 p/c/m The null and alternative hypotheses are:

H_0: The population percentages buying snacks are the same for all three movies

H_1: The population percentages buying snacks are not the same for all three movies

In the Minitab printout below, the p-value is 0.047. Because p-value = 0.047 is not less than the $\alpha = 0.025$ level of significance for the test, we do not reject H_0. At this level, the population percentages buying snacks could be the same for all three movies.

```
Tabulated Statistics: SnackCat, Movie

 Rows: SnackCat       Columns: Movie

            1         2         3       All

 1         28        35        24        87
 2         22        65        36       123
 All       50       100        60       210

Chi-Square = 6.129, DF = 2, P-Value = 0.047

    Cell Contents --
              Count
```

13.55 d/p/e The assumption of normality is required because the test for a population variance relies on transforming the equation for the sampling distribution of chi-square. This distribution is based on random samples from a normal population.

13.57 c/a/d

The lower limit of the 95% confidence interval is: $(30 - 1)\dfrac{23.8}{45.722} = 15.096$

The upper limit of the 95% confidence interval is: $(30 - 1)\dfrac{23.8}{16.047} = 43.011$

Therefore, the 95% confidence interval is (15.096, 43.011).

13.59 c/a/d Using the Estimators.xls workbook that accompanies Data Analysis Plus, we obtain the 95% confidence interval for the population variance as (2.620, 9.664).

	A	B	C	D
1	Chi-Squared Estimate of a Variance			
2				
3	Sample variance	4.53	Confidence Interval Estimate	
4	Sample size	20	Lower confidence limit	2.620
5	Confidence level	0.95	Upper confidence limit	9.664

13.61 p/a/d The sample variance is $s^2 = 0.18^2 = 0.0324$.

The lower limit for the 90% confidence interval for the variance is: $(12-1)\dfrac{0.0324}{19.675} = 0.01811$

The upper limit for the 90% confidence interval for the variance is: $(12-1)\dfrac{0.0324}{4.575} = 0.07790$

Thus, the 90% confidence interval for the variance is (0.01811, 0.07790).
Taking the square root of the variance to determine the standard deviation yields a 90% confidence interval of (0.1346, 0.2791) for σ.

Given the summary information in the exercise, we can also use the Estimators.xls workbook that accompanies Data Analysis Plus. The results, expressed in terms of the variance limits, are shown below:

	A	B	C	D
1	Chi-Squared Estimate of a Variance			
2				
3	Sample variance	0.0324	Confidence Interval Estimate	
4	Sample size	12	Lower confidence limit	0.0181
5	Confidence level	0.90	Upper confidence limit	0.0779

13.63 c/a/d The degrees of freedom for this problem are 30 - 1 = 29. This is a two-tail test; we will reject the null hypothesis if the calculated value is <16.047 or the calculated value > 45.722. The calculated value is $(30-1)\dfrac{41.5}{29.0} = 41.50$. Since the calculated value falls between the critical values, we fail to reject the null hypothesis. There is no evidence, at the 0.05 level, that $\sigma^2 \neq 29.0$.

Given the summary information in the exercise, we can also use the Test Statistics.xls workbook that accompanies Data Analysis Plus. The p-value for this two-tail test (0.1245) is not less than the 0.05 level of significance for the test, and we are not able to reject the null hypothesis that the population variance is 29.0. (In this solution, note that the calculated chi-square and the sample variance just happen to have the same numerical value.)

	A	B	C	D
1	Chi-squared Test of a Variance			
2				
3	Sample variance	41.5	Chi-squared Stat	41.50
4	Sample size	30	P(CHI<=chi) one-tail	0.0622
5	Hypothesized variance	29	chi-squared Critical one-tail	42.557
6	Alpha	0.05	P(CHI<=chi) two-tail	0.1245
7			chi-squared Critical two-tail	16.047
8				45.722

13.65 c/a/d In terms of the population variance, the null and alternative hypotheses are:

H_0: $\sigma^2 \leq 0.000009$ and H_1: $\sigma^2 > 0.000009$

The degrees of freedom for this problem are 25 - 1 = 24. This is an upper tail test. The critical value of chi-square at the 0.025 level is 39.364. The calculated chi-square is $(25-1)\dfrac{0.00001369}{0.000009} = 36.507$.

Since the calculated value is less than the critical value, we fail to reject the null hypothesis. There is no evidence, at the 0.025 level, to suggest that the population variance is greater than 0.000009 (or that the population standard deviation is greater than 0.003 inches). Using the chi-square table in the appendix, the most accurate statement that we can make about the p-value for this problem is that it is between 0.05 and 0.025.

Given the summary information in the exercise, we can also use the Test Statistics.xls workbook that accompanies Data Analysis Plus. The p-value for this one-tail test (0.0490) is not less than the 0.025 level of significance for the test, and we are not able to reject the null hypothesis.

	A	B	C	D
1	Chi-squared Test of a Variance			
2				
3	Sample variance	0.00001369	Chi-squared Stat	36.507
4	Sample size	25	P(CHI<=chi) one-tail	0.0490
5	Hypothesized variance	0.000009	chi-squared Critical one-tail	39.364
6	Alpha	0.025	P(CHI<=chi) two-tail	0.0979
7			chi-squared Critical two-tail	11.202
8				42.124

13.67 p/c/d Using Excel and Data Analysis Plus, the 98% confidence interval for the variance is shown in the printout below. We are 98% confident that the population variance is between 0.0329 and 0.0775.

	A	B
1	Chi Squared Estimate: Variance	
2		
3		Ounces
4	Sample Variance	0.0484
5	df	60
6	LCL	0.0329
7	UCL	0.0775

CHAPTER EXERCISES

13.69 p/a/d

a. The null and alternative hypotheses are:

H_0: The data are drawn from a population that is binomial with $\pi = 0.1$

H_1: The data are not drawn from a population that is binomial with $\pi = 0.1$

The degrees of freedom for this problem are 5 - 1 - 0 = 4, and the critical value of chi-square at the 0.05 level is 9.488. The final three categories had to be combined to achieve a high enough expected cell count. We will use Excel worksheet template tmchifit.xls, on the disk supplied with the text. The calculated value exceeds the critical value at the 0.05 level, and we reject H_0. At this level, there is evidence to suggest that the data are not drawn from a population that follows a binomial distribution with $\pi = 0.1$. Alternatively, using the p-value approach, we reject H_0 because p-value = 0.000 is $< \alpha = 0.05$ level of significance for the test.

	A	B	C	D	E
1	Chi-Square Goodness-of-Fit Test			no. of cells, k =	5
2	Cell Frequencies:			no. of parameters estimated, m =	0
3	Observed (Oj):	Expected (Ej):	(Oj-Ej)^2/Ej:	df = k - 1 - m =	4
4	17	6.08	19.613	calculated chi-square =	25.313
5	15	13.51	0.164	p-value =	0.000
6	9	14.26	1.940		
7	4	9.505	3.188		
8	5	6.645	0.407		

b. The null and alternative hypotheses are:

H_0: The data are drawn from a population that is Poisson with $\lambda = 1.4$

H_1: The data are not drawn from a population that is Poisson with $\lambda = 1.4$

The degrees of freedom for this problem are 4 - 1 - 0 = 3, and the critical value of chi-square at the 0.05 level is 7.815. The final four categories had to be combined to achieve a high enough expected cell count. We will use Excel worksheet template tmchifit.xls, on the disk supplied with the text. The calculated value does not exceed the critical value at the 0.05 level, and we are not able to reject H_0. At this level, there is no evidence to suggest that the data are not drawn from a population that follows a Poisson distribution with $\lambda = 1.4$. Alternatively, using the p-value approach, we are unable to reject H_0 because p-value = 0.406 is not $< \alpha = 0.05$ level of significance for the test.

	A	B	C	D	E
1	Chi-Square Goodness-of-Fit Test			no. of cells, k =	4
2	Cell Frequencies:			no. of parameters estimated, m =	0
3	Observed (Oj):	Expected (Ej):	(Oj-Ej)^2/Ej:	df = k - 1 - m =	3
4	17	12.330	1.769	calculated chi-square =	2.907
5	15	17.260	0.296	p-value =	0.406
6	9	12.085	0.788		
7	9	8.325	0.055		

13.71 p/a/d The null and alternative hypotheses are:

H_0: The data are drawn from a population that is Poisson-distributed

H_1: The data are not drawn from a population that is Poisson-distributed

The degrees of freedom for this problem are 5 - 1 - 1 = 3. The sample mean (180/100 = 1.8) is used to estimate the population mean (λ). The critical value of chi-square at the 0.05 level is 7.815. The upper two categories are combined so that no expected cell is < 5. We will use Excel worksheet template tmchifit.xls, on the disk supplied with the text. The calculated value exceeds the critical value at the 0.05 level, and we reject H_0. At this level, there is evidence to suggest that the data are not drawn from a population that follows a Poisson distribution. Alternatively, using the p-value approach, we reject H_0 because p-value = 0.0264 is $< \alpha = 0.05$ level of significance for the test.

	A	B	C	D	E
1	**Chi-Square Goodness-of-Fit Test**			no. of cells, k =	5
2	Cell Frequencies:			no. of parameters estimated, m =	1
3	Observed (Oj):	Expected (Ej):	(Oj-Ej)^2/Ej:	df = k - 1 - m =	3
4	24	16.530	3.376	calculated chi-square =	9.228
5	24	29.750	1.111	p-value =	0.0264
6	18	26.780	2.879		
7	20	16.070	0.961		
8	14	10.870	0.901		

13.73 p/a/d The null and alternative hypotheses are:

H_0: The amounts for ordinary life insurance policies sold within the state follow the national distribution

H_1: The amounts for ordinary life insurance policies sold within the state do not follow the national distribution

The degrees of freedom for this problem are 7 - 1 - 0 = 6, and the critical value of chi-square at the 0.05 level is 12.592. We will use Excel worksheet template tmchifit.xls, on the disk supplied with the text. The calculated value does not exceed the critical value at the 0.05 level, and we do not reject H_0. At this level, there is no evidence to suggest that the amounts for ordinary life insurance policies sold within the state do not follow the national distribution. Using the p-value approach, we do not reject H_0 because p-value = 0.114 is not < α = 0.05 level of significance for the test.

	A	B	C	D	E
1	**Chi-Square Goodness-of-Fit Test**			no. of cells, k =	7
2	Cell Frequencies:			no. of parameters estimated, m =	0
3	Observed (Oj):	Expected (Ej):	(Oj-Ej)^2/Ej:	df = k - 1 - m =	6
4	34	27.00	1.815	calculated chi-square =	10.273
5	39	27.00	5.333	p-value =	0.114
6	17	21.00	0.762		
7	38	39.00	0.026		
8	56	57.00	0.018		
9	105	120.00	1.875		
10	11	9.00	0.444		

13.75 p/a/m The null and alternative hypotheses are:

H_0: The regional distribution for the magazine's subscribers follows the regional distribution for persons who are 65 or over

H_1: The regional distribution for the magazine's subscribers does not follow the regional distribution for persons who are 65 or over

The degrees of freedom for this problem are 4 - 1 - 0 = 3, and the critical value of chi-square at the 0.05 level is 7.815. We will use Excel worksheet template tmchifit.xls, on the disk supplied with the text. The calculated value exceeds the critical value at the 0.05 level, and we reject H_0. At this level, we conclude that the regional distribution for the magazine's subscribers does not follow the regional distribution for persons who are 65 or over. Alternatively, using the p-value approach, we reject H_0 because p-value = 0.022 is < α = 0.05 level of significance for the test.

	A	B	C	D	E
1	**Chi-Square Goodness-of-Fit Test**			no. of cells, k =	4
2	Cell Frequencies:			no. of parameters estimated, m =	0
3	Observed (Oj):	Expected (Ej):	(Oj-Ej)^2/Ej:	df = k - 1 - m =	3
4	11	21.8	5.350	calculated chi-square =	9.609
5	32	24.1	2.590	p-value =	0.022
6	41	34.8	1.105		
7	16	19.3	0.564		

13.77 p/a/m The null and alternative hypotheses are:

H_0: The employees have been selected at random

H_1: The employees have not been selected at random

Note that a random selection should tend to follow the overall distribution of employees. The degrees of freedom for this problem are 3 - 1 - 0 = 2, and the critical value of chi-square at the 0.025 level is 7.378. We will use Excel worksheet template tmchifit.xls, on the disk supplied with the text. The calculated value does not exceed the critical value at the 0.025 level, and we do not reject H_0. At this level, we conclude that the employees could have been selected at random. Using the table in the appendix, the most accurate statement that we can make about the p-value for this problem is that it is between 0.10 and 0.05. Alternatively, using Excel and the p-value approach, we do not reject H_0 because p-value = 0.064 is not $< \alpha = 0.025$ level of significance for the test.

	A	B	C	D	E
1	Chi-Square Goodness-of-Fit Test			no. of cells, k =	3
2	Cell Frequencies:			no. of parameters estimated, m =	0
3	Observed (Oj):	Expected (Ej):	(Oj-Ej)^2/Ej:	df = k - 1 - m =	2
4	3	5	0.800	calculated chi-square =	5.496
5	5	8	1.125	p-value =	0.064
6	12	7	3.571		

13.79 p/a/m The null and alternative hypotheses are:

H_0: The constituency of the marching band is similar to that for the entire university

H_1: The constituency of the marching band is not similar to that for the entire university

The degrees of freedom for this problem are 4 - 1 - 0 = 3, and the critical value of chi-square at the 0.05 level is 7.815. We will use Excel worksheet template tmchifit.xls, on the disk supplied with the text. The calculated value does not exceed the critical value at the 0.05 level, and we do not reject H_0. At this level, we conclude that the constituency of the marching band could be similar to that for the entire university. Using the table in the appendix, the most accurate statement that we can make about the p-value for this problem is that it is between 0.10 and 0.05. Alternatively, using Excel and the p-value approach, we do not reject H_0 because p-value = 0.070 is not $< \alpha = 0.05$ level of significance for the test.

	A	B	C	D	E
1	Chi-Square Goodness-of-Fit Test			no. of cells, k =	4
2	Cell Frequencies:			no. of parameters estimated, m =	0
3	Observed (Oj):	Expected (Ej):	(Oj-Ej)^2/Ej:	df = k - 1 - m =	3
4	24	30	1.200	calculated chi-square =	7.050
5	28	25	0.360	p-value =	0.070
6	19	25	1.440		
7	29	20	4.050		

13.81 p/a/d The population variance of interest is 0.03^2, or 0.0009.

There are $12 - 1 = 11$ degrees of freedom for this problem.

The lower limit for the 98% confidence interval for σ^2 is $(12-1)\dfrac{0.0009}{24.725} = 0.0004004$.

The upper limit for the 98% confidence interval for σ^2 is $(12-1)\dfrac{0.0009}{3.053} = 0.0032427$.

Therefore, the 98% confidence interval for σ is $(0.02001, 0.05694)$.

Given the summary information in the exercise, we can also use the Estimators.xls workbook that accompanies Data Analysis Plus. The results, expressed in terms of the variance limits, are shown below:

	A	B	C	D
1	Chi-Squared Estimate of a Variance			
2				
3	Sample variance	0.0009	Confidence Interval Estimate	
4	Sample size	12	Lower confidence limit	0.00040
5	Confidence level	0.98	Upper confidence limit	0.00324

13.83 p/a/m The null and alternative hypotheses are:

H_0: The population proportions are equal for men in the four age groups
H_1: The population proportions are not equal for men in the four age groups

With d.f. $= (2 - 1)(4 - 1) = 3$, the critical value of chi-square at the 0.05 level is 7.815. We will use Excel worksheet template tmchipro.xls, on the disk supplied with the text. The calculated chi-square exceeds the critical value, and we reject H_0. At this level, men in the four age groups are not equally likely to keep their paper money in order of denomination. Alternatively, because p-value $= 0.000$ is $< \alpha = 0.05$ level of significance for the test, we reject H_0.

	A	B	C	D	E	F	G	H
1	Chi-Square Test							
2	for Comparing Proportions							
3	Observed Freqs.:							
4		21-34	35-44	45-54	55+			
5	in order	90	61	80	80	311		
6	not in order	10	39	20	20	89		
7		100	100	100	100	400		
8	Expected Freqs.:							
9		21-34	35-44	45-54	55+			
10	in order	77.75	77.75	77.75	77.75	311.00		
11	not in order	22.25	22.25	22.25	22.25	89.00		
12		100.00	100.00	100.00	100.00	400.00		
13							no. rows	2
14							no. cols.	4
15							d.f.	3
16							calc. chi-square	25.478
17							p-value	0.000

13.85 p/a/m The null and alternative hypotheses are:

H_0: The population percentages are the same for the four income groups

H_1: The population percentages are not the same for the four income groups

With d.f. = (2 - 1)(4 - 1) = 3, the critical value of chi-square at the 0.10 level is 6.251.

We will use Excel worksheet template tmchipro.xls, on the disk supplied with the text. The calculated chi-square exceeds the critical value, and we reject H_0. At this level, the population percentages for participation in adult education are not the same for the four income groups. Alternatively, because p-value = 0.000 < α = 0.10 level of significance for the test, we reject H_0.

	A	B	C	D	E	F	G	H
1	Chi-Square Test							
2	for Comparing Proportions							
3	Observed Freqs.:							
4		>20 to 30	>30 to 40	>40 to 50	>50 to 75			
5	participated	349	427	468	520	1764		
6	didn't	651	573	532	480	2236		
7		1000	1000	1000	1000	4000		
8	Expected Freqs.:							
9		>20 to 30	>30 to 40	>40 to 50	>50 to 75			
10	participated	441.00	441.00	441.00	441.00	1764.00		
11	didn't	559.00	559.00	559.00	559.00	2236.00		
12		1000.00	1000.00	1000.00	1000.00	4000.00		
13							no. rows	2
14							no. cols.	4
15							d.f.	3
16							calc. chi-square	63.403
17							p-value	0.000

13.87 p/a/m The null and alternative hypotheses are:

H_0: The population proportions are the same for the three departments

H_1: The population proportions are not the same for the three departments

With d.f. = (2 - 1)(3 - 1) = 2, the critical value of chi-square at the 0.025 level is 7.378. We will use Minitab. The calculated chi-square does not exceed the critical value, and we do not reject H_0. At this level, the population proportions for the three departments could be the same. Alternatively, because p-value = 0.045 is not < α = 0.025 level of significance for the test, we do not reject H_0.

```
Chi-Square Test: C1, C2, C3

Expected counts are printed below observed counts
              C1        C2        C3     Total
    1         20        30        25        75
           18.75     25.00     31.25
    2         10        10        25        45
           11.25     15.00     18.75
Total         30        40        50       120

Chi-Sq =   0.083 +   1.000 +   1.250 +
           0.139 +   1.667 +   2.083 =   6.222      DF = 2, P-Value = 0.045
```

13.89 p/a/m The null and alternative hypotheses are:

H_0: The population percentages are the same for the four colleges

H_1: The population percentages are not the same for the four colleges

With d.f. = (2 - 1)(4 - 1) = 3, the critical value of chi-square at the 0.05 level is 7.815. We will use Excel worksheet template tmchipro.xls, on the disk supplied with the text. The calculated chi-square exceeds the critical value, and we reject H_0. At this level, the population percentages for support of the football team's move to division I are not the same for the four colleges.

Alternatively, because p-value = 0.009 < α = 0.05 level of significance for the test, we reject H_0.

	A	B	C	D	E	F	G	H
1	Chi-Square Test							
2	for Comparing Proportions							
3	Observed Freqs.:							
4		Business	Fine Arts	Sci/Tech	Hum/SS			
5	support	47	62	45	39	193		
6	don't	53	38	55	61	207		
7		100	100	100	100	400		
8	Expected Freqs.:							
9		Business	Fine Arts	Sci/Tech	Hum/SS			
10	support	48.25	48.25	48.25	48.25	193.00		
11	don't	51.75	51.75	51.75	51.75	207.00		
12		100.00	100.00	100.00	100.00	400.00		
13							no. rows	2
14							no. cols.	4
15							d.f.	3
16							calc. chi-square	11.484
17							p-value	0.009

13.91 p/c/m The null and alternative hypotheses are:

H_0: Customer size and accuracy are independent

H_1: Customer size and accuracy are not independent

With d.f. = (3 - 1)(3 - 1) = 4, the critical value of chi-square at the 0.025 level is 11.143. We will use Minitab. The calculated chi-square exceeds the critical value and we reject H_0. Alternatively, because p-value = 0.023 is less than α = 0.025 level of significance for the test, we reject H_0. At this level, customer size and accuracy are not independent.

```
Tabulated Statistics: ESTvsUSE, CompSize

   Rows: ESTvsUSE     Columns: CompSize

              1         2         3       All

  1          53       112        46       211
  2          19        31        34        84
  3          32        57        35       124
  All       104       200       115       419

Chi-Square = 11.302, DF = 4, P-Value = 0.023


    Cell Contents --
               Count
```

13.93 p/c/d The null and alternative hypotheses are:

H_0: The number of TV sets per household is Poisson distributed
H_1: The number of TV sets per household is not Poisson distributed

First, we use Minitab to determine the sample mean -- this value is 2.300 televisions per household. We then use Minitab to tally the frequency with which each value occurs and to describe the individual and cumulative probability distributions for a Poisson distribution having a mean of 2.300.

```
Descriptive Statistics: NumbSets

Variable          N        Mean     Median     TrMean     StDev     SE Mean
NumbSets        100       2.300      2.000      2.244     1.474       0.147

Variable    Minimum     Maximum         Q1         Q3
NumbSets      0.000       6.000      1.000      3.000
```

```
Tally for Discrete Variables: NumbSets

NumbSets    Count
       0        7
       1       27
       2       28
       3       18
       4       10
       5        7
       6        3
      N=      100
```

```
Probability Density Function          Cumulative Distribution Function

Poisson with mu = 2.30000             Poisson with mu = 2.30000

        x      P( X = x )                     x      P( X <= x )
     0.00        0.1003                     0.00        0.1003
     1.00        0.2306                     1.00        0.3309
     2.00        0.2652                     2.00        0.5960
     3.00        0.2033                     3.00        0.7993
     4.00        0.1169                     4.00        0.9162
     5.00        0.0538                     5.00        0.9700
```

Next, we determine the expected frequencies for each x value. For example, the expected frequency for x = 1 is 0.2306(100) = 23.06. We have combined the two highest categories in order that each expected count will be $\geq$ 5. As a result, the expected frequency for x $\geq$ 5 TV sets is (1 - 0.9162)(300) = 8.38.

We now use Excel worksheet template tmchifit.xls, on the disk accompanying the text. Note that the d.f. value is calculated as d.f. = k - 1 - m, or d.f. = 5 - 1 - 1 = 3. The "m" term is 1 because we have used the sample mean (2.300) to estimate its (unknown) population counterpart.

The calculated value does not exceed the critical value at the 0.025 level, and we do not reject H_0. Alternatively, using the p-value approach, we do not reject H_0 because p-value = 0.645 is not less than α = 0.025 level of significance for the test. At this level, we conclude that the number of TV sets per household could be Poisson distributed.

	A	B	C	D	E
1	Chi-Square Goodness-of-Fit Test			no. of cells, k =	6
2	Cell Frequencies:			no. of parameters estimated, m =	1
3	Observed (Oj):	Expected (Ej):	(Oj-Ej)^2/Ej:	df = k - 1 - m =	4
4	7	10.03	0.915	calculated chi-square =	2.496
5	27	23.06	0.673	p-value =	0.645
6	28	26.52	0.083		
7	18	20.33	0.267		
8	10	11.69	0.244		
9	10	8.38	0.313		

13.95 p/a/d The null and alternative hypotheses are:

H_0: $\sigma^2 \le 25$ and H_1: $\sigma^2 > 25$

The degrees of freedom for this problem are 51 - 1 = 50. This is an upper tail test, so the critical value of chi-square at the 0.025 level is 71.420, as shown in the printout below that was generated using Excel and Data Analysis Plus. The calculated chi-square is 74.420, and this exceeds the critical value of 71.420. Since the calculated value exceeds the critical value, we reject H_0. From a p-value standpoint, the p-value in this one-tail test is 0.0141, which is less than the 0.025 level of significance, so we reject H_0. At this level, the researchers would appear to be correct in their assertion that too much variability exists in the speeds of vehicles passing this location.

	A	B	C	D
1	**Chi Squared Test: Variance**			
2				
3				*mph*
4	Sample Variance			37.210
5	Hypothesized Variance			25
6	df			50
7	chi-squared Stat			74.420
8	P (CHI<=chi) one-tail			0.0141
9	chi-squared Critical one tail	Left-tail		32.357
10		Right-tail		71.420
11	P (CHI<=chi) two-tail			0.0282
12	chi-squared Critical two tail	Left-tail		30.307
13		Right-tail		75.039

CHAPTER 14
NONPARAMETRIC METHODS

SECTION EXERCISES

14.1 d/p/e Nonparametric tests make no assumptions about the specific shape of the population from which a sample was drawn; most of the tests seen in previous chapters assumed a normally distributed population.

14.3 d/p/m A nonparametric test should be used instead of its parametric counterpart whenever:
(1) data are of the nominal or ordinal scales of measurement, or (2) data are of the interval or ratio scales of measurement but one or more other assumptions, such as the normality of the underlying population distribution, have not been met.

14.5 d/p/m The parametric counterpart to the Wilcoxon signed rank test is the one-sample t-test.
The one sample t-test assumes that the underlying population is normally distributed and that the data are continuous on the ratio or interval scale. The Wilcoxon signed rank test assumes that the underlying population is approximately symmetric and that the data are continuous on the ratio or interval scale.

14.7 c/a/m The null and alternative hypotheses are H_0: M $\leq$ 37.0 and H_1: M > 37.0.
In this right-tail test, there are 14 observations and none is equal to the hypothesized median. At the 0.01 level, we will reject H_0 if the calculated test statistic (W = the sum of the R+ ranks) is greater than 89. The Minitab printout is shown below.

```
Wilcoxon Signed Rank Test: x

Test of median = 37.00 versus median > 37.00
                N for   Wilcoxon              Estimated
        N       Test    Statistic       P     Median
x       14      14        94.5        0.005    41.90
```

Since W is greater than 89, we reject H_0. There is evidence to suggest that the median is greater than 37.0. Using the table of critical values for the Wilcoxon signed rank test, we find that the p-value for the test is less than 0.005. Using the printout and the p-value approach, we reject H_0 because Minitab's p-value = 0.005 is < α = 0.01 level of significance for the test.

14.9 p/a/m The null and alternative hypotheses are H_0: M $\leq$ 15 minutes and H_1: M > 15 minutes. This is a right-tail test with 22 observations, but only 21 that differ from the hypothesized median. Because n is large, we can use the normal approximation to the Wilcoxon signed rank test. The calculated test statistic is determined as shown below, and the critical z in this right-tail test is 1.96.

$$z = \frac{176.5 - \dfrac{(21)(22)}{4}}{\sqrt{\dfrac{(21)(22)(43)}{24}}} = 2.12$$ exceeds the critical value of 1.96, and we reject H_0.

With z = 2.12, and using the normal distribution table in the text, we can estimate the p-value as being 0.5 - 0.4830 = 0.0170. We can also apply the normal approximation using Excel and Data Analysis Plus. The results are comparable to those shown above.

	A	B	C	D
1	Wilcoxon Signed Rank Sum Test			
2				
3	Difference		minutes - Median_0	
4				
5	T+		176.5	
6	T-		54.5	
7	Observations (for test)		21	
8	z Stat		2.12	
9	P(Z<=z) one-tail		0.017	
10	z Critical one-tail		1.96	
11	P(Z<=z) two-tail		0.034	
12	z Critical two-tail		2.24	

14.11 p/c/m The null and alternative hypotheses are H_0: M = 37.4 years and H_1: M ≠ 37.4 years. There are 24 observations and none is equal to the hypothesized median. The Minitab printout is shown below. Because p-value = 0.700 is not < α = 0.05 level of significance for the test, we do not reject H_0. There is no evidence to suggest that the median age at which married males in this region are divorced differs from that for the nation as a whole.

```
Wilcoxon Signed Rank Test: Years

Test of median = 37.40 versus median not = 37.40
                   N for    Wilcoxon              Estimated
             N     Test    Statistic        P       Median
Years       24      24       164.0      0.700        38.25
```

Since we have ≥ 20 nonzero differences, we can also apply Data Analysis Plus and the z-test approximation to the Wilcoxon signed rank test. Using this method, we obtain the comparable results shown below.

	A	B	C	D
1	Wilcoxon Signed Rank Sum Test			
2				
3	Difference		Years - Median_0	
4				
5	T+		164	
6	T-		136	
7	Observations (for test)		24	
8	z Stat		0.4	
9	P(Z<=z) one-tail		0.3446	
10	z Critical one-tail		1.960	
11	P(Z<=z) two-tail		0.6892	
12	z Critical two-tail		2.241	

14.13 c/a/m With d = x_1 - x_2 for each pair, the null and alternative hypotheses are: H_0: $m_d \leq 0$ and H_1: $m_d > 0$. This is a one-tail test with 9 observations, and we will not lose any when we calculate the differences. At the 0.10 level, we will reject H_0 if the calculated test statistic (W = the sum of the R+ ranks) is greater than 34. The Minitab printout is shown below.

```
Wilcoxon Signed Rank Test: diff

Test of median = 0.000000 versus median > 0.000000

               N for    Wilcoxon              Estimated
          N    Test    Statistic        P       Median
diff      9      9        34.5      0.087       0.8500
```

Since W is greater than 34, we reject H_0 at the 0.10 level. There is evidence to suggest that the median difference is greater than 0. Alternatively, using Minitab and the p-value approach, because p-value = 0.087 is < α = 0.10 level of significance for the test, we reject H_0.

14.15 p/a/m With d = x_1 - x_2 (spam messages before filter minus spam messages after filter) for each pair, the null and alternative hypotheses are H_0: $m_d \leq 0$ and H_1: $m_d > 0$. This is a one-tail test with 9 observations, and we will not lose any when we calculate the differences. At the 0.05 level, we will reject H_0 if the calculated test statistic (W = the sum of the R+ ranks) is greater than 36. The Minitab printout is shown below.

```
Wilcoxon Signed Rank Test: Bef-Aft

Test of median = 0.000000 versus median  >  0.000000
                  N for    Wilcoxon                Estimated
            N     Test    Statistic        P        Median
Bef-Aft     9       9         39.0      0.029         3.000
```

Since W = 39.0 is greater than 36, we reject H_0 at the 0.05 level. There is evidence to suggest that the median difference is positive and that the spam filter is effective. Alternatively, using Minitab and the p-value approach, because p-value = 0.029 is < α = 0.05 level of significance for the test, we reject H_0.

14.17 p/a/m With d = x_1 - x_2 for each pair, the null and alternative hypotheses are:
H_0: $m_d \leq 0$ and H_1: $m_d > 0$. (Note: If the procedure increases efficiency, the times should decrease and d should be greater than zero.) This is a right-tail test with 5 observations, and we will not lose any when we take the differences. At the 0.025 level, we will reject H_0 if the calculated test statistic (W = the sum of the R+ ranks) is greater than 15. This will not occur, since the maximum possible R+ sum is actually 15, but we will carry out the analysis anyway and get its conclusion and p-value. The Minitab printout is shown below.

```
Wilcoxon Signed Rank Test: differ

Test of median = 0.000000 versus median  >  0.000000
                  N for    Wilcoxon                Estimated
            N     Test    Statistic        P        Median
differ      5       5         14.0      0.053         5.250
```

As anticipated, W (14.0) does not exceed 15, and we do not reject H_0. Alternatively, because p-value = 0.053 is not < α = 0.025 level of significance for the test, we do not reject H_0. At the 0.025 level, we conclude that the proposed method could be no faster than the current method. Using only the table of critical values for the Wilcoxon signed rank test, we would be able to determine that the p-value for this test is between 0.05 and 0.10.

14.19 p/c/m With d = x_1 - x_2 for each pair, the null and alternative hypotheses are:
H_0: $m_d \geq 0$ and H_1: $m_d < 0$. (Note: If the speed is really increased, d should be less than 0.)
This is a left-tail test with 5 observations, and one observation is lost when we take the differences. The Minitab printout is shown below.

```
Wilcoxon Signed Rank Test: diff

Test of median = 0.000000 versus median  <  0.000000

                  N for    Wilcoxon                Estimated
            N     Test    Statistic        P        Median
diff        5       4          1.5      0.137        -75.00
```

The p-value = 0.137 is not < α = 0.01 level of significance for the test, and we do not reject H_0.
At the 0.01 level, there is no evidence to indicate that the program is effective.

14.21 d/p/m The parametric counterpart to the Wilcoxon rank sum test is the two-sample pooled-variances t-test for independent samples. The t-test assumes data on the interval or ratio scale, independent random samples, normally-distributed populations, and equal variances. The Wilcoxon rank sum test assumes data on the ordinal, interval, or ratio scale and that the samples are independent random samples from populations with similar shapes.

14.23 c/a/m The null and alternative hypotheses are H_0: $m_1 \geq m_2$ and H_1: $m_1 < m_2$. In this left-tail test at the 0.025 level, the critical value of W is 54. The Minitab printout is shown below.

```
Mann-Whitney Test and CI: Sample1, Sample2

Sample1    N =   8    Median =        47.40
Sample2    N =  10    Median =        56.10
Point estimate for ETA1-ETA2 is       -5.90
95.4 Percent CI for ETA1-ETA2 is (-12.20,1.29)
W = 55.0
Test of ETA1 = ETA2  vs  ETA1 < ETA2 is significant at 0.0343
```

Since W = 55.0 is not less than 54, we do not reject H_0 at the 0.025 level. Alternatively, we do not reject H_0 because p-value = 0.0343 is not $< \alpha = 0.025$ level of significance for the test.

14.25 p/c/m With sample 1 as "MasterCard" and sample 2 as "Visa", the null and alternative hypotheses are H_0: $m_1 = m_2$ and H_1: $m_1 \neq m_2$. In this two-tail test at the 0.05 level, the critical values of W are 41 and 78. The Minitab printout is shown below.

```
Mann-Whitney Test and CI: MCard, Visa

MCard      N =   7    Median =        46.00
Visa       N =   9    Median =        76.00
Point estimate for ETA1-ETA2 is      -24.00
95.6 Percent CI for ETA1-ETA2 is (-52.99,20.00)
W = 48.0
Test of ETA1 = ETA2  vs  ETA1 not = ETA2 is significant at 0.2443

Cannot reject at alpha = 0.05
```

Since W = 48.0 is between 41 and 78, we do not reject H_0 at the 0.05 level. Alternatively, we do not reject H_0 because p-value = 0.2443 is not $< \alpha = 0.05$ level of significance for the test. At the 0.05 level, there is no evidence to suggest the shop's median sales differ for the two cards.

14.27 d/p/m The one-way analysis of variance test assumes that samples have been drawn from normally distributed populations with equal variances. In addition, data must be of the interval or ratio scale. The Kruskal-Wallis test requires neither normal populations nor equal variances. The data must be at least ordinal, and the samples are assumed to be randomly selected. The one-way ANOVA test is preferred whenever the necessary assumptions are met.

14.29 c/a/m The null and alternative hypotheses are H_0: the population medians are equal and H_1: the population medians are not equal. For this problem, d.f. = 4 - 1 = 3, and the critical value of H (using the chi-square table) at the 0.10 level is 6.251. The Minitab printout is shown below.

```
Kruskal-Wallis Test: x versus sample

Kruskal-Wallis Test on x

sample     N     Median    Ave Rank          Z
1          6      246.5        11.5       0.00
2          5      316.0        12.2       0.27
3          5      197.0         5.4      -2.39
4          6      336.0        16.0       1.99
Overall   22                   11.5

H = 7.35   DF = 3   P = 0.061
H = 7.36   DF = 3   P = 0.061 (adjusted for ties)
```

Since the adjusted H = 7.36 exceeds the critical value, we reject H_0. Alternatively, we reject H_0 because p-value = 0.061 is $< \alpha = 0.10$ level of significance for the test. At this level, there is evidence to suggest that at least one population median differs from the others.

14.31 p/a/m The null and alternative hypotheses are H_0: the three compounds have the same median tread life and H_1: at least one compound differs. For d.f. = 3 - 1 = 2, the critical value of H (using the chi-square table) at the 0.05 level is 5.991. The Minitab printout is shown below.

```
Kruskal-Wallis Test: distance versus Design

Kruskal-Wallis Test on distance

Design     N     Median    Ave Rank          Z
1          5      33.00         3.9      -2.51
2          5      43.00        11.1       1.90
3          5      39.00         9.0       0.61
Overall   15                    8.0

H = 6.86   DF = 2   P = 0.032
H = 6.90   DF = 2   P = 0.032 (adjusted for ties)
```

Since the adjusted H exceeds the critical value, we reject H_0. Alternatively, we reject H_0 because p-value = 0.032 is $< \alpha = 0.05$ level of significance for the test. At this level, there is evidence to suggest that the median tread life for at least one of the compounds differs from the others.

Using Data Analysis Plus, we obtain the comparable results shown below. Note that the H statistic and p-value have not been adjusted for ties.

	A	B	C
1	**Kruskal-Wallis Test**		
2			
3	Group	Rank Sum	Observations
4	*Design 1:*	19.5	5
5	*Design 2:*	55.5	5
6	*Design 3:*	45	5
7			
8	H Stat		6.855
9	df		2
10	p-value		0.033
11	chi-squared Critical		5.99

14.33 p/c/m The null and alternative hypotheses are H_0: the absences are equal across age groups and H_1: at least one group exhibits a different level of absences. For this problem, d.f. = 3 - 1 = 2, and the critical value of H (using the chi-square table) at the 0.10 level is 4.605. The Minitab printout is shown below.

```
Kruskal-Wallis Test: Absences versus AgeGrp

Kruskal-Wallis Test on Absences

AgeGrp      N     Median    Ave Rank          Z
1           7      3.000         7.0      -1.58
2           5      8.000        14.1       2.27
3           6      2.500         8.6      -0.52
Overall    18                    9.5

H = 5.42  DF = 2   P = 0.066
H = 5.68  DF = 2   P = 0.058 (adjusted for ties)
```

Since the adjusted H exceeds the critical value, we reject H_0. Alternatively, we reject H_0 because p-value = 0.058 is $< \alpha = 0.10$ level of significance for the test. At this level, there is evidence to suggest that at least one age group exhibits a different level of absences.

Using Data Analysis Plus, we obtain the comparable results shown below. Note that the H statistic and p-value have not been adjusted for ties.

	A	B	C
1	**Kruskal-Wallis Test**		
2			
3	Group	Rank Sum	Observations
4	18to30yr	49	7
5	31to50yr	70.5	5
6	51to65yr	51.5	6
7			
8	H Stat		5.424
9	df		2
10	p-value		0.066
11	chi-squared Critical		4.605

14.35 d/p/m

a. The null and alternative hypotheses for the Friedman test and the randomized block ANOVA are almost identical except that the Friedman test deals with medians and the ANOVA test deals with means.

b. Randomized block ANOVA requires that observations be from normal populations with equal variances and that the observations be measured on the interval or ratio scale. The Friedman test makes no assumptions about the underlying populations, and the observations may be measured on the ordinal, interval, or ratio scale.

14.37 p/a/m The null and alternative hypotheses are H_0: the median salaries for the three positions are equal and H_1: at least one median differs from the others. For this problem, d.f. = 3 - 1 = 2, and (from the chi-square table) the critical value for F_r at the 0.10 level is 4.605. The table below shows the ranks within each block and is used to obtain the calculated F_r, which is 3.58.

	Mayor	Police Chief	Fire Chief
Atlanta	3	1	2
Chicago	2	1	3
Denver	3	1	2
Jacks'vle	3	1.5	1.5
Newark	3	1	2
San Diego	1	3	2
	sum = 15	sum = 8.5	sum = 12.5

The Minitab printout is shown below.

```
Friedman Test: Salary versus Position, City

Friedman test for Salary by Position blocked by City

S = 3.58  DF = 2  P = 0.167
S = 3.74  DF = 2  P = 0.154 (adjusted for ties)

                     Est     Sum of
Position      N   Median      Ranks
1             6    75593       15.0
2             6    62275        8.5
3             6    69095       12.5

Grand median   =    68987
```

Since the adjusted F_r (shown in the printout as S = 3.74) does not exceed the critical value, we do not reject H_0. Alternatively, we do not reject H_0 because p-value = 0.154 is not $< \alpha = 0.10$ level of significance for the test. At this level, there is no evidence to suggest a difference in the effectiveness of the treatments. Using only the chi-square table, we would find the p-value to be between 0.10 and 0.90.

Using Data Analysis Plus, we obtain comparable results, although not adjusted for ties.

	A	B	C
1	**Friedman Test**		
2			
3	Group		Rank Sum
4	Mayor		15.0
5	Police		8.5
6	Fire		12.5
7			
8	Fr Stat		3.583
9	df		2
10	p-value		0.167
11	chi-squared Critical		4.605

14.39 p/c/m The null and alternative hypotheses are H_0: the median ratings for the three critics are equal and H_1: the median rating given by at least one critic differs from the others. The three critics are considered to be the treatments, and the movies are the blocks. The Minitab printout is shown below.

```
Friedman Test: Rating versus Critic, Movie

Friedman test for Rating by Critic blocked by Movie

S = 1.75  DF = 2  P = 0.417
S = 1.87  DF = 2  P = 0.393 (adjusted for ties)

                    Est     Sum of
Critic        N   Median    Ranks
1             8    7.833     17.0
2             8    8.167     18.0
3             8    6.500     13.0

Grand median  =   7.500
```

The adjusted F_r is shown in the printout as $S = 1.87$. Because p-value = 0.393 is not $< \alpha = 0.025$ level of significance for the test., we do not reject H_0. At this level, there is no evidence to suggest a difference in the ratings and/or value systems among the critics.

Using Data Analysis Plus, we obtain comparable results, although not adjusted for ties.

	A	B	C
1	**Friedman Test**		
2			
3	Group		Rank Sum
4	CriticA		17
5	CriticB		18
6	CriticC		13
7			
8	Fr Stat		1.75
9	df		2
10	p-value		0.417
11	chi-squared Critical		7.378

14.41 d/p/e The sign test relies on the fact that, if the true population (of differences) median is equal to zero, there is a 0.50 probability of obtaining a positive difference. Thus, this is a binomial experiment and a difference will be either positive or negative (ties are ignored). When the number of observations (non-zero differences) is at least ten, the normal approximation may be used.

14.43 c/a/m This is a two-tail test. Using the cumulative binomial table in the appendix, we see that the probability of finding 5 or more positive differences when n = 8 is (1 - 0.6367) = 0.3633. Since this is a two-tail test, we multiply this probability by 2, yielding a p-value of 0.7266. Since p-value = 0.7266 is not $< \alpha = 0.05$ level of significance for the test, we do not reject H_0. There is no evidence to suggest that the median is not zero.

14.45 p/a/m Given the data in exercise 14.13, the variable of interest is $d = x_1 - x_2$. In the earlier exercise, the null and alternative hypotheses are H_0: $m_d \leq 0$ and H_1: $m_d > 0$. In applying the sign test to these data, we have 9 nonzero differences and 7 of them are positive. We will use Excel worksheet template tmsign.xls, on the disk supplied with the text. This worksheet template can be used regardless of the number of nonzero differences, because it relies on the binomial distribution, and is not an approximation.

	A	B	C	D	E
1	Sign Test for Testing One Sample or for Comparing Paired Samples				
2					
3	n, the Number of NonZero Differences:	9	*Calculated Values:*		
4	T, the Number of Differences with di > 0:	7	*p-value if the test is:*		
5			Left-Tail	Two-Tail	**Right-Tail**
6			0.9805	0.1797	**0.0898**

The p-value for this right-tail test is 0.0898, which is $< \alpha = 0.10$ level of significance for the test, and we reject H_0. This conclusion is the same as in exercise 14.13, although the p-value for the sign test is not quite as low as that for the Wilcoxon signed rank test (0.087) performed earlier. Because these data are at least interval in nature, the Wilcoxon signed rank test (which uses more of the information contained in the data) is the more powerful of the two tests. The corresponding Minitab printout is shown below.

```
Sign Test for Median: diff

Sign test of median = 0.00000 versus  >  0.00000

              N  Below  Equal  Above         P    Median
diff          9      2      0      7    0.0898    0.5000
```

14.47 d/p/e A "run" is a sequence of like events or the consecutive appearance of one or more observations that are similar.

14.49 p/a/m The null and alternative hypotheses are H_0: the sequence is random and H_1: the sequence is not random. The data can be divided into two categories, above 1.5 and below 1.5.
The Minitab printout is shown below.

```
Runs Test: NamCode

    NamCode

    K =    1.5000

    The observed number of runs =  18
    The expected number of runs =  17.6111
    13 Observations above K    23 below
            The test is significant at  0.8864
            Cannot reject at alpha = 0.05
```

Since p-value = 0.8864 is not $< \alpha = 0.10$ level of significance for this test, we do not reject H_0. There is no evidence to suggest that the sequence is not random. For this problem, we can also use Excel worksheet template tmruns.xls, on the disk supplied with the text. There are 18 runs in the 36-year series, with 13 observations above 1.5 and 23 below 1.5. The p-value in this approximation is, to four decimal places, the same as the value generated by Minitab, above.

	A	B	C	D	E
1	**Z-Test Approximation , Runs Test for Randomness**				
2					
3	Number of Runs, T:	18	*Calculated Values:*		
4	Number of Observs. of Type 1, n1:	13	Expected No. of Runs =	17.6	
5	Number of Observs. of Type 2, n2:	23	z =	0.143	
6	Total Number of Observations, n:	36	*p-Value If the Test Is:*		
7			Left-Tail	**Two-Tail**	Right-Tail
8			0.5568	**0.8864**	0.4432

14.51 p/a/m The null and alternative hypotheses are H_0: the data are from a normal distribution and H_1: the data are not from a normal distribution. For the sample values, the mean and standard deviation are 152.5 and 10.88905, respectively. As an example of the calculations involved, the z value corresponding to the test score of 145 can be calculated as z = (145 - 152.5)/10.88905, or z = -0.69. The expected cumulative relative frequencies can then be determined using the standard normal distribution table. For example, the expected cumulative relative frequency associated with the test score of 145 would be 0.5000 - 0.2549, or 0.2451. Because there are 8 observations and we are using the 0.10 level of significance, the critical value of D is 0.261. Therefore, we will reject H_0 if $D_{max} > 0.261$.

Data	z	rel. freq.	cum. freq.	expected	D
142	-0.96	0.125	0.125	0.1685	0.0435
142	-0.96	0.125	0.250	0.1685	0.0815
145	-0.69	0.125	0.375	0.2451	0.1299
147	-0.51	0.125	0.500	0.3050	0.1950
149	-0.32	0.125	0.625	0.3745	0.2505
161	0.78	0.125	0.750	0.7823	0.0323
164	1.06	0.125	0.875	0.8554	0.0196
170	1.61	0.125	1.000	0.9463	0.0537

Since the largest absolute value of D (0.2505) is less than the critical value (0.261), we do not reject H_0. At the 0.10 level, there is no evidence to suggest that the data are not from a normal distribution. Using the table of critical values for the Lilliefors test, we can determine that the p-value for this test is between 0.10 and 0.15.

The corresponding results using Minitab and Data Analysis Plus are shown below. As discussed in the text, Minitab performs the test in reponse to the Kolmogorov-Smirnov menu selection, and Data Analysis Plus automatically provides the critical value for a test at the 0.05 level of significance.

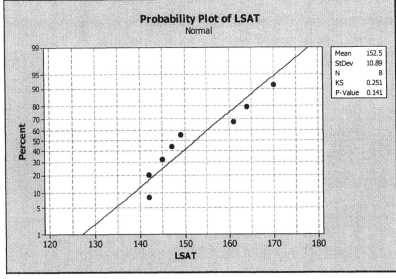

204

	A	B
1	**Lilliefors Test**	
2		
3	*LSAT*	
4	D Stat	0.2511
5	D Critical	0.285

14.53 p/a/m This test can be performed with pocket calculator, formulas and tables, as we showed in the solution to exercise 14.51, but we will use Minitab and Data Analysis Plus. The null and alternative hypotheses are H_0: the data are from a normal distribution and H_1: the data are not from a normal distribution. Because there are 7 observations and we are using the 0.10 level of significance, the critical value of D is 0.276. Therefore, we will reject H_0 if $D_{max} > 0.276$. Referring to the printouts below, $D_{max} = 0.2239$ is less than the critical value of 0.276, and we do not reject H_0. At this level, the data could have come from a normal distribution. Minitab shows the approximate p-value as greater than 0.15. Data Analysis Plus has automatically provided a critical value corresponding to the 0.05 level of significance.

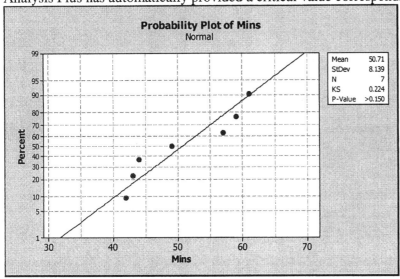

	A	B
1	**Lilliefors Test**	
2		
3	*Mins*	
4	D Stat	0.2239
5	D Critical	0.300

14.55 p/c/m The null and alternative hypotheses are H_0: $M_d \geq 8$ and H_1: $M_d < 8$. The Minitab and Data Analysis Plus results are shown below. Note that Data Analysis Plus has used a normal approximation to the sign test. For this left-tail test, p-value = 0.0287 is less than the 0.05 level of significance for the test, and we reject H_0. We conclude that the median age for cars driven to work by the company's executives is less than 8 years.

Sign Test for Median: Age

Sign test of median = 8.000 versus < 8.000

```
            N   Below  Equal  Above       P    Median
Age        15     11      1      3   0.0287    5.000
```

	A	B	C	D
1	**Sign Test**			
2				
3	Difference			*Age - Median_0*
4				
5	Positive Differences			3
6	Negative Differences			11
7	Zero Differences			1
8	z Stat			-2.138
9	P(Z<=z) one-tail			0.0163
10	z Critical one-tail			1.645
11	P(Z<=z) two-tail			0.0326
12	z Critical two-tail			1.960

14.57 p/c/m The null and alternative hypotheses are H_0: the sequence is random and H_1: the sequence is not random. The Minitab printout is shown below and the default cutoff is the mean winning margin (16.2162 points). Because p-value = 0.8710 is not less than the 0.10 level of significance for the test, we do not reject H_0. At this level, the sequence of winning margins could be random.

Runs Test: Margin

```
    Margin

    K =    16.2162

    The observed number of runs =  19
    The expected number of runs =  19.4865
    18 Observations above K   19 below
            The test is significant at  0.8710
            Cannot reject at alpha = 0.05
```

For this problem, we can also use Excel worksheet template tmruns.xls, on the disk supplied with the text. There are 19 runs in the 37-year series, with 18 observations above 16.2162 and 19 below 16.2162. The p-value in this approximation is (to four decimal places) the same as the value generated by Minitab, above.

	A	B	C	D	E
1	**Z-Test Approximation , Runs Test for Randomness**				
2					
3	Number of Runs, T:	19	*Calculated Values:*		
4	Number of Observs. of Type 1, n1:	18	Expected No. of Runs =	19.5	
5	Number of Observs. of Type 2, n2:	19	z =	-0.162	
6	Total Number of Observations, n:	37	*p-value if the test is:*		
7			Left-Tail	**Two-Tail**	Right-Tail
8			0.4355	**0.8710**	0.5645

206

14.59 p/c/m The null and alternative hypotheses are H_0: the data are from a normal distribution and H_1: the data are not from a normal distribution. For 25 observations, we refer to the appendix table and identify the critical value of D for the 0.10 level as 0.158. Because $D_{max} = 0.136$ is less than the critical value, we do not reject H_0. At this level, the rates of return could have come from a normal distribution. Minitab lists the approximate p-value as >0.15. Note that Data Analysis Plus has listed the critical value for a test at the 0.05 level.

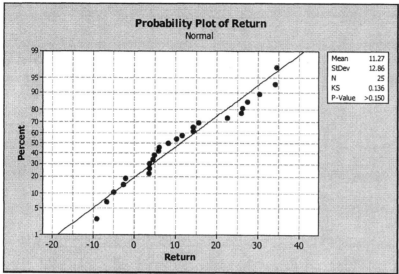

	A	B
1	**Lilliefors Test**	
2		
3	*Return*	
4	D Stat	0.136
5	D Critical	0.173

CHAPTER EXERCISES

14.61 p/a/m The Wilcoxon rank sum test is appropriate for this problem. The null and alternative hypotheses are H_0: $M_A \leq M_B$ and H_1: $M_A > M_B$. The Minitab printout is shown below.

```
Mann-Whitney Test and CI: BrandA, BrandB

BrandA     N = 12     Median =        92.95
BrandB     N = 10     Median =        81.60
Point estimate for ETA1-ETA2 is        16.10
95.6 Percent CI for ETA1-ETA2 is (0.90,28.80)
W = 173.0
Test of ETA1 = ETA2  vs  ETA1 > ETA2 is significant at 0.0115
The test is significant at 0.0114 (adjusted for ties)
```

For this test, p-value = 0.0114 is < α = 0.05 level of significance for the test, and we reject H_0. At this level, evidence suggests that brand A is superior to brand B. Because n is large, we can also use Data Analysis Plus and the normal approximation to the Wilcoxon rank sum test. The results are comparable and the printout is shown below.

	A	B	C	D
1	Wilcoxon Rank Sum Test			
2				
3			Rank Sum	Observations
4	*BrandA*		173	12
5	*BrandB*		80	10
6	z Stat		2.31	
7	P(Z<=z) one-tail		0.0105	
8	z Critical one-tail		1.645	
9	P(Z<=z) two-tail		0.0210	
10	z Critical two-tail		1.96	

14.63 p/a/m The Wilcoxon signed rank test is appropriate here. The null and alternative hypotheses are H_0: the median pulse rate is unchanged and H_1: the median pulse rate has changed. This is a two-tail test. With n = 6, critical values of W at the 0.05 level are 1 and 20. Considering the difference between before and after pulse rates as the variable of interest, the Minitab printout is shown below.

```
Wilcoxon Signed Rank Test: diff

Test of median = 0.000000 versus median not = 0.000000
                  N for   Wilcoxon              Estimated
             N    Test    Statistic       P      Median
diff         6     6         19.5       0.075     4.000
```

Since the calculated value of W falls between the critical values, we do not reject H_0. At the 0.05 level, there is no evidence to suggest that the exercise program has affected resting pulse rates. Alternatively, since p-value = 0.075 is not $< \alpha = 0.05$ level of significance for the test, we do not reject H_0. Using only the table of critical values for the Wilcoxon signed rank test, we could determine the p-value to be between 0.05 and 0.10.

14.65 p/a/m The Friedman test is appropriate for this problem. The null and alternative hypotheses are H_0: the medians are equal for each of the compactors and H_1: at least one median differs.
Using the chi-square approximation for F_R, the appropriate degrees of freedom are 5 - 1 = 4. At the 0.025 level, the critical value of F_R is 11.143. The table below shows the basic calculations leading to F_R.

	comp 1	comp 2	comp 3	comp 4	comp 5
cans & bottles	1	2	3	5	4
cardboard boxes	1.5	1.5	4.5	4.5	3
newspapers & mags	1	2	3	5	4
	sum = 3.5	sum = 5.5	sum = 10.5	sum = 14.5	sum = 11.0

The calculated $F_R = 10.53$ is less than the critical value, and we do not reject H_0. At the 0.025 level, there is no evidence to suggest that the compactors are not equally effective. Using only the chi-square table, we could find the p-value for this test to be between 0.025 and 0.05. The Minitab printout is shown below and the p-value (adjusted for ties) is 0.028. Since p-value = 0.028 is not $< \alpha = 0.025$ level of significance for the test, we do not reject H_0. The Minitab printout is shown below.

```
Friedman Test: CubeFt versus Comp, Block

Friedman test for CubeFt by Comp blocked by Block

S = 10.53   DF = 4   P = 0.032
S = 10.90   DF = 4   P = 0.028 (adjusted for ties)

                    Est       Sum of
Comp        N     Median      Ranks
1           3     1.4000       3.5
2           3     1.5000       5.5
3           3     1.6600      10.5
4           3     1.8200      14.5
5           3     1.7200      11.0

Grand median  =   1.6200
```

Using Data Analysis Plus, we obtain the comparable results (without adjustment for ties) shown below.

	A	B	C
1	Friedman Test		
2			
3	Group		Rank Sum
4	Comp1		3.5
5	Comp2		5.5
6	Comp3		10.5
7	Comp4		14.5
8	Comp5		11
9			
10	Fr Stat		10.53
11	df		4
12	p-value		0.032
13	chi-squared Critical		11.143

14.67 p/a/m The Friedman test is appropriate for this problem. The null and alternative hypotheses are H_0: the shapes are equally effective and H_1: at least one shape differs in effectiveness.
Using the chi-square approximation for F_R, d.f. $= 3 - 1 = 2$ and the critical value of F_R at the 0.025 level is 7.378. The table below shows the basic calculations leading to F_R.

	o-ring 1	o-ring 2	o-ring 3
86 degrees	3	2	1
75 degrees	2	3	1
65 degrees	2	3	1
55 degrees	3	2	1
45 degrees	3	2	1
33 degrees	3	1	2
	sum = 16	sum = 13	sum = 7

The calculated $F_R = 7.00$ is less than the critical value, and we do not reject H_0. At the 0.025 level, there is no evidence to suggest that the o-ring shapes are not equally effective. Using only the chi-square table, we could find the p-value for this test to be between 0.025 and 0.05. The Minitab printout is shown below. Since p-value = 0.030 is not $< \alpha = 0.025$ level of significance for the test, we do not reject H_0.

```
Friedman Test: Pressure versus Shape, Temp

Friedman test for Pressure by Shape blocked by Temp

S = 7.00   DF = 2   P = 0.030

                   Est    Sum of
Shape        N   Median    Ranks
1            6   12.892     16.0
2            6   12.525     13.0
3            6   10.908      7.0

Grand median  =   12.108
```

209

Using Data Analysis Plus, we obtain the comparable results shown below.

	A	B	C
1	**Friedman Test**		
2			
3	Group		Rank Sum
4	*Shape1*		16
5	*Shape2*		13
6	*Shape3*		7
7			
8	Fr Stat		7.00
9	df		2
10	p-value		0.030
11	chi-squared Critical		7.378

14.69 p/a/m The answers will vary widely with the data collected for this particular exercise. It might also be interesting to compare data for the same location at different times of the day or for different days of the week.

14.71 p/a/m Applying the Wilcoxon signed rank test to this problem, the null and alternative hypotheses will be H_0: M ≤ 20 and H_1: M > 20. For this right-tail test with n = 10 and using the 0.05 level, the critical value of W is 44. The Minitab printout is shown below.

```
Wilcoxon Signed Rank Test

Test of median = 20.00 versus median  >  20.00
                  N for   Wilcoxon              Estimated
           N      Test   Statistic         P     Median
time      10        10        45.0     0.042      22.20
```

Since the calculated value of W exceeds the critical value, we reject H_0. At the 0.05 level, there is evidence to suggest that the median travel time is more than twenty minutes. Alternatively, since p-value = 0.042 is < α = 0.05 level of significance for the test, we reject H_0. Using only the table of critical values for the Wilcoxon signed rank test, we could determine that the p-value is between 0.05 and 0.025.

14.73 p/a/m The null and alternative hypotheses are H_0: the data are from a normal distribution and H_1: the data are not from a normal distribution. There are n = 8 observations, so the critical value of D for the 0.05 level of significance is 0.285. Referring to the Data Analysis Plus printout below, the observed value of D_{max} is 0.1925. The observed D_{max} does not exceed the critical value for the 0.05 level of signficance and we do not reject H_0. At this level, the insurance amounts could have come from a normal distribution. Minitab lists the approximate p-value as >0.15.

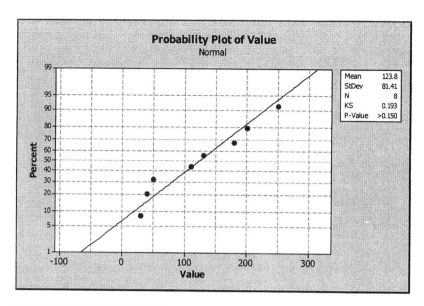

Probability Plot of Value
Normal

Mean	123.8
StDev	81.41
N	8
KS	0.193
P-Value	>0.150

	A	B
1	**Lilllefors Test**	
2		
3	*Value*	
4	D Stat	0.1925
5	D Critical	0.285

14.75 p/c/m Applying the Wilcoxon signed rank test, the null and alternative hypotheses are H_0: $M_d \geq 0$ and H_1: $M_d < 0$. The Minitab printout is shown below.

```
Wilcoxon Signed Rank Test: diff

Test of median = 0.000000 versus median   <  0.000000
                     N for   Wilcoxon              Estimated
               N     Test    Statistic      P       Median
diff           5      5         5.5      0.343      -2.000
```

Since p-value = 0.343 is not < α = 0.05 level of significance for the test, we do not reject H_0. At the 0.05 level, there is no evidence to suggest that the incentive policy has improved performance.

14.77 p/c/m Applying the Wilcoxon rank sum test, the null and alternative hypotheses are H_0: the two routes are equally efficient and H_1: the routes are not equally efficient. The Minitab printout is shown below.

```
Mann-Whitney Test and CI: RouteA, RouteB

RouteA      N =   5      Median =      19.000
RouteB      N =   5      Median =      21.100
Point estimate for ETA1-ETA2 is     -2.700
96.3 Percent CI for ETA1-ETA2 is (-7.502,1.600)
W = 21.0
Test of ETA1 = ETA2  vs  ETA1 not = ETA2 is significant at 0.2101

Cannot reject at alpha = 0.05
```

Since p-value = 0.2101 is not < α = 0.05 level of significance for the test, we do not reject H_0. At the 0.05 level, there is no evidence to suggest that the two routes are not equally efficient.

211

14.79 p/c/m Applying the Friedman test, the null and alternative hypotheses are H_0: the rankings are equal across price ranges and H_1: the rankings for at least one price range differ. The Minitab printout is shown below.

```
Friedman Test: Rating versus Price, Taster

Friedman test for Rating by Price blocked by Taster

S = 7.90  DF = 2  P = 0.019
S = 8.32  DF = 2  P = 0.016 (adjusted for ties)

                    Est     Sum of
Price        N    Median     Ranks
1            5     4.000       5.0
2            5     8.333      11.5
3            5     9.667      13.5

Grand median  =    7.333
```

Since p-value = 0.016 is < α = 0.05 level of significance for the test, we reject H_0. At the 0.05 level, there is evidence to suggest that the rankings do differ across price ranges -- i.e., it appears the tasters can tell the difference between cheaper and more expensive wines. The corresponding Data Analysis Plus printout (without adjustment for ties) is shown below.

	A	B	C
1	**Friedman Test**		
2			
3	Group		Rank Sum
4	*Elderberry*		5.0
5	*Rhine*		11.5
6	*Sauvignon*		13.5
7			
8	Fr Stat		7.90
9	df		2
10	p-value		0.0193
11	chi-squared Critical		5.99

14.81 p/c/m Applying the runs test, the null and alternative hypotheses are H_0: the sequence is random and H_1: the sequence is not random. The median for the data is 37.6. To ensure that the median value is included in the "median or above" category, we will have Minitab categorize on the basis of a 37.599 cutoff. The printout is shown below.

```
Runs Test: Ounces

    Ounces

    K =     37.5990

    The observed number of runs =    9
    The expected number of runs =   13.0000
    12 Observations above K    12 below
            The test is significant at  0.0950
            Cannot reject at alpha = 0.05
```

Since p-value = 0.0950 is < α = 0.10 level of significance for this test, we reject H_0. At this level, there is evidence to suggest that the sequence of weights is not random. For this problem, we can also use Excel worksheet template tmruns.xls, on the disk supplied with the text. There are 9 runs in the 24-item series, with 12 observations above 37.599 and 12 below 37.599. The p-value in this approximation is comparable to the value generated by Minitab, above.

	A	B	C	D	E
1	**Z-Test Approximation , Runs Test for Randomness**				
2					
3	Number of Runs, T:	9	*Calculated Values:*		
4	Number of Observs. of Type 1, n1:	12	Expected No. of Runs =	13.0	
5	Number of Observs. of Type 2, n2:	12	z =	-1.670	
6	Total Number of Observations, n:	24	*p-Value If the Test Is:*		
7			Left-Tail	**Two-Tail**	Right-Tail
8			0.0475	**0.0950**	0.9525

14.83 p/c/m The null and alternative hypotheses are H_0: the data are from a normal distribution and H_1: the data are not from a normal distribution. There are n = 7 observations, so the critical value of D for the 0.10 level of significance is 0.276. Referring to the Data Analysis Plus printout, the observed value of D_{max} is 0.1919. (Note: The critical value listed by Data Analysis Plus corresponds to its default level of significance for the test, 0.05.) The observed $D_{max} = 0.1919$ does not exceed the critical value of 0.276 for the 0.10 level of significance and we do not reject H_0. At this level, the cereal box weights could have come from a normal distribution. Minitab shows the p-value as >0.15.

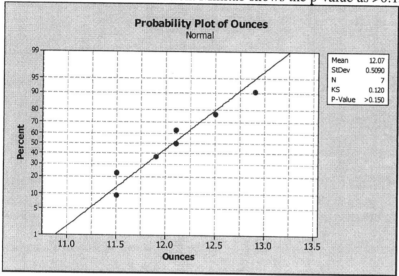

	A	B
1	**Lilliefors Test**	
2		
3	*Ounces*	
4	D Stat	0.1919
5	D Critical	0.300

SECTION EXERCISES

15.1 d/p/m y_i = is a value of the dependent variable, and x_i is a value of the independent variable. β_0 is the y-intercept of the regression line, and β_1 is the slope of the regression line. Finally, ε_i is the random error term for the given value of x.

15.3 d/p/e It is "y-hat" because it is an estimated value for the dependent variable given a value of x.

15.5 c/a/m The tables below show the calculations for $(y - \hat{y})^2$ for the two lines.

Let $\hat{y} = 10 + x$:

x	y	$\hat{y}$	$(y - \hat{y})$	$(y - \hat{y})^2$
2	10	12	-2	4
3	12	13	-1	1
4	20	14	6	36
5	16	15	1	1
				sum = 42

Let $\hat{y} = 8 + 2x$:

x	y	$\hat{y}$	$(y - \hat{y})$	$(y - \hat{y})^2$
2	10	12	-2	4
3	12	14	-2	4
4	20	16	4	16
5	16	18	-2	4
				sum = 28

Using the least squares criterion, the second regression line fits the data better.

15.7 c/a/m The tables below show the sum of squares value for each line.

Let $\hat{y} = 5 + 3x$:

x	y	$\hat{y}$	$(y - \hat{y})$	$(y - \hat{y})^2$
3	8	14	-6	36
5	18	20	-2	4
7	30	26	4	16
9	32	32	0	0
				sum = 56

Let $\hat{y} = -2 + 4x$:

x	y	$\hat{y}$	$(y - \hat{y})$	$(y - \hat{y})^2$
3	8	10	-2	4
5	18	18	0	0
7	30	26	4	16
9	32	34	-2	4
				sum = 24

Using the least squares criterion, the second line fits the data better.

An important note to readers using pocket calculators for some of the solutions in this chapter:
To get calculator-based results that are as close as possible to those of Minitab, Excel and other computer packages, we have sometimes had to carry along many more significant digits than are shown in the arithmetic expressions themselves. For example, you may see a regression coefficient expressed as 0.00333 when in fact we are still carrying it along as the more exact 0.003328198623 in our calculations. If your calculator carries 10 significant digits, and you carry as many digits as possible through your sequence of calculations in any given exercise, you should arrive at the same results we obtained on the calculator and be extremely close to those generated by the computer.

15.9 p/a/m This exercise can be solved using a pocket calculator and the method shown in the solution to exercise 15.8.

Method 1. Using the regression formulas and a calculator:

x	y	xy	x^2
6.00	300.00	1800.00	36.00
12.00	408.00	4896.00	144.00
14.00	560.00	7840.00	196.00
6.00	252.00	1512.00	36.00
9.00	288.00	2592.00	81.00
13.00	650.00	8450.00	169.00
15.00	630.00	9450.00	225.00
9.00	522.00	4698.00	81.00
$\Sigma x = 84.00$	$\Sigma y = 3610.00$	$\Sigma xy = 41{,}238.00$	$\Sigma x^2 = 968.00$

$n = 8$, $\bar{x} = 10.5$ and $\bar{y} = 451.25$

slope, $b_1 = \dfrac{\sum x_i y_i - n\bar{x}\bar{y}}{\sum x_i^2 - n\bar{x}^2} = \dfrac{41{,}238.00 - 8(10.5)(451.25)}{968.00 - 8(10.5)^2} = \dfrac{3333}{86} = 38.756$

y-intercept, $b_0 = \bar{y} - b_1\bar{x} = 451.25 - (38.756)(10.5) = 44.314$

The regression equation is $\hat{y} = 44.314 + 38.756x$

Method 2. Finding the regression equation using Minitab:
In generating the Minitab printout shown below, we have specified a prediction for y when x is 10 years.

```
Regression Analysis: Shares versus Years

The regression equation is Shares = 44 + 38.8 Years

Predictor       Coef        StDev          T         P
Constant        44.3        108.5       0.41     0.697
Years         38.756        9.864       3.93     0.008

S = 91.48        R-Sq = 72.0%      R-Sq(adj) = 67.3%

Analysis of Variance
Source        DF           SS           MS          F         P
Regression     1       129173       129173      15.44     0.008
Error          6        50210         8368
Total          7       179383

Predicted Values for New Observations
    Fit   StDev Fit        95.0% CI                95.0% PI
  431.9        32.7  (   351.8,    511.9)  (   194.1,    669.7)

Values of Predictors for New Observations
New Obs      Years
1               10
```

a. To the greatest number of decimal places in the printout, the regression equation is Shares = 44.3 + 38.756*Years. Since the slope is positive, we can deduce that the number of shares held increases as the number of years with the firm increases. Also, the value of the slope implies that for each additional year with the firm, the number of shares held increases by nearly 39 (38.756) shares.

b. Substituting Years = 10 into the equation in part a, the predicted value of Shares will be 431.9, as shown in the "Fit" column of the printout.

Using the Chart Wizard procedure described in textbook chapter 2, in Computer Solutions 2.9, we generate the corresponding Excel plot and equation shown below. Note that the default Excel equation shows the slope term and the intercept in reverse order compared to Minitab and the textbook. Also note

216

that when using the Excel Chart Wizard, the data should be arranged so that the x variable is at the left and the y variable is at the right within the field, as shown here.

	A	B	C	D	E	F	G	H	I
1	Years	Shares							
2	6	300							
3	12	408							
4	14	560							
5	6	252							
6	9	288							
7	13	650							
8	15	630							
9	9	522							
10									
11									
12									
13									
14									
15									
16									

$y = 38.756x + 44.314$

The Complete Excel regression printout.

	C	D	E	F	G	H	I
1	SUMMARY OUTPUT						
2							
3	*Regression Statistics*						
4	Multiple R	0.849					
5	R Square	0.720					
6	Adjusted R Square	0.673					
7	Standard Error	91.479					
8	Observations	8					
9							
10	ANOVA						
11		*df*	*SS*	*MS*	*F*	*Significance F*	
12	Regression	1	129173.13	129173.13	15.436	0.008	
13	Residual	6	50210.37	8368.40			
14	Total	7	179383.50				
15							
16		*Coefficients*	*Standard Error*	*t Stat*	*P-value*	*Lower 95%*	*Upper 95%*
17	Intercept	44.3140	108.51	0.408	0.697	-221.197	309.825
18	Years	38.7558	9.86	3.929	0.008	14.618	62.893

15.11 p/a/m This exercise can be solved using a pocket calculator and the method shown in the solution to exercise 15.9. We will use Minitab. In generating the printout shown below, we have specified that a prediction be made for total gross sales when sales after two weeks = $100 million.

```
Regression Analysis: totgross versus 2weeks

The regression equation is totgross = 106 + 1.47 2weeks

Predictor        Coef      SE Coef        T        P
Constant       106.28       92.31      1.15    0.333
2weeks          1.474        1.246      1.18    0.322

S = 57.88       R-Sq = 31.8%     R-Sq(adj) = 9.1%

Analysis of Variance
Source           DF          SS         MS        F        P
Regression        1        4690       4690     1.40    0.322
Residual Error    3       10051       3350
Total             4       14740

Predicted Values for New Observations
New Obs    Fit      SE Fit       95.0% CI              95.0% PI
1        253.7       44.3    ( 112.7,   394.6)  (   21.7,    485.6)

Values of Predictors for New Observations
New Obs    2weeks
1            100
```

The regression equation is totgross = 106.28 + 1.474*2weeks. Substituting 2weeks = $100 million into the equation, we obtain $253.7 million as the estimated total gross sales for an animated film that had $100 million in ticket sales during the first two weeks of its run.

Applying the procedure described in textbook chapter 2, Computer Solutions 2.9, we obtain the corresponding Excel plot and equation shown below. A reminder: When using the Excel Chart Wizard, the data should be arranged so that the x variable is at the left and the y variable is at the right within the field, as shown here.

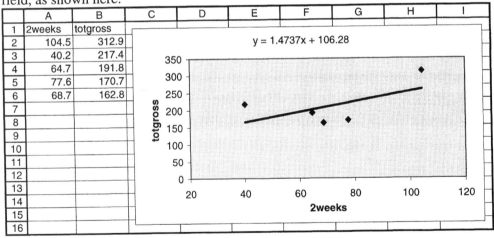

The complete Excel printout.

	C	D	E	F	G	H	I
1	SUMMARY OUTPUT						
2							
3	*Regression Statistics*						
4	Multiple R	0.564					
5	R Square	0.318					
6	Adjusted R Square	0.091					
7	Standard Error	57.881					
8	Observations	5					
9							
10	ANOVA						
11		*df*	*SS*	*MS*	*F*	*Significance F*	
12	Regression	1	4689.850	4689.850	1.400	0.3220	
13	Residual	3	10050.618	3350.206			
14	Total	4	14740.468				
15							
16		*Coefficients*	*Standard Error*	*t Stat*	*P-value*	*Lower 95%*	*Upper 95%*
17	Intercept	106.278	92.315	1.151	0.3331	-187.509	400.066
18	2weeks	1.474	1.246	1.183	0.3220	-2.490	5.438

15.13 p/c/m In generating the Minitab printout shown below, we have specified that a prediction be made for the number of acres burned during a January-May period having 18 inches of rainfall.

Regression Analysis: AcresBurned versus RainInches

The regression equation is AcresBurned = 349550 - 7851 RainInches

```
Predictor       Coef     SE Coef          T        P
Constant      349550      179579       1.95    0.078
RainInch       -7851        9902      -0.79    0.445

S = 185098      R-Sq = 5.4%      R-Sq(adj) = 0.0%

Analysis of Variance
Source          DF          SS          MS         F        P
Regression       1  21539746092  21539746092      0.63    0.445
Residual Error  11  3.76873E+11  34261159213
Total           12  3.98412E+11

Unusual Observations
Obs   RainInch   AcresBur         Fit     SE Fit     Residual     St Resid
  2       12.1     645331      254861      73546       390470        2.30R
  4       30.1      86948      113142     136122       -26194       -0.21 X

R denotes an observation with a large standardized residual
X denotes an observation whose X value gives it large influence.

Predicted Values for New Observations
New Obs      Fit      SE Fit        95.0% CI              95.0% PI
1         208223       51704   (  94423,  322024)   ( -214770,  631217)

Values of Predictors for New Observations
New Obs   RainInch
1            18.0
```

The regression equation is AcresBurned = 349550 - 7851 RainInches. Substituting RainInches = 18 into the equation, we obtain an estimate of 208,223 for AcresBurned. The corresponding Excel plot and equation are shown below.

	A	B	C	D	E	F	G	H	I
1	Year	RainInches	AcresBurned						
2	1988	17.51	193881						
3	1989	12.06	645331						
4	1990	14.03	249912						
5	1991	30.11	86948						
6	1992	16.03	82230						
7	1993	19.61	80484						
8	1994	18.15	180048						
9	1995	16.34	45586						
10	1996	20.43	93849						
11	1997	18.46	146122						
12	1998	22.24	506970						
13	1999	12.70	340124						
14	2000	8.25	118860						
15									
16									

Chart: $y = -7851.5x + 349550$; Acres Burned vs Rain Inches

15.15 d/p/m The standard error of the estimate is a measure which describes the dispersion of the data points above and below the regression line. The standard error may be used as a measure of how well the regression line fits the data. In addition, the standard error is used in calculating confidence intervals for the mean value of y given a specific value of x, and in calculating the prediction interval for an individual y observation.

15.17 d/p/d A confidence interval provides a range of possible values for the mean value of y given a specific value of x. The point estimate for the mean will fall on the regression line. A prediction interval provides a range of possible values for an individual observation of y given a value of x.

15.19 d/p/d The prediction interval for y gets wider as the x value on which the interval estimate is based gets farther away from the mean of x because there is less error in making interval estimates based on x values that are closer to the mean. This can be seen in the formula; the numerator of the fraction under the square root includes $(x - \bar{x})^2$. This value, of course, increases as the x value moves away from the mean.

15.21 c/a/m

a. To determine the least squares regression line, we must calculate the slope and y-intercept.

$$\text{slope, } b_1 = \frac{\sum x_i y_i - n\bar{x}\bar{y}}{\sum x_i^2 - n\bar{x}^2} = \frac{400 - 6(6.67)(12.67)}{346 - 6(6.67)^2} = -1.354$$

y-intercept, $b_0 = \bar{y} - b_1\bar{x} = 12.67 - (-1.354)(6.67) = 21.701$

The regression equation is $\hat{y} = 21.701 - 1.354x$

b. The standard error of the estimate is

$$s_{y.x} = \sqrt{\frac{\sum y_i^2 - b_0\sum y_i - b_1\sum x_i y_i}{n-2}} \quad \sqrt{\frac{1160 - 21.701(76) - (-1.354)(400)}{6-2}} = \sqrt{13.081} = 3.617$$

c. To calculate the confidence interval for the mean of y, we must first calculate the estimated value of y:

$\hat{y} = 21.701 - 1.354(7) = 12.223$.

We also need the t-value with 6 - 2 = 4 degrees of freedom for a 95% interval; this value is 2.776. Therefore the 95% confidence interval for the mean value of y when x = 7 is:

$$\hat{y} \pm t s_{y.x} \sqrt{\frac{1}{n} + \frac{(x-\bar{x})^2}{(\sum x_i^2) - \frac{(\sum x_i)^2}{n}}} = 12.223 \pm 2.776(3.617)\sqrt{\frac{1}{6} + \frac{(7-6.67)^2}{(346 - \frac{40^2}{6})}}$$

$$= 12.223 \pm 4.116 = (8.107, \ 16.339)$$

d. To calculate the confidence interval for the mean of y, we must first calculate the estimated value of y:
$\hat{y} = 21.701 - 1.354(9) = 9.515$.

We also need the t-value with $6 - 2 = 4$ degrees of freedom for a 95% interval; this value is 2.776. Therefore the 95% confidence interval for the mean value of y when $x = 9$ is:

$$\hat{y} \pm ts_{y.x} \sqrt{\frac{1}{n} + \frac{(x - \bar{x})^2}{(\sum x_i^2) - \frac{(\sum x_i)^2}{n}}} = 9.515 \pm 2.776(3.617) \sqrt{\frac{1}{6} + \frac{(9 - 6.67)^2}{(346 - \frac{40^2}{6})}}$$

$$= 9.515 \pm 4.868 = (4.647, \; 14.383)$$

e. The confidence interval in d is wider because 9 is farther from the mean of x than 7.

15.23 p/a/m One way to solve this exercise is with the regression formulas and a pocket calculator:

x = TD%	y = Rating	xy	x^2	y^2
5.6	96.8	542.08	31.36	9370.24
5.1	92.3	470.73	26.01	8519.29
5.4	87.1	470.34	29.16	7586.41
5.0	86.4	432.00	25.00	7464.96
4.0	85.4	341.60	16.00	7293.16
5.0	84.4	422.00	25.00	7123.36
5.2	83.4	433.68	27.04	6955.56
$\sum x = 35.3$	$\sum y = 615.8$	$\sum xy = 3112.43$	$\sum x^2 = 179.57$	$\sum y^2 = 54{,}312.98$

$n = 7$, $\bar{x} = 5.043$ and $\bar{y} = 87.971$

$$\text{slope, } b_1 = \frac{\sum x_i y_i - n\bar{x}\bar{y}}{\sum x_i^2 - n\bar{x}^2} = \frac{3112.43 - 7(5.043)(87.971)}{179.57 - 7(5.043)^2} = 4.52$$

y-intercept, $b_0 = \bar{y} - b_1\bar{x} = 87.971 - (4.52)(5.043) = 65.18$

The regression equation is $\hat{y} = 65.18 + 4.52x$

In generating the Minitab printout below, we have specified point and 95% interval estimates for Rating when TD% = 5.0.

```
Regression Analysis: Rating versus TD%

The regression equation is Rating = 65.2 + 4.52 TD%

Predictor        Coef     SE Coef        T          P
Constant        65.18       18.90     3.45      0.018
TD%             4.520        3.731     1.21      0.280

S = 4.655      R-Sq = 22.7%      R-Sq(adj) = 7.2%

Analysis of Variance
Source          DF          SS          MS        F         P
Regression       1       31.82       31.82     1.47      0.280
Residual Error   5      108.36       21.67
Total            6      140.17

Predicted Values for New Observations
New Obs    Fit     SE Fit        95.0% CI              95.0% PI
1        87.78       1.77    ( 83.24,   92.32)   ( 74.98,  100.58)

Values of Predictors for New Observations
New Obs       TD%
1            5.00
```

a. To the greatest number of decimal places in the Minitab printout, the regression equation is Rating = 65.18 + 4.520*TD%.

b. Substituting TD% = 5.0 into the regression equation, the estimated value of Rating will be 87.78, as shown in the "Fit" column of the printout.

c. The Minitab printout shows the standard error of the estimate to be 4.655 rating points. Using a calculator and the formula in the text:

$$s_{y.x} = \sqrt{\frac{\sum y_i^2 - b_0 \sum y_i - b_1 \sum x_i y_i}{n-2}} = \sqrt{\frac{54,312.98 - 65.18(615.8) - 4.520(3112.43)}{7-2}} = 4.655$$

(Please note once again the calculator-solution commentary on the first page of this chapter's solution manual materials. As mentioned in that discussion, we are carrying more decimal places than practicality allows us to show.)

d. For a quarterback with TD% = 5.0, the 95% prediction interval for his quarterback rating will be from 74.98 to 100.58, as shown in the "PI" column of the Minitab printout. Using a calculator and the formula in the text:

When x = 5.0, $\hat{y}$ = 65.18 + 4.520(5.0) = 87.78, and d.f. = n - 2 = 5, then t = 2.571 for a 95% prediction interval:

$$\hat{y} \pm ts_{y.x}\sqrt{1 + \frac{1}{n} + \frac{(x - \bar{x})^2}{(\sum x_i^2) - \frac{(\sum x_i)^2}{n}}} = 87.78 \pm 2.571(4.655)\sqrt{1 + \frac{1}{7} + \frac{(5 - 5.043)^2}{(179.57 - \frac{35.3^2}{7})}}$$

or from 74.98 to 100.58.

e. For all quarterbacks with TD% = 5.0, the 95% confidence interval for their mean rating will be from 83.24 to 92.32, as shown in the "CI" column of the Minitab printout. Using a calculator and the formula in the text:

$$\hat{y} \pm ts_{y.x}\sqrt{\frac{1}{n} + \frac{(x - \bar{x})^2}{(\sum x_i^2) - \frac{(\sum x_i)^2}{n}}} = 87.78 \pm 2.571(4.655)\sqrt{\frac{1}{7} + \frac{(5 - 5.043)^2}{(179.57 - \frac{35.3^2}{7})}}$$

or from 83.24 to 92.32

For this exercise, the detailed Excel regression printout is shown below. Although it does not provide confidence and prediction intervals, it does provide the other items shown earlier. The Excel printout provides a 95% confidence interval for the intercept and the slope of the population regression equation. The square of the standard error of the estimate is shown in the "Residual" row, "MS" column as 21.6717. Its positive square root (4.6553) is listed as the standard error of the estimate.

	E	F	G	H	I	J	K
1	SUMMARY OUTPUT						
2							
3	*Regression Statistics*						
4	Multiple R	0.4764					
5	R Square	0.2270					
6	Adjusted R Square	0.0724					
7	Standard Error	4.6553					
8	Observations	7					
9							
10	ANOVA						
11		*df*	*SS*	*MS*	*F*	*Significance F*	
12	Regression	1	31.8156	31.8156	1.4681	0.2798	
13	Residual	5	108.3587	21.6717			
14	Total	6	140.1743				
15							
16		*Coefficients*	*Standard Error*	*t Stat*	*P-value*	*Lower 95%*	*Upper 95%*
17	Intercept	65.1768	18.8952	3.4494	0.0182	16.605	113.748
18	TD%	4.5202	3.7306	1.2116	0.2798	-5.070	14.110

We can also use Data Analysis Plus for the point estimate and the 95% PI and CI estimates in parts b, d, and e. The values are comparable to those from Minitab, and the printout is shown below.

	A	B	C
1	Prediction Interval		
2			
3			Rating
4			
5	Predicted value		87.78
6			
7	Prediction Interval		
8	Lower limit		74.98
9	Upper limit		100.58
10			
11	Interval Estimate of Expected Value		
12	Lower limit		83.24
13	Upper limit		92.32

15.25 p/a/m Shown below are relevant preliminary calculations for pocket-calculator solutions to this exercise. We have included y^2 values with those presented earlier, in the solution to exercise 15.9.

x	y	xy	x^2	y^2
6.00	300.00	1800.00	36.00	90,000.00
12.00	408.00	4896.00	144.00	166,464.00
14.00	560.00	7840.00	196.00	313,600.00
6.00	252.00	1512.00	36.00	63,504.00
9.00	288.00	2592.00	81.00	82,944.00
13.00	650.00	8450.00	169.00	422,500.00
15.00	630.00	9450.00	225.00	396,900.00
9.00	522.00	4698.00	81.00	272,484.00
$\Sigma x = 84.00$	$\Sigma y = 3610.00$	$\Sigma xy = 41,238.00$	$\Sigma x^2 = 968.00$	$\Sigma y^2 = 1,808,396.00$

$n = 8$, $\bar{x} = 10.5$ and $\bar{y} = 451.25$

$$\text{slope, } b_1 = \frac{\sum x_i y_i - n\bar{x}\bar{y}}{\sum x_i^2 - n\bar{x}^2} = \frac{41,238.00 - 8(10.5)(451.25)}{968.00 - 8(10.5)^2} = \frac{3333}{86} = 38.756$$

y-intercept, $b_0 = \bar{y} - b_1\bar{x} = 451.25 - (38.756)(10.5) = 44.314$

The regression equation is $\hat{y} = 44.314 + 38.756x$

$$s_{y.x} = \sqrt{\frac{\sum y_i^2 - b_0\sum y_i - b_1\sum x_i y_i}{n-2}} = \sqrt{\frac{1,808,396.0 - (44.314)(3610) - (38.756)(41,238)}{8-2}} = 91.48$$

The prediction interval for the amount of stock owned when x = 5:
When x = 5, $\hat{y} = 44.314 + 38.756(5) = 238.09$. With n = 8, d.f. = n - 2 = 6, the t-value for the 95% prediction interval is 2.447, and the interval can be calculated as:

$$\hat{y} \pm t s_{y.x}\sqrt{1 + \frac{1}{n} + \frac{(x - \bar{x})^2}{(\sum x_i^2) - \frac{(\sum x_i)^2}{n}}} = 238.09 \pm 2.447(91.48)\sqrt{1 + \frac{1}{8} + \frac{(5 - 10.5)^2}{(968 - \frac{84^2}{8})}}$$

or from -34.0 to 510.2

In generating the Minitab printout below, we have specified point and 95% interval estimates for Shares when Years = 5.0.

```
Regression Analysis: Shares versus Years

The regression equation is Shares = 44 + 38.8 Years

Predictor        Coef     SE Coef         T        P
Constant         44.3       108.5      0.41    0.697
Years          38.756       9.864      3.93    0.008

S = 91.48       R-Sq = 72.0%    R-Sq(adj) = 67.3%

Analysis of Variance
Source              DF          SS          MS         F        P
Regression           1      129173      129173     15.44    0.008
Residual Error       6       50210        8368
Total                7      179384

Predicted Values for New Observations
New Obs     Fit      SE Fit  (       95.0% CI          95.0% PI
1         238.1        63.2  (     83.5,   392.7)  (   -34.0,    510.2)

Values of Predictors for New Observations
New Obs     Years
1            5.00
```

Referring to the Minitab printout, the regression equation is Shares = 44.3 + 38.756*Years and the standard error of estimate is 91.48 shares of stock. For an employee with Years = 5, the estimated value of Shares is 238.1 (see the "Fit" column in the printout). For this individual employee, the 95% prediction interval for the number of shares of stock he or she owns is shown as -34.0 to 510.2. (See the "PI" column of the printout.) Mathematically, the 95% prediction interval would have -34.0 as its lower limit. It could be argued that this would revert to 0 because of the impossibility of owning negative shares of stock (aside from short sales in the financial market, which involve selling stock which you have not yet purchased, in hopes that the price will be going down).

We can also use Data Analysis Plus for the point estimate and the 95% PI and CI estimates. The results are comparable to those reported by Minitab and the printout is shown below.

	A	B	C
1	**Prediction Interval**		
2			
3			Shares
4			
5	Predicted value		238.1
6			
7	Prediction Interval		
8	Lower limit		-33.9
9	Upper limit		510.1
10			
11	Interval Estimate of Expected Value		
12	Lower limit		83.5
13	Upper limit		392.6

15.27 p/c/m The Minitab printout is shown below. We have specified point and interval estimates for total revenue associated with a dealership group consisting of 100 dealers.

```
Regression Analysis: $GroupRevenue versus NumDealrs

The regression equation is $GroupRevenue = -1.66E+08 +64076803 NumDealrs

Predictor        Coef     SE Coef        T        P
Constant    -166227473    72426532    -2.30    0.024
NumDealr      64076803     2030145    31.56    0.000

S = 641757544   R-Sq = 91.2%     R-Sq(adj) = 91.1%

Analysis of Variance
Source          DF         SS           MS         F        P
Regression       1  4.10289E+20  4.10289E+20   996.20    0.000
Residual Error  96  3.95379E+19  4.11853E+17
Total           97  4.49827E+20

Unusual Observations
Obs   NumDealr  $GroupRe         Fit      SE Fit    Residual   St Resid
  1       290  2.0926e+10  18416045305  560209649  2510154695     8.02RX
  4       114  3100030000   7138528033  209426763 -4038498033    -6.66RX
  9        45  1242660000   2717228647   87696854 -1474568647    -2.32R

R denotes an observation with a large standardized residual
X denotes an observation whose X value gives it large influence.

Predicted Values for New Observations
New Obs    Fit      SE Fit          95.0% CI               95.0% PI
  1    6.24E+09  182612760   (5.88E+09,6.60E+09)   (4.92E+09,7.57E+09) X
X  denotes a row with X values away from the center

Values of Predictors for New Observations
New Obs  NumDealr
  1         100
```

For a dealership group consisting of 100 dealers, the point estimate for total revenue is $GroupRevenue = -1.66E+08 +64,076,803*100 = $6.24 billion. (Note the scientific notation in the printout. For example, -1.66E+08 is $-1.66*10^8$.)

For this dealership group, the 95% prediction interval for total revenue is from $4.92E+09 to $7.57E+09, or from $4.92 billion to $7.57 billion.

For all dealership groups like this one, the 95% confidence interval for their mean total revenue is from $5.88E+09 to $6.60E+09, or from $5.88 billion to $6.60 billion.

We can also use Data Analysis Plus for the point estimate and the 95% PI and CI estimates. The results are comparable to those reported by Minitab and the printout is shown below.

	A	B	C
1	**Prediction Interval**		
2			
3			$GroupRevenue
4			
5	Predicted value		6241452795
6			
7	Prediction Interval		
8	Lower limit		4917004315
9	Upper limit		7565901275
10			
11	Interval Estimate of Expected Value		
12	Lower limit		5878969054
13	Upper limit		6603936536

15.29 p/c/m The Minitab printout is shown below. We have specified point and interval estimates for burned acreage when the January-May rainfall is 20 inches.

```
Regression Analysis: AcresBurned versus RainInches

The regression equation is AcresBurned = 349550 - 7851 RainInches

Predictor       Coef    SE Coef        T        P
Constant      349550     179579     1.95    0.078
RainInch       -7851       9902    -0.79    0.445

S = 185098    R-Sq = 5.4%    R-Sq(adj) = 0.0%

Analysis of Variance
Source          DF          SS           MS        F       P
Regression       1  21539746092  21539746092     0.63    0.445
Residual Error  11  3.76873E+11  34261159213
Total           12  3.98412E+11

Unusual Observations
Obs  RainInch  AcresBur      Fit   SE Fit  Residual  St Resid
  2      12.1    645331   254861    73546    390470     2.30R
  4      30.1     86948   113142   136122    -26194    -0.21 X

R denotes an observation with a large standardized residual
X denotes an observation whose X value gives it large influence.

Predicted Values for New Observations
New Obs    Fit   SE Fit       90.0% CI             90.0% PI
  1     192521    57527  (  89209,  295832)  ( -155578,  540619)

Values of Predictors for New Observations
New Obs  RainInch
  1          20.0
```

The point estimate for burned acreage for a year during which January-May rainfall is 20 inches is AcresBurned = 349,550 - 7851*20 = 192,521 acres.

For a year during which January-May rainfall is 20 inches, the 90% prediction interval for the number of acres burned is from -155,578 to 540,619 acres.

For all years during which January-May rainfall is 20 inches, the 90% confidence interval for the mean number of acres burned is from 89,209 to 295,832 acres.

We can also use Data Analysis Plus for the point estimate and the 90% PI and CI estimates. The results are comparable to those reported by Minitab and the printout is shown below.

	A	B	C
1	**Prediction Interval**		
2			
3			AcresBurned
4			
5	Predicted value		192521
6			
7	Prediction Interval		
8	Lower limit		-155578
9	Upper limit		540619
10			
11	Interval Estimate of Expected Value		
12	Lower limit		89209
13	Upper limit		295832

15.31 d/p/m The coefficient of correlation (r) describes both the direction and the strength of the linear relationship between two variables. The coefficient of determination (r^2) expresses the proportion of the variation in the dependent variable (y) that is explained by the regression line, $\hat{y} = b_0 + b_1x$, but it does not indicate the direction of the relationship.

15.33 c/a/e The coefficient of determination is 0.81 (or the coefficient of correlation squared). This means that 81% of the variation in y is explained by x.

15.35 d/p/m Variation explained by the regression line (SSR) and variation not explained by the regression line (SSE) are the two components of the total variation in y (SST). The coefficient of determination can be expressed as: $1 - \dfrac{SSE}{SST} = \dfrac{SSR}{SST}$

15.37 c/a/e The coefficients of correlation and determination could be found using the formula shown in the text and a pocket calculator. With x = theft rating and y = collision rating, the preliminary calculations would reveal the following summary information:

$\Sigma x = 1445.0$, $\Sigma y = 1291.0$, $\Sigma xy = 166{,}185.0$, $\Sigma x^2 = 283{,}593.0$, $\Sigma y^2 = 141{,}167.0$
$n = 12$, $\bar{x} = 120.417$, $\bar{y} = 107.583$

a. Determining the regression line:

$$\text{slope, } b_1 = \frac{\sum x_i y_i - n\bar{x}\bar{y}}{\sum x_i^2 - n\bar{x}^2} = \frac{166{,}185 - 12(120.417)(107.583)}{283{,}593 - 12(120.417)^2} = 0.09788$$

y-intercept, $b_0 = \bar{y} - b_1\bar{x} = 107.583 - (0.09788)(120.417) = 95.797$
The regression equation is $\hat{y} = 95.797 + 0.09788x$

b. Determining the values of r and r^2:

$$r = \frac{n(\sum x_i y_i) - (\sum x_i)(\sum y_i)}{\sqrt{n(\sum x_i^2) - (\sum x_i)^2} * \sqrt{n(\sum y_i^2) - (\sum y_i)^2}}$$

$$= \frac{12(166{,}185) - (1445)(1291)}{\sqrt{(12*283{,}593 - 1445^2)*(12*141{,}167 - 1291^2)}} = 0.679$$

and the coefficient of determination is $r^2 = 0.679^2 = 0.461$

The Minitab and Excel printouts for the coefficient of correlation are shown below.

Correlations: Collision, Theft

Pearson correlation of Collision and Theft = 0.679

		E	F	G
1		Collision	Theft	
2	Collision	1		
3	Theft	0.679	1	

c. The Minitab printout, including point and interval estimates when Theft = 110, is shown below.

```
Regression Analysis: Collision versus Theft

The regression equation is Collision = 95.8 + 0.0979 Theft

Predictor         Coef      SE Coef          T         P
Constant        95.797        5.144      18.62     0.000
Theft          0.09788      0.03346       2.93     0.015

S = 11.08       R-Sq = 46.1%     R-Sq(adj) = 40.7%

Analysis of Variance
Source           DF          SS          MS         F         P
Regression        1      1050.0      1050.0      8.56     0.015
Residual Error   10      1226.9       122.7
Total            11      2276.9

Unusual Observations
Obs      Theft    Collisio         Fit      SE Fit      Residual      St Resid
  5         90      127.00      104.61        3.36         22.39         2.12R
  8        425      139.00      137.40       10.68          1.60         0.55 X

R denotes an observation with a large standardized residual
X denotes an observation whose X value gives it large influence.

Predicted Values for New Observations
New Obs      Fit      SE Fit          95.0% CI              95.0% PI
  1       106.56        3.22    (  99.40,  113.73)    (  80.86,  132.26)

Values of Predictors for New Observations
New Obs      Theft
  1            110
```

The Excel printout and regression equation.

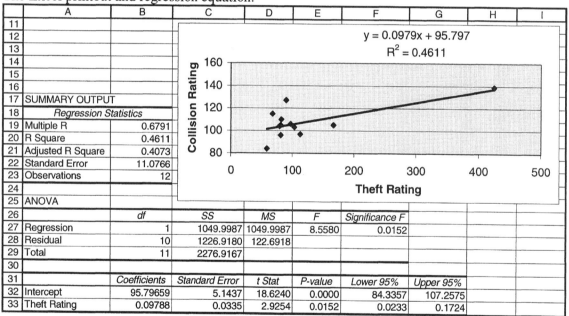

	A	B	C	D	E	F	G	H	I
11									
12									
13									
14									
15									
16									
17	SUMMARY OUTPUT								
18	Regression Statistics								
19	Multiple R	0.6791							
20	R Square	0.4611							
21	Adjusted R Square	0.4073							
22	Standard Error	11.0766							
23	Observations	12							
24									
25	ANOVA								
26		df	SS	MS	F	Significance F			
27	Regression	1	1049.9987	1049.9987	8.5580	0.0152			
28	Residual	10	1226.9180	122.6918					
29	Total	11	2276.9167						
30									
31		Coefficients	Standard Error	t Stat	P-value	Lower 95%	Upper 95%		
32	Intercept	95.79659	5.1437	18.6240	0.0000	84.3357	107.2575		
33	Theft Rating	0.09788	0.0335	2.9254	0.0152	0.0233	0.1724		

15.39 p/a/m Use of the formulas and pocket calculator will be facilitated by the following summary of preliminary calculations for this exercise, with x = U.S. population (millions) and y = shipments of socks (millions of pairs):

$\Sigma x = 1339.70$, $\Sigma y = 15,976.60$, $\Sigma xy = 4,284,916.52$, $\Sigma x^2 = 359,051.49$, $\Sigma y^2 = 51,256,363.78$
$n = 5$, $\bar{x} = 267.94$, $\bar{y} = 3195.32$. First, determine the regression line:

$$\text{slope, } b_1 = \frac{\sum x_i y_i - n\bar{x}\bar{y}}{\sum x_i^2 - n\bar{x}^2} = \frac{4,284,916.52 - 5(267.94)(3195.32)}{359,051.49 - 5(267.94)^2} = 44.936$$

y-intercept, $b_0 = \bar{y} - b_1\bar{x} = 3195.32 - (44.936)(267.94) = -8845$

The regression equation is $\hat{y} = -8845 + 44.936x$

Next, determine the values of r and r^2:

$$r = \frac{n(\sum x_i y_i) - (\sum x_i)(\sum y_i)}{\sqrt{n(\sum x_i^2) - (\sum x_i)^2} * \sqrt{n(\sum y_i^2) - (\sum y_i)^2}}$$

$$= \frac{5(4,284,916.52) - (1339.70)(15,976.60)}{\sqrt{(5*359,051.49 - 1339.70^2)*(5*51,256,363.78 - 15,976.60^2)}} = 0.951, r^2 = 0.951^2 = 0.904$$

The Minitab printout, with point and interval estimates when population = 280.0 million, is shown below. Also shown is the Minitab printout for the coefficient of correlation.

```
Correlations: population, millsocks

Pearson correlation of population and millsocks = 0.951

Regression Analysis: millsocks versus population

The regression equation is millsocks = - 8845 + 44.9 population

Predictor        Coef      SE Coef          T        P
Constant        -8845         2260      -3.91    0.030
populati       44.936        8.435       5.33    0.013

S = 81.03       R-Sq = 90.4%      R-Sq(adj) = 87.3%

Analysis of Variance
Source            DF            SS           MS        F        P
Regression         1        186318       186318    28.38    0.013
Residual Error     3         19696         6565
Total              4        206014

Predicted Values for New Observations
New Obs     Fit      SE Fit         95.0% CI              95.0% PI
1        3737.2       108.0  ( 3393.6,   4080.9) (  3307.6,   4166.9) XX
X  denotes a row with X values away from the center
XX denotes a row with very extreme X values

Values of Predictors for New Observations
New Obs   populati
1              280
```

a. To the greatest number of decimal places expressed in the printout, the regression equation is: millsocks = -8845 + 44.936*population. Prior to our regression analysis, Minitab reported the coefficient of correlation to be r = 0.951.

b. The "R-Sq" portion of the printout provides us with the coefficient of determination (0.904, or 90.4%). This indicates that 90.4% of the variation in annual shipments of socks is explained by the size of the population.

c. During a year in which population = 280.0 million, we would predict the shipment of 3737.2 million pairs of socks. This can be obtained by substitution into the regression equation or by referring to the "Fit" column of the printout.

The Excel printout and plot are shown below.

	A	B	C	D	E	F	G	H	I
1	population	millsocks	SUMMARY OUTPUT						
2	260.7	2927.2							
3	266.1	3002.4	*Regression Statistics*						
4	268.7	3217.3	Multiple R	0.951					
5	271.3	3410.5	R Square	0.904					
6	272.9	3419.2	Adjusted R Square	0.873					
7			Standard Error	81.027					
8			Observations	5					
9									
10			ANOVA						
11				*df*	*SS*	*MS*	*F*	*Significance F*	
12			Regression	1	186318.02	186318.02	28.38	0.013	
13			Residual	3	19696.25	6565.42			
14			Total	4	206014.27				
15									
16				*Coefficients*	*Standard Error*	*t Stat*	*P-value*	*Lower 95%*	*Upper 95%*
17			Intercept	-8844.778	2260.422	-3.913	0.030	-16038.456	-1651.100
18			population	44.936	8.435	5.327	0.013	18.091	71.780

Chart: $y = 44.936x - 8844.8$, $R^2 = 0.9044$ (millsocks vs population)

15.41 p/c/m The Minitab and Excel printouts are shown below.

Correlations: Profits/Partner, No. of Partners

Pearson correlation of Profits/Partner and No. of Partners = -0.279

	D	E	F
1		*Profits/Partner*	*No. of Partners*
2	Profits/Partner	1	
3	No. of Partners	-0.2791	1

The coefficient of correlation is r = -0.2791. The coefficient of determination is $r^2 = (-0.2791)^2 = 0.0779$. The number of partners explains only 7.79% of the variation in profits per partner.

15.43 p/c/m The Minitab and Excel printouts for the coefficient of correlation are shown below.

Correlations: DomGross, ForGross

Pearson correlation of DomGross and ForGross = 0.736

	D	E	F
1		*DomGross*	*ForGross*
2	DomGross	1	
3	ForGross	0.7356	1

The coefficient of correlation is r = 0.7356. The coefficient of determination is $r^2 = 0.7356^2 = 0.5411$. Domestic gross explains 54.11% of the variation in foreign gross.

To the number of decimal places shown in the Excel printout, the regression equation is ForGross = - 69.382 + 1.5266*DomGross. The Minitab and Excel printouts are shown below.

Regression Analysis: ForGross versus DomGross

The regression equation is ForGross = - 69 + 1.53 DomGross

Predictor	Coef	SE Coef	T	P
Constant	-69.4	105.2	-0.66	0.518
DomGross	1.5266	0.3314	4.61	0.000

S = 152.5   R-Sq = 54.1%   R-Sq(adj) = 51.6%

Analysis of Variance

Source	DF	SS	MS	F	P
Regression	1	493846	493846	21.23	0.000
Residual Error	18	418805	23267		
Total	19	912651			

Unusual Observations

Obs	DomGross	ForGross	Fit	SE Fit	Residual	St Resid
1	601	1233.4	847.8	105.3	385.6	3.49RX
5	461	324.7	634.4	63.3	-309.7	-2.23R

R denotes an observation with a large standardized residual
X denotes an observation whose X value gives it large influence.

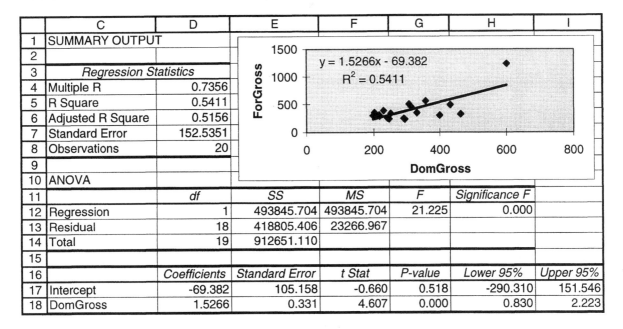

	C	D	E	F	G	H	I
1	SUMMARY OUTPUT						
2							
3	*Regression Statistics*						
4	Multiple R	0.7356					
5	R Square	0.5411					
6	Adjusted R Square	0.5156					
7	Standard Error	152.5351					
8	Observations	20					
9							
10	ANOVA						
11		*df*	*SS*	*MS*	*F*	*Significance F*	
12	Regression	1	493845.704	493845.704	21.225	0.000	
13	Residual	18	418805.406	23266.967			
14	Total	19	912651.110				
15							
16		*Coefficients*	*Standard Error*	*t Stat*	*P-value*	*Lower 95%*	*Upper 95%*
17	Intercept	-69.382	105.158	-0.660	0.518	-290.310	151.546
18	DomGross	1.5266	0.331	4.607	0.000	0.830	2.223

15.45 c/a/m This calls for a hypothesis test with H_0: $\rho = 0$ (there is no linear relationship) and H_1: $\rho \neq 0$ (there is a linear relationship). This is a two-tail t-test with $\alpha = 0.05$ and d.f. = 15 - 2 = 13. The critical values of t are -2.160 and +2.160. The calculated value of the test statistic is:

$$t = \frac{r}{\sqrt{\frac{1-r^2}{n-2}}} = \frac{0.9}{\sqrt{(1-0.81)/(15-2)}} = 7.445$$

Since the calculated test statistic falls outside the critical values, we reject H_0. There is evidence to suggest that a linear relationship exists between x and y.

15.47 d/p/d At the 0.10 level of significance, we would not reject the null hypothesis that the true slope is equal to zero. This implies that the p-value is greater than 0.10. Since the 90% confidence interval contains 0, the p-value must be greater than 1 - 0.90; that is, it must be greater than 0.10.

15.49 p/a/m Referring to the printout shown in the solution to exercise 15.37, the coefficient of correlation is $r = 0.679$ and the slope of the regression equation is $b_1 = 0.09788$.

a. For testing the coefficient of correlation, the null and alternative hypotheses are:

H_0: $\rho = 0$ (there is no linear relationship) and H_1: $\rho \neq 0$ (there is a linear relationship)

In the computer printout, the p-value for the two-tail test of the coefficient of correlation is 0.015. Since p-value = 0.015 is $< \alpha = 0.05$ level of significance for the test, we reject H_0. At this level, the coefficient of correlation differs significantly from zero.

If the computer is not used, the coefficient of correlation can be obtained using preliminary calculation results, a pocket calculator, and the formula:

$$r = \frac{n(\sum x_i y_i) - (\sum x_i)(\sum y_i)}{\sqrt{n(\sum x_i^2) - (\sum x_i)^2} * \sqrt{n(\sum y_i^2) - (\sum y_i)^2}} = 0.679$$

With $n = 12$ pairs of observations, d.f. = 12 - 2 = 10, and for the 0.05 level of significance the critical values of the test statistic are $t = -2.228$ and $t = +2.228$. The calculated value of the test statistic is:

$$t = \frac{r}{\sqrt{\dfrac{1-r^2}{n-2}}} = \frac{0.679}{\sqrt{(1-0.679^2)/(12-2)}} = 2.92$$

Since the calculated value (2.92) falls outside the critical values, we reject the null hypothesis at the 0.05 level. At this level of significance, the population coefficient of correlation (ρ) is some value other than zero.

b. The null and alternative hypotheses are H_0: $\beta_1 = 0$ and H_1: $\beta_1 \neq 0$. In the computer printout shown in exercise 15.37, the p-value for the two-tail test of the slope of the regression equation is 0.015. Since p-value = 0.015 is $< \alpha = 0.05$ level of significance for the test, we reject the null hypothesis that the population slope (β_1) could be zero.

If the computer is not used, the standard error of estimate must be calculated, then we must compute the standard deviation of the slope:

$$s_{y.x} = \sqrt{\frac{\sum y_i^2 - b_0 \sum y_i - b_1 \sum x_i y_i}{n-2}} = 11.08 \text{ and } s_{b_1} = \frac{s_{y.x}}{\sqrt{\sum x_i^2 - n\overline{x}^2}} = 0.03346$$

With $n = 12$ pairs of observations, d.f. = 12 - 2 = 10. For the 0.05 level of significance, the critical values of the test statistic are $t = -2.228$ and $t = +2.228$. The calculated value of the test statistic is:

$$t = \frac{b_1 - 0}{s_{b_1}} = \frac{0.09788 - 0}{0.03346} = 2.93$$

Since the calculated value (2.93) falls outside the critical values, we reject the null hypothesis at the 0.05 level. At this level of significance, the slope of the population regression equation is some value other than zero. (Note: The calculated values, $t = 2.92$ in part a and $t = 2.93$ in part b, differ only due to rounding, as the two tests are equivalent. The more exact of the two values is $t = 2.93$, as reported by Minitab.)

c. The standard deviation of the slope is $s_{b_1} = 0.03346$, which appears in the denominator of the calculated t statistic in part b. With d.f. = 12 - 2 = 10, the 95% confidence interval for the slope of the population regression line is:

$b_1 \pm t\,s_{b_1} = 0.09788 \pm 2.228(0.03346)$, or from 0.0233 to 0.1724.

Minitab does not directly provide any confidence interval for the slope of the population regression line, but a confidence interval can be obtained by using the appropriate value of t, along with the sample slope and the standard deviation of the slope, as shown in the preceding calculation.

Excel does provide lower and upper confidence limits for the population slope (β_1). The default is a 95% confidence interval. For the regression line in this exercise, Excel's 95% confidence interval for the slope is from 0.023 to 0.172. See the Excel printout that is included in the exercise 15.37 solution.

15.51 c/a/m
a. We can use this hypothesis test:

H_0: $\beta_1 = 0$ (the true slope is 0) and H_1: $\beta_1 \neq 0$ (the true slope is not zero)

This is a two-tail t-test with alpha = 0.05 and d.f. = 30 - 2 = 28. For the 0.05 level, the critical values of t are t = -2.048 and t = +2.048. The calculated value of the test statistic is:

$t = \dfrac{b_1 - 0}{s_{b_1}} = \dfrac{5.0 - 0}{2.25} = 2.222$

Since the calculated value of the test statistic falls outside the critical values, we reject H_0. There is evidence to suggest that there is a linear relationship between the x and y values.

b. Since we rejected the null hypothesis (of no linear relationship) in testing the slope of the regression equation, we will also reject the null hypothesis at the same level of significance in testing the coefficient of correlation. Testing either one of these is equivalent to testing both.

c. The 95% confidence interval for the true slope of the regression line is:

$b_1 \pm t\,s_{b_1} = 5.0 \pm 2.048(2.25) = 5.0 \pm 4.608$, or from 0.392 to 9.608

15.53 c/a/e We know that the coefficient of determination is $r^2 = 1 - (SSE/SST)$.
Therefore: 0.49 = 1 - (SSE/120), 0.51 = SSE/120, and SSE = 61.2
Since SSE = 61.2, SSR must be 58.8. (SST = SSR + SSE.)

15.55 c/a/m The table below details the calculations for SSE and SST.
Recall that SSR = SST - SSE. The regression equation is $\hat{y} = 20.66 + 4.02x$, and the mean of the y values is $\overline{y} = 51.2$.

x	y	$\hat{y}$	$(y - \hat{y})$	$(y - \hat{y})^2$	$(y - \overline{y})$	$(y - \overline{y})^2$
5	34	40.76	-6.76	45.6976	-17.2	295.84
4	44	36.74	7.26	52.7076	-7.2	51.84
10	65	60.86	4.14	17.1396	13.8	190.44
9	47	56.84	-9.84	96.8256	-4.2	17.64
10	66	60.86	5.14	26.4196	14.8	219.04
				sum = 238.7900		sum = 774.80

SST = 774.80 and SSE = 238.79.
Therefore, SSR = SST - SSE = 536.01 and $r^2 = 536.01/774.80 = 0.692$.
The appropriate null and alternative hypotheses are:
H_0: There is no linear relationship between x and y, and H_1: There is a linear relationship

The F statistic will have 1 and 3 degrees of freedom. For $\alpha = 0.05$ level of significance, the critical value of F is 10.13. The calculated value of F is: $F = \dfrac{SSR/1}{SSE/(n-2)} = \dfrac{536.01/1}{238.79/(5-2)} = 6.734$

Since the calculated F does not exceed the critical value, we do not reject H_0. At this level, there is no evidence to suggest that a linear relationship exists between x and y. Therefore, we cannot say that r^2 is significantly different from 0.

15.57 p/c/m The Minitab and Excel printouts are shown below.

```
Regression Analysis: Gallons versus Hours

The regression equation is Gallons = 5921561 + 10.4 Hours

Predictor        Coef      SE Coef         T        P
Constant      5921561     10790160      0.55    0.612
Hours         10.4481       0.6132     17.04    0.000

S = 1056418     R-Sq = 98.6%     R-Sq(adj) = 98.3%

Analysis of Variance
Source          DF          SS          MS         F        P
Regression       1  3.24029E+14  3.24029E+14    290.34    0.000
Residual Error   4  4.46408E+12  1.11602E+12
Total            5  3.28493E+14
```

	E	F	G	H	I	J	K
1	SUMMARY OUTPUT						
2							
3	*Regression Statistics*						
4	Multiple R	0.993					
5	R Square	0.986					
6	Adjusted R Square	0.983					
7	Standard Error	1056418.05					
8	Observations	6					
9							
10	ANOVA						
11		*df*	*SS*	*MS*	*F*	*Significance F*	
12	Regression	1	3.240E+14	3.240E+14	2.903E+02	6.957E-05	
13	Residual	4	4.464E+12	1.116E+12			
14	Total	5	3.285E+14				
15							
16		*Coefficients*	*Standard Error*	*t Stat*	*P-value*	*Lower 95%*	*Upper 95%*
17	Intercept	5921560.921	10790159.777	0.549	0.612	-24036787.427	35879909.269
18	Hours	10.44806	0.613	17.039	0.000	8.746	12.150

The regression equation is Gallons = 5921560.92 + 10.44806*Hours.
The slope is 10.44806 -- on average, one additional hour of flying time will require 10.44806 additional gallons of fuel. The coefficients of correlation and determination are 0.993 and 0.986, respectively. The flying hours variable explains 98.6% of the variation in fuel consumed. To three decimal places, the p-value is 0.000, and this is less than 0.05. At this level of significance, the population slope and the population coefficient of correlation could not be zero. The lower and upper 95% confidence limits for the population slope are 8.746 and 12.150, respectively.

15.59 p/c/m The Minitab and Excel printouts are shown below.

```
Regression Analysis: Net Income versus Revenue

The regression equation is Net Income = 59.8 + 0.0382 Revenue
Predictor        Coef     SE Coef        T        P
Constant        59.80       21.23     2.82    0.006
Revenue      0.038208    0.003413    11.20    0.000

S = 201.3      R-Sq = 56.1%     R-Sq(adj) = 55.7%

Analysis of Variance
Source            DF          SS          MS        F        P
Regression         1     5078439     5078439   125.36    0.000
Residual Error    98     3970099       40511
Total             99     9048539

Unusual Observations
Obs   Revenue   Net Inco       Fit    SE Fit    Residual    St Resid
 29     45625     1109.0    1803.0     150.3      -694.0     -5.18RX
 48     27008     1757.0    1091.7      87.8       665.3      3.67RX
 50      3621      667.6     198.1      20.9       469.5      2.35R
 59      7055     1120.2     329.4      26.6       790.8      3.96R
 76     16894     1825.3     705.3      54.7      1120.0      5.78RX
 86     20779      265.1     853.7      67.2      -588.6     -3.10RX
 95      2489      577.0     154.9      20.2       422.1      2.11R

R denotes an observation with a large standardized residual
X denotes an observation whose X value gives it large influence.
```

	D	E	F	G	H	I	J
1	SUMMARY OUTPUT						
2							
3	*Regression Statistics*						
4	Multiple R	0.7492					
5	R Square	0.5612					
6	Adjusted R Square	0.5568					
7	Standard Error	201.2740					
8	Observations	100					
9							
10	ANOVA						
11		*df*	*SS*	*MS*	*F*	*Significance F*	
12	Regression	1	5078439	5078439	125.36	0.0000	
13	Residual	98	3970099	40511			
14	Total	99	9048539				
15							
16		*Coefficients*	*Standard Error*	*t Stat*	*P-value*	*Lower 95%*	*Upper 95%*
17	Intercept	59.8006	21.228	2.817	0.006	17.675	101.926
18	Revenue	0.0382	0.003	11.196	3.13E-19	0.031	0.045

The regression equation is Net Income = 59.8006 + 0.0382*Revenue.

The slope is 0.0382 -- on average, each additional dollar of revenue is accompanied by an additional $0.0382 in net income. The coefficients of correlation and determination are 0.7492 and 0.5612, respectively. The amount of revenue explains 56.12% of the variation in net income. To four decimal places, the p-value is 0.0000, which is less than 0.05. At the 0.05 level of significance, the population slope and the population coefficient of correlation could not be zero. The lower and upper 95% confidence limits for the population slope are 0.031 and 0.045, respectively.

15.61 d/p/m Residual analysis refers to a broad class of tests which can be performed using the error terms from a regression line to test the assumptions of regression analysis. Residual analysis can test the assumption of normal distribution of the error terms, the assumption of equal standard deviations of the error terms, and the assumption that the error terms are independent from one another.

In residual analysis, we may also "flag" error values that are especially large.

15.63 d/p/d Correlation does not necessarily imply causation of any kind. Perhaps sales have increased as a result of better products which came from more money invested in research and development. On the other hand, perhaps the R&D budget is set based on past or anticipated sales.

It is also very possible that both of these variables have simply increased over time.

15.65 c/c/m We will use Minitab to obtain the regression equation and to analyze the residuals.

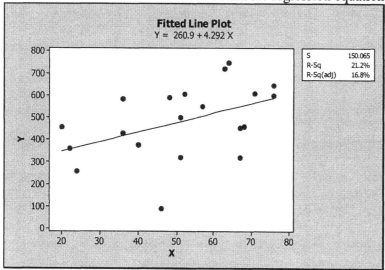

a. The histogram of residuals looks fairly normal.

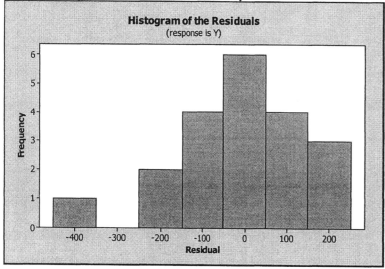

b. The Lilliefors test. As noted in the text, Minitab does the Lilliefors test as a "Kolmogorov-Smirnov" procedure. The approximate p-value is > 0.15. At the 0.05 level of significance, we would conclude that the residuals could have come from a normal population.

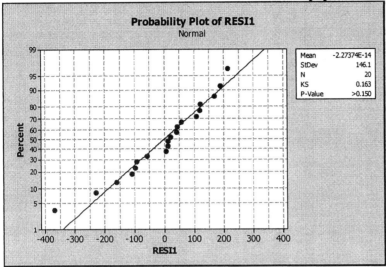

c. Plot of residuals versus the x values. The residuals appear to be relatively randomly scattered above and below the "0" line.

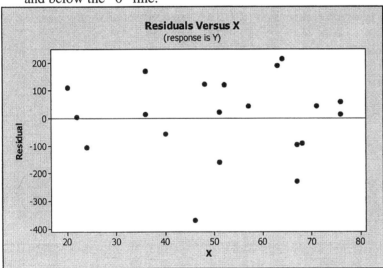

d. Plot of residuals versus the order of the data. Six of the first 8 residuals are positive, 4 of the next 6 are negative, and 5 of the last 6 are positive. There may be a nonrandom pattern here.

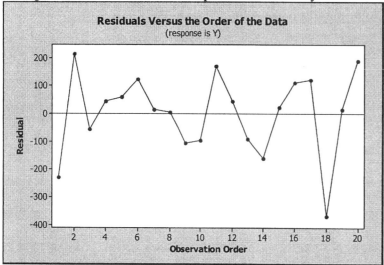

15.67 p/c/m We will use Minitab to obtain the regression equation and to analyze the residuals. In the scatter diagram below, it would appear that none of the years was associated with especially high or low fuel consumption for the number of flying hours during the year.

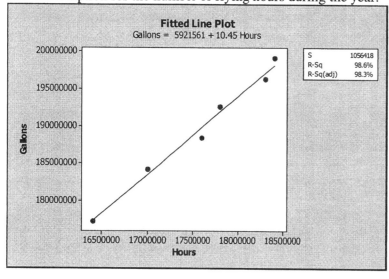

a. The histogram of residuals does not appear normal. However, this is based on visual examination and judgement, so we will rely on the Lilliefors test in part (b).

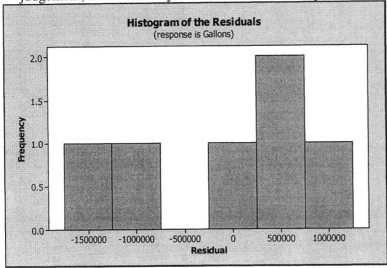

b. The Lilliefors test. As noted in the text, Minitab does the Lilliefors test as a "Kolmogorov-Smirnov" procedure. The approximate p-value is > 0.15. At the 0.05 level of significance, we would conclude that the residuals could have come from a normal population.

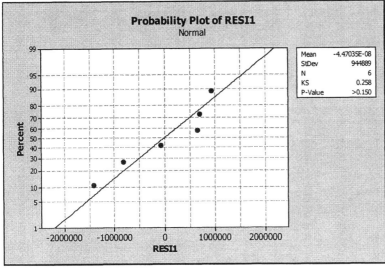

c. Plot of residuals versus the x values. The points in this plot seem to be rather random.

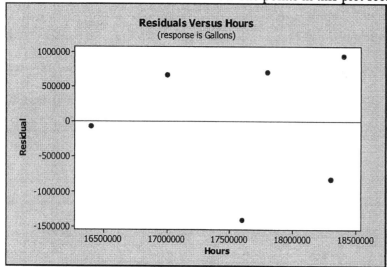

d. Plot of residuals versus the order of the data. There seems to be a downward tendency, but the number of points is very small.

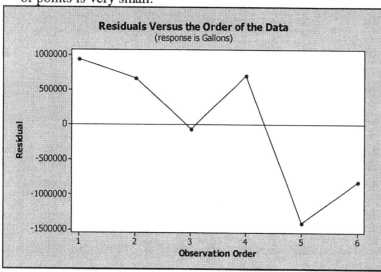

CHAPTER EXERCISES

15.69 d/p/m

a. The monthly mortgage payment would likely be directly related to the market value of the house, the interest rate, the size of the house, or the monthly taxes and insurance, among other variables.

b. The monthly mortgage payment would likely be inversely related to the age of the house, among other variables.

c. The monthly mortgage payment would likely be unrelated to the amount of chocolate consumed by the owners, and a wide variety of other variables.

15.71 p/a/m The least-squares equation is NetIncome = 1.81 - 0.0183*TotRev. Curiously, the slope indicates an inverse relationship between net income and total revenue. For each additional $billion of total revenue, the equation would estimate a decrease of $0.0183 billion in the company's net income. Note that the p-value is not very impressive, 0.794, and the slope of the population regression equation could very easily be 0, at least based on this sample of years. For a year in which total revenues are $10 billion, the equation would estimate net income as $1.632 billion. The Minitab printout is shown below.

```
Regression Analysis: NetIncome versus TotRev

The regression equation is NetIncome = 1.81 - 0.0183 TotRev

Predictor        Coef      SE Coef         T         P
Constant       1.8149       0.8686      2.09     0.082
TotRev        -0.01834      0.06732     -0.27     0.794

S = 0.3621      R-Sq = 1.2%      R-Sq(adj) = 0.0%

Analysis of Variance
Source             DF          SS          MS        F        P
Regression          1      0.0097      0.0097     0.07    0.794
Residual Error      6      0.7867      0.1311
Total               7      0.7964

Unusual Observations
Obs     TotRev   NetIncom         Fit      SE Fit    Residual     St Resid
  8       15.4      0.893       1.532       0.219      -0.639        -2.22R

R denotes an observation with a large standardized residual

Predicted Values for New Observations
New Obs     Fit      SE Fit         95.0% CI              95.0% PI
1         1.632       0.226    (  1.079,   2.184)   (  0.587,   2.676)

Values of Predictors for New Observations
New Obs    TotRev
1            10.0
```

15.73 p/a/m

a. The least squares equation and slope interpretation.

To facilitate pocket calculator computations, the following preliminary calculation results are listed here: $\Sigma x = 42.531$, $\Sigma y = 42.363$, $\Sigma xy = 359.056$, $\Sigma x^2 = 362.224$, $\Sigma y^2 = 369.105$ $n = 5$, $\bar{x} = 8.506$, $\bar{y} = 8.473$. The regression line can be determined as shown below.

$$\text{slope, } b_1 = \frac{\sum x_i y_i - n\bar{x}\bar{y}}{\sum x_i^2 - n\bar{x}^2} = \frac{359.056 - 5(8.506)(8.473)}{362.224 - 5(8.506)^2} = -2.893$$

y-intercept, $b_0 = \bar{y} - b_1\bar{x} = 8.473 - (-2.893)(8.506) = 33.08$

The regression equation is $\hat{y} = 33.08 - 2.893x$

The least squares regression analysis below is generated using Minitab:

```
Regression Analysis: Retired versus New

The regression equation is Retired = 33.1 - 2.89 New

Predictor         Coef      SE Coef           T        P
Constant         33.08        18.67        1.77    0.174
New             -2.893        2.193       -1.32    0.279

S = 1.466      R-Sq = 36.7%     R-Sq(adj) = 15.6%

Analysis of Variance
Source             DF          SS          MS        F        P
Regression          1       3.737       3.737     1.74    0.279
Residual Error      3       6.443       2.148
Total               4      10.180

Predicted Values for New Observations
New Obs     Fit      SE Fit          95.0% CI              95.0% PI
1        12.830       3.368    (  2.112,  23.549)   (  1.141,  24.520) XX
2         7.044       1.266    (  3.015,  11.073)   (  0.881,  13.207)
X   denotes a row with X values away from the center
XX  denotes a row with very extreme X values

Values of Predictors for New Observations

New Obs        New
1             7.00
2             9.00
```

To the maximum number of decimal places shown in the printout, the regression equation is Retired = 33.08 - 2.893*New. The slope is -2.893 -- on average, an increase of 1 million new cars registered tends to be accompanied by a decrease of 2.893 million in the number of cars retired from service. The slope is actually the reverse of what we might expect -- i.e., that more new cars would lead to more existing cars being retired. However, any relationship based on these data would appear to be very weak. Note that the p-value is not very impressive, 0.279, and the slope of the population regression equation could very easily be 0.

b. For a year in which x = 7.0 million new passenger cars are registered, we would predict that 33.08 - 2.893(7.0) = 12.83 million cars would be retired from service during that year. See the first row of the "Fit" column in the Minitab printout.

c. For a year in which x = 9.0 million new passenger cars are registered, we would predict that 33.08 - 2.893(9.0) = 7.044 million cars would be retired from service during that year. See the second row of the "Fit" column in the Minitab printout.

The corresponding Excel printout and plot are shown below.

	C	D	E	F	G	H	I
1	SUMMARY OUTPUT						
2							
3	*Regression Statistics*						
4	Multiple R	0.6059					
5	R Square	0.3671					
6	Adjusted R Square	0.1561					
7	Standard Error	1.4655					
8	Observations	5					
9							
10	ANOVA						
11		*df*	*SS*	*MS*	*F*	*Significance F*	
12	Regression	1	3.737	3.737	1.740	0.279	
13	Residual	3	6.443	2.148			
14	Total	4	10.180				
15							
16		*Coefficients*	*Standard Error*	*t Stat*	*P-value*	*Lower 95%*	*Upper 95%*
17	Intercept	33.084	18.668	1.772	0.174	-26.327	92.494
18	New	-2.893	2.193	-1.319	0.279	-9.873	4.087

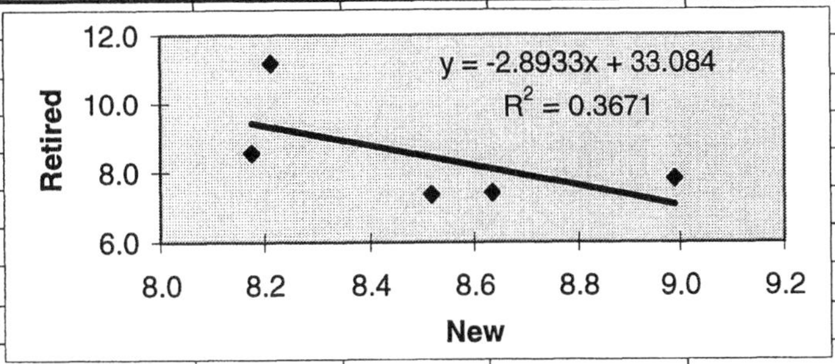

Chart (rows 1–9, columns E–I):
$y = -2.8933x + 33.084$
$R^2 = 0.3671$
Y-axis: Retired (6.0 to 12.0)
X-axis: New (8.0 to 9.2)

15.75 p/a/m

To facilitate pocket calculator computations, the following preliminary calculation results are listed here:
$\Sigma x = 77.990$, $\Sigma y = 2.840$, $\Sigma xy = 29.901$, $\Sigma x^2 = 800.946$, $\Sigma y^2 = 1.958$, $n = 8$, $\bar{x} = 9.749$,
$\bar{y} = 0.355$. The regression line can be determined as shown below.

$$\text{slope, } b_1 = \frac{\sum x_i y_i - n\bar{x}\bar{y}}{\sum x_i^2 - n\bar{x}^2} = \frac{29.901 - 8(9.749)(0.355)}{800.946 - 8(9.749)^2} = 0.0551$$

$$\text{y-intercept, } b_0 = \bar{y} - b_1\bar{x} = 0.355 - (0.0551)(9.749) = -0.1821$$

The regression equation is $\hat{y} = -0.1821 + 0.0551x$

The least squares regression analysis below is generated using Minitab:

```
Regression Analysis: NetIncome versus Revenue

The regression equation is NetIncome = - 0.182 + 0.0551 Revenue

Predictor        Coef     SE Coef           T        P
Constant      -0.1821      0.5861       -0.31    0.767
Revenue       0.05510     0.05859        0.94    0.383

S = 0.3714      R-Sq = 12.8%     R-Sq(adj) = 0.0%

Analysis of Variance
Source           DF         SS          MS        F        P
Regression        1     0.1220      0.1220     0.88    0.383
Residual Error    6     0.8278      0.1380
Total             7     0.9498

Unusual Observations
Obs    Revenue   NetIncom       Fit     SE Fit   Residual    St Resid
  3       10.0     -0.390     0.368      0.132     -0.758      -2.18R
R denotes an observation with a large standardized residual


Predicted Values for New Observations
New Obs     Fit    SE Fit       95.0% CI            95.0% PI
  1       0.369     0.132   ( 0.045,  0.692)   ( -0.596,  1.334)

Values of Predictors for New Observations
New Obs   Revenue
  1          10.0
```

243

To the maximum number of decimal places shown in the printout, the regression equation is NetIncome = - 0.1821 + 0.0551*Revenue. For every increase of $1 million in revenue, the company would anticipate an additional $0.0551 million in net income. Alternatively, the company might view each additional dollar of revenue as generating about $0.0551 in additional net income.

For a year in which x = $10 million in revenue, the company would predict net income as -0.1821 + 0.0551*(10) = $0.369 million. See the "Fit" column in the Minitab printout.

The corresponding Excel printout and plot are shown below.

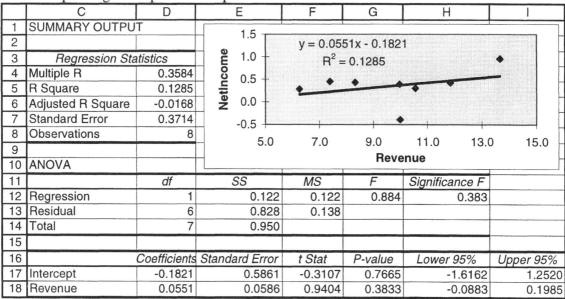

	C	D	E	F	G	H	I
1	SUMMARY OUTPUT						
2							
3	*Regression Statistics*						
4	Multiple R	0.3584					
5	R Square	0.1285					
6	Adjusted R Square	-0.0168					
7	Standard Error	0.3714					
8	Observations	8					
9							
10	ANOVA						
11		*df*	*SS*	*MS*	*F*	*Significance F*	
12	Regression	1	0.122	0.122	0.884	0.383	
13	Residual	6	0.828	0.138			
14	Total	7	0.950				
15							
16		*Coefficients*	*Standard Error*	*t Stat*	*P-value*	*Lower 95%*	*Upper 95%*
17	Intercept	-0.1821	0.5861	-0.3107	0.7665	-1.6162	1.2520
18	Revenue	0.0551	0.0586	0.9404	0.3833	-0.0883	0.1985

15.77 p/a/m

a. The least squares equation.

To facilitate pocket calculator computations, the following preliminary calculation results are listed here for x = flying time (millions of hours) and y = fuel consumed (millions of gallons): $\Sigma x = 8.900$, $\Sigma y = 725.500$, $\Sigma xy = 1106.220$, $\Sigma x^2 = 13.590$, $\Sigma y^2 = 90,076.190$, n = 6, $\bar{x} = 1.483$, $\bar{y} = 120.917$. The regression line can be determined as shown below.

$$\text{slope, } b_1 = \frac{\sum x_i y_i - n\bar{x}\bar{y}}{\sum x_i^2 - n\bar{x}^2} = \frac{1106.220 - 6(1.483)(120.917)}{13.590 - 6(1.483)^2} = 77.412$$

$$\text{y-intercept, } b_0 = \bar{y} - b_1\bar{x} = 120.917 - (77.412)(1.483) = 6.089$$

The regression equation is $\hat{y} = 6.089 + 77.412x$

The least squares regression analysis below is generated using Minitab:

```
Regression Analysis: Fuel versus Hours

The regression equation is Fuel = 6.09 + 77.4 Hours

Predictor       Coef     SE Coef          T         P
Constant       6.089       5.917       1.03     0.362
Hours         77.412       3.932      19.69     0.000

S = 2.450      R-Sq = 99.0%      R-Sq(adj) = 98.7%
```

```
Analysis of Variance
Source              DF          SS          MS          F          P
Regression          1        2327.1      2327.1      387.63     0.000
Residual Error      4          24.0         6.0
Total               5        2351.1

Predicted Values for New Observations
New Obs    Fit      SE Fit       95.0% CI            95.0% PI
1        160.91      2.26   ( 154.63,  167.20)  ( 151.65,  170.18)

Values of Predictors for New Observations
New Obs    Hours
1          2.00
```

b. Interpreting the Minitab printout, the coefficient of determination is 0.99. Note: Using either the formula below or the Excel printout that follows, we can express both r and r^2 using several more decimal places. Doing so, we find that that $r = 0.995$ and $r^2 = 0.990$. Using the latter value, we can say that 99.0% of the variation in fuel consumption for is explained by the number of flying hours. There is obviously an extremely strong direct linear relationship between fuel consumption and flying hours for these aircraft.

To find the coefficient of correlation with a calculator, we can use the formula below.

$$r = \frac{n(\sum x_i y_i) - (\sum x_i)(\sum y_i)}{\sqrt{n(\sum x_i^2) - (\sum x_i)^2} * \sqrt{n(\sum y_i^2) - (\sum y_i)^2}}$$

$$= \frac{6(1106.220) - (8.900)(725.500)}{\sqrt{6(13.590) - 8.900^2} * \sqrt{6(90,076.190) - 725.500^2}} = 0.995$$

c. During a year when these aircraft fly for 2.0 million hours, the estimate for fuel consumption would be $\hat{y} = 6.089 + 77.412(2) = 160.91$ million gallons of fuel consumed. See the "Fit" column in the Minitab printout.

The corresponding Excel printout and plot are shown below.

	C	D	E	F	G	H	I
1	SUMMARY OUTPUT						
2							
3	*Regression Statistics*						
4	Multiple R	0.99488001					
5	R Square	0.98978623					
6	Adjusted R Square	0.98723279					
7	Standard Error	2.45020801					
8	Observations	6					
9							
10	ANOVA						
11		*df*	*SS*	*MS*	*F*	*Significance F*	
12	Regression	1	2327.134	2327.134	387.628	0.00004	
13	Residual	4	24.014	6.004			
14	Total	5	2351.148				
15							
16		*Coefficients*	*Standard Error*	*t Stat*	*P-value*	*Lower 90.0%*	*Upper 90.0%*
17	Intercept	6.089	5.917	1.029	0.36165	-6.53	18.70
18	Hours	77.412	3.932	19.688	0.00004	69.03	85.79

Chart (rows 1-9, columns E-I): Fuel vs Hours scatterplot with fitted line. $y = 77.412x + 6.0888$, $R^2 = 0.9898$. Y-axis "Fuel" ranges 70 to 150; X-axis "Hours" ranges 1.0 to 2.0.

245

15.79 p/a/m The Minitab printout and least-squares equation are shown below.

Regression Analysis: Fuel versus Miles

The regression equation is Fuel = 15.1 + 0.0367 Miles

```
Predictor        Coef      SE Coef           T        P
Constant       15.084        5.475        2.75    0.070
Miles        0.036739     0.003634       10.11    0.002

S = 0.3950      R-Sq = 97.1%     R-Sq(adj) = 96.2%
```

Analysis of Variance

```
Source            DF           SS          MS          F        P
Regression         1       15.942      15.942     102.19    0.002
Residual Error     3        0.468       0.156
Total              4       16.410
```

Predicted Values for New Observations
```
New Obs     Fit      SE Fit         95.0% CI              95.0% PI
1        73.131       0.322   ( 72.105,  74.158)   ( 71.509,  74.754)
```

Values of Predictors for New Observations
```
New Obs      Miles
1             1580
```

a. The regression equation is Fuel = 15.1 + 0.0367*Miles
b. The coefficient of correlation is the square root of 0.971, or 0.985. It is positive and indicates a direct relationship between the variables. The coefficient of determination is 0.971, and the number of miles traveled explains 97.1% of the variation in fuel consumption.
c. If passenger car travel during a year were 1580 billion miles, the equation would predict a fuel consumption level of 73.131 billion gallons.

15.81 d/p/m The Minitab printout and least-squares equation are shown below.

Regression Analysis: Rear versus Front

The regression equation is Rear = 1856 - 0.306 Front

```
Predictor        Coef      SE Coef           T        P
Constant       1855.9        776.0        2.39    0.040
Front         -0.3057       0.9729       -0.31    0.760

S = 1114        R-Sq = 1.1%      R-Sq(adj) = 0.0%
```

Analysis of Variance

```
Source            DF           SS          MS          F        P
Regression         1       122500      122500       0.10    0.760
Residual Error     9     11162919     1240324
Total             10     11285419
```

Unusual Observations
```
Obs       Front        Rear         Fit      SE Fit     Residual     St Resid
 1          213        3683        1791         596         1892         2.01R
```

R denotes an observation with a large standardized residual

Predicted Values for New Observations
```
New Obs     Fit      SE Fit  (     95.0% CI            95.0% PI
1          1550         433  (      571,    2530)  (   -1153,    4253)
```

Values of Predictors for New Observations
```
New Obs      Front
1             1000
```

Correlations: Front, Rear

Pearson correlation of Front and Rear = -0.104

a. The regression equation is Rear = 1856 - 0.306*Front.
b. We have also used Minitab to separately determine the coefficient of correlation, which is shown above as -0.104. It is negative and indicates an inverse relationship between the variables.

The coefficient of determination is 0.011, and the amount of damage in a rear crash explains just 1.1% of the variation in front crash damage.

c. If a sport utility vehicle were to incur $1000 in damages in the front-crash test, the predicted repair bill for damage in a rear-crash test would be $1550, as shown in the printout above.

15.83 p/a/m

a. The least squares equation.

To facilitate pocket calculator computations, the following preliminary calculation results are listed here for x = age and y = shares held (thousands):

$\Sigma x = 1153.00$, $\Sigma y = 759.90$, $\Sigma xy = 44,293.00$, $\Sigma x^2 = 70,937.00$, $\Sigma y^2 = 52,374.01$, $n = 19$, $\bar{x} = 60.68421$, $\bar{y} = 39.99474$. The regression line can be determined as shown below.

$$\text{slope, } b_1 = \frac{\sum x_i y_i - n\bar{x}\bar{y}}{\sum x_i^2 - n\bar{x}^2} = \frac{44,293.0 - 19(60.68421)(39.99474)}{70,937.0 - 19(60.68421)^2} = -1.881$$

$$\text{y-intercept, } b_0 = \bar{y} - b_1\bar{x} = 39.99474 - (-1.881)(60.68421) = 154.14$$

The regression equation is $\hat{y} = 154.14 - 1.881x$

The least squares regression analysis below is generated using Minitab:

```
Regression Analysis: Shares versus Age

The regression equation is Shares = 154 - 1.88 Age

Predictor        Coef      SE Coef         T       P
Constant       154.14        64.88     ·2.38   0.030
Age            -1.881         1.062     -1.77   0.094

S = 33.04        R-Sq = 15.6%      R-Sq(adj) = 10.6%

Analysis of Variance
Source           DF          SS         MS       F       P
Regression        1        3425       3425    3.14   0.094
Residual Error   17       18557       1092
Total            18       21982

Unusual Observations
Obs       Age      Shares        Fit    SE Fit    Residual    St Resid
  8      62.0      121.10      37.52      7.71       83.58        2.60R

R denotes an observation with a large standardized residual

Predicted Values for New Observations
New Obs      Fit      SE Fit        95.0% CI             95.0% PI
1          33.76        8.36   (  16.13,   51.39)   ( -38.14,  105.66)

Values of Predictors for New Observations
New Obs          Age
1               64.0
```

To the greatest number of decimal places shown in the printout, the regression equation is Shares = 154.14 - 1.881*Age.

b. Interpreting the Minitab printout, the coefficient of determination is (to three decimal places) 0.156. Given this information, we would estimate r as the negative square root of 0.156, or r = -0.395. We select the negative square root because the slope of the equation is negative, making the relationship inverse between number of shares owned and age of the stock owner. Based on $r^2 = 0.156$, we can say that just 15.6% of the variation in shares owned is explained by the age of the stock owner.

To find the coefficient of correlation with a calculator, we can use the formula below.

$$r = \frac{n(\sum x_i y_i) - (\sum x_i)(\sum y_i)}{\sqrt{n(\sum x_i^2) - (\sum x_i)^2} * \sqrt{n(\sum y_i^2) - (\sum y_i)^2}}$$

$$= \frac{19(44,293.0) - (1153.0)(759.9)}{\sqrt{19(70,937.0) - (1153.0)^2} * \sqrt{19(52,374.01) - 759.9^2}} = -0.395$$

c. A board member who is 64 years old would be expected to hold $\hat{y}$ = 154.14 - 1.881(64) = 33.76 thousand shares. See the "Fit" column in the Minitab printout.

The corresponding Excel printout and plot are shown below.

	C	D	E	F	G	H	I
1	SUMMARY OUTPUT						
2							
3	*Regression Statistics*						
4	Multiple R	0.3947					
5	R Square	0.1558					
6	Adjusted R Square	0.1062					
7	Standard Error	33.0392					
8	Observations	19					
9							
10	ANOVA						
11		*df*	*SS*	*MS*	*F*	*Significance F*	
12	Regression	1	3425.032	3425.032	3.138	0.094	
13	Residual	17	18556.977	1091.587			
14	Total	18	21982.009				
15							
16		*Coefficients*	*Standard Error*	*t Stat*	*P-value*	*Lower 90.0%*	*Upper 90.0%*
17	Intercept	154.137	64.882	2.376	0.030	41.27	267.01
18	Age	-1.881	1.062	-1.771	0.094	-3.73	-0.03

Plot (rows 1–10, columns E–I): Scatter plot of Shares vs Age with regression line, equation y = -1.8809x + 154.14, R^2 = 0.1558.

15.85 p/a/m

a. The regression equation is: STRENGTH = 60.02 + 10.507 TEMPTIME

b. 53% of the variation in strength is explained by the regression. (See R-sq.)

c. The slope of the line differs from zero at the 0.017 level of significance. Minitab used a two-tail t-test to reach this conclusion.

d. The coefficient of correlation differs from zero at the 0.017 level (using the output from the ANOVA table). This is the same level seen in part c. They will always be the same because they are testing essentially the same thing.

e. The 95% confidence interval for the slope of the regression line is:

$b_1 \pm t\, s_{b_1}$ = 10.507 ± 2.306(3.496) = 10.507 ± 8.062, or from 2.445 to 18.569.

15.87 p/a/m
a. The least squares regression line is: ROLRESIS = 9.450 - 0.08113 PSI.
b. 23.9% of the variation in rolling resistance is explained by the regression line. (See R-sq.)
c. The slope of the line differs from zero at the 0.029 level of significance. Minitab used a two-tail t-test to reach this conclusion.
d. The coefficient of correlation differs from zero at the 0.029 level of significance. (See the ANOVA table.) This is the same level found in part c. These levels will always be the same because the tests are equivalent.
e. The 95% confidence interval for the slope of the population regression line is:
$$b_1 \pm t \, s_{b_1} = -0.08113 \pm 2.101(0.03416) = -0.08113 \pm 0.07177, \text{ or from } -0.15290 \text{ to } -0.00936.$$

15.89 p/a/m The solution could be obtained with formulas and a calculator, but we will use Minitab:

```
Regression Analysis: Sold versus Score

The regression equation is Sold = - 7.62 + 0.436 Score

Predictor         Coef      SE Coef          T        P
Constant        -7.616        8.421      -0.90    0.432
Score           0.4363       0.1275       3.42    0.042

S = 7.233       R-Sq = 79.6%     R-Sq(adj) = 72.8%

Analysis of Variance
Source            DF           SS          MS        F        P
Regression         1       613.05      613.05    11.72    0.042
Residual Error     3       156.95       52.32
Total              4       770.00

Predicted Values for New Observations
New Obs     Fit      SE Fit        95.0% CI            95.0% PI
1         18.56        3.24    (  8.26,   28.87)  (  -6.66,   43.78)

Values of Predictors for New Observations
New Obs    Score
1           60.0
```

The 95% confidence interval is from 8.26 to 28.87. For all individuals who score 60 on the sales aptitude test, we are 95% confident that their mean sales performance will be between 8.26 and 28.87 units. The 95% prediction interval is from -6.66 to 43.78. For an individual scoring 60 on the sales aptitude test, we are 95% confident that this person's sales performance will be between -6.66 and 43.78 units. As a practical matter (barring returns), the lower limit can be considered to be 0, since the individual cannot sell a negative number of units.

15.91 p/a/m The solution could be obtained with formulas and a calculator, but we will use Minitab:

```
Regression Analysis: RugCarp versus HouStart

The regression equation is RugCarp = 3.64 + 4.35 HouStart

Predictor        Coef      SE Coef          T        P
Constant        3.638        1.586       2.29    0.083
HouStart        4.353        1.150       3.79    0.019

S = 0.2949      R-Sq = 78.2%     R-Sq(adj) = 72.7%

Analysis of Variance
Source           DF          SS          MS        F        P
Regression        1      1.2458      1.2458    14.33    0.019
Residual Error    4      0.3477      0.0869
Total             5      1.5935

Predicted Values for New Observations
New Obs     Fit     SE Fit         95.0% CI              95.0% PI
1         9.732      0.124   (  9.389,  10.076)   (  8.844,  10.620)

Values of Predictors for New Observations
New Obs  HouStart
1            1.40
```

The 95% confidence interval is from 9.389 to 10.076. For all years in which there are 1.4 million housing starts, we are 95% confident that the mean value of rugs and carpeting shipped will be between $9.389 billion and $10.076 billion.

The 95% prediction interval is from 8.844 to 10.620. For any given year in which there are 1.4 million housing starts, we are 95% confident that the value of rugs and carpeting shipped during that year will be between $8.844 billion and $10.620 billion.

15.93 p/c/m The Minitab printout is shown below.

```
Regression Analysis: Fr GPA versus SAT Total

The regression equation is Fr GPA = - 0.696 + 0.00333 SAT Total

Predictor        Coef      SE Coef          T        P
Constant      -0.6964       0.9938      -0.70    0.510
SAT Tota    0.0033282    0.0008991       3.70    0.010

S = 0.3339      R-Sq = 69.5%     R-Sq(adj) = 64.5%

Analysis of Variance
Source           DF          SS          MS        F        P
Regression        1      1.5281      1.5281    13.70    0.010
Residual Error    6      0.6691      0.1115
Total             7      2.1972

Predicted Values for New Observations
New Obs     Fit     SE Fit         99.0% CI              99.0% PI
1         2.965      0.118   (  2.527,   3.402)   (  1.651,   4.278)

Values of Predictors for New Observations
New Obs  SAT Tota
1            1100
```

a. The regression equation is Fr GPA = - 0.6964 + 0.0033282*SAT Total. The predicted freshman grade point average for an applicant who has scored a total of 1100 on his SAT examination is -0.6964 + 0.0033282(1100) = 2.965. See the "Fit" column of the Minitab printout.

b. The SAT scores explain 69.5% of the variation in the freshman grade point averages.

c. The 99% confidence interval is from 2.527 to 3.402. For all those scoring 1100 on their SAT exam, we are 99% confident that their mean freshman grade point average will be between 2.527 and 3.402. The 99% prediction interval is from 1.651 to 4.278. For an individual who has scored 1100 on her SAT exam, we are 99% confident that her freshman grade point average will be between 1.651 and 4.278.

The corresponding Excel printout and plot are shown below.

	D	E	F	G	H	I	J
1	SUMMARY OUTPUT						
2							
3	*Regression Statistics*						
4	Multiple R	0.8339					
5	R Square	0.6955					
6	Adjusted R Square	0.6447					
7	Standard Error	0.3339					
8	Observations	8					
9							
10	ANOVA						
11		*df*	*SS*	*MS*	*F*	*Significance F*	
12	Regression	1	1.528	1.528	13.702	0.010	
13	Residual	6	0.669	0.112			
14	Total	7	2.197				
15							
16		*Coefficients*	*Standard Error*	*t Stat*	*P-value*	*Lower 95%*	*Upper 95%*
17	Intercept	-0.69645	0.99382	-0.7008	0.5097	-3.1282	1.7354
18	SAT Total	0.00333	0.00090	3.7016	0.0101	0.0011	0.0055

Chart (rows 2–10, columns F–J): Fr GPA vs SAT Total scatterplot with fitted line $y = 0.0033x - 0.6964$, $R^2 = 0.6955$.

15.95 p/c/m The Minitab printout is shown below.

```
Regression Analysis: Pay% versus Rate%

The regression equation is Pay% = 7.02 + 1.52 Rate%

Predictor        Coef      SE Coef         T         P
Constant       7.0202       0.6124     11.46     0.000
Rate%          1.51597      0.07629    19.87     0.000

S = 0.2200      R-Sq = 98.0%     R-Sq(adj) = 97.8%

Analysis of Variance
Source           DF          SS           MS          F         P
Regression        1      19.102       19.102     394.83     0.000
Residual Error    8       0.387        0.048
Total             9      19.489

Unusual Observations
Obs    Rate%     Pay%         Fit     SE Fit    Residual     St Resid
  1     10.0  22.0000     22.2405     0.1722     -0.2405        -1.76 X

X denotes an observation whose X value gives it large influence.

Predicted Values for New Observations
New Obs     Fit    SE Fit        95.0% CI              95.0% PI
  1     19.1479    0.0696   ( 18.9874, 19.3084)   ( 18.6159, 19.6799)

Values of Predictors for New Observations
New Obs    Rate%
  1         8.00
```

a. The regression equation is Pay% = 7.0202 + 1.51597*Rate%. If the average mortgage rate is 8.0%, the predicted average monthly mortgage payment as a percentage of median household income will be 7.0202 + 1.51597(8) = 19.1479%. See the "Fit" column of the Minitab printout.

b. The mortgage rate percentages explain 98.0% of the variation in mortgage affordability. The p-value (0.000) for the test of the slope is less than the 0.05 level of significance. At this level, we conclude that the slope of the population regression equation is not zero.

c. The 95% confidence interval is from 18.9874 to 19.3084. For all years in which the average mortgage rate is 8.0%, we are 95% confident that the mean mortgage affordability for these years will be between 18.9874% and 19.3084%.

The 95% prediction interval is from 18.6159 to 19.6799. For a given year in which the average mortgage rate is 8.0%, we are 95% confident that the mortgage affordability for this year will be between 18.6159% and 19.6799%.

The corresponding Excel printout and plot are shown below.

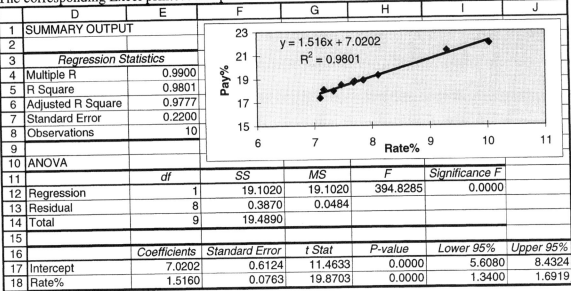

	D	E	F	G	H	I	J
1	SUMMARY OUTPUT						
2							
3	*Regression Statistics*						
4	Multiple R	0.9900					
5	R Square	0.9801					
6	Adjusted R Square	0.9777					
7	Standard Error	0.2200					
8	Observations	10					
9							
10	ANOVA						
11		*df*	*SS*	*MS*	*F*	*Significance F*	
12	Regression	1	19.1020	19.1020	394.8285	0.0000	
13	Residual	8	0.3870	0.0484			
14	Total	9	19.4890				
15							
16		*Coefficients*	*Standard Error*	*t Stat*	*P-value*	*Lower 95%*	*Upper 95%*
17	Intercept	7.0202	0.6124	11.4633	0.0000	5.6080	8.4324
18	Rate%	1.5160	0.0763	19.8703	0.0000	1.3400	1.6919

15.97 p/c/m The Minitab printout is shown below.

```
Regression Analysis: Est P/E Ratio versus Revenue%Growth

The regression equation is Est P/E Ratio = 52.6 - 0.096 Revenue%Growth

96 cases used 4 cases contain missing values

Predictor      Coef     SE Coef        T        P
Constant      52.56       15.01     3.50    0.001
Revenue%     -0.0959     0.2132    -0.45    0.654

S = 63.97      R-Sq = 0.2%     R-Sq(adj) = 0.0%

Analysis of Variance
Source           DF         SS        MS       F        P
Regression        1        828       828    0.20    0.654
Residual Error   94     384660      4092
Total            95     385488

Unusual Observations
Obs    Revenue%    Est P/E       Fit    SE Fit    Residual    St Resid
  3        117      245.00     41.33     13.17      203.67       3.25R
  4         80      286.00     44.88      7.43      241.12       3.79R
  8         71      363.00     45.75      6.73      317.25       4.99R
 12        132       10.00     39.90     16.03      -29.90      -0.48 X
 13         55      176.00     47.28      6.77      128.72       2.02R
 14        221        3.00     31.36     34.24      -28.36      -0.52 X
 21        134       17.00     39.70     16.42      -22.70      -0.37 X
 63         33      309.00     49.39      9.20      259.61       4.10R

R denotes an observation with a large standardized residual
X denotes an observation whose X value gives it large influence.

Predicted Values for New Observations
New Obs      Fit     SE Fit         95.0% CI              95.0% PI
  1        38.17      19.59    ( -0.74,   77.07)   ( -94.67,  171.01) X
X  denotes a row with X values away from the center

Values of Predictors for New Observations
New Obs    Revenue%
  1             150
```

a. The regression equation is Est P/E Ratio = 52.56 - 0.0959*Revenue%Growth. A company whose annual revenue growth has been 150% would have a predicted P/E ratio of 52.56 - 0.0959(150) = 38.17. See the "Fit" column of the Minitab printout.

b. Annual revenue growth percentage explains only 0.2% of the variation in the P/E estimates. The p-value (0.654) for the test of the slope is not less than the 0.05 level of significance. At this level, we conclude that the slope of the population regression equation could be zero.

c. The 95% confidence interval is from -0.74 to 77.07. For all companies having an annual revenue growth of 150%, we are 95% confident that the mean P/E estimate for these companies will be between -0.74 and 77.07.
The 95% prediction interval is from -94.67 to 171.01. For an individual company having an annual revenue growth of 150%, we are 95% confident that this company's P/E estimate will be between -94.67 and 171.01.

The corresponding Excel printout and plot are shown below. Note: Before carrying out the Excel regression analysis, it will be necessary to delete the four cases for which there is missing data on these variables.

SUMMARY OUTPUT						
Regression Statistics						
Multiple R	0.0463					
R Square	0.0021					
Adjusted R Square	-0.0085					
Standard Error	63.9697					
Observations	96					
ANOVA						
	df	*SS*	*MS*	*F*	*Significance F*	
Regression	1	827.9321	827.9321	0.2023	0.6539	
Residual	94	384660.0262	4092.1279			
Total	95	385487.9583				
	Coefficients	*Standard Error*	*t Stat*	*P-value*	*Lower 95%*	*Upper 95%*
Intercept	52.5569	15.0067	3.5022	0.0007	22.7608	82.3529
Revenue%Growth	-0.0959	0.2132	-0.4498	0.6539	-0.5193	0.3275

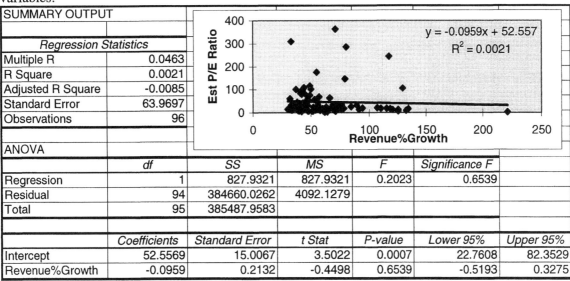

SECTION EXERCISES

16.1 d/p/e Simple linear regression involves only one independent variable; multiple regression involves two or more independent variables. Multiple regression analysis is preferred whenever two or more variables impact upon the dependent variable.

16.3 d/p/m Many variables could affect the annual household expenditure for auto maintenance and repair: the number of cars owned, the number of miles driven each year, the age(s) of the car(s), the make(s) of the car(s). These are just a few of the many variables that could have a notable effect.

16.5 d/p/e The multiple regression model is:

$y_i = \beta_0 + \beta_1 x_{1i} + \beta_2 x_{2i} + ... + \beta_k x_{ki} + \varepsilon_i$ where

y_i = a value of the dependent variable, y

β_0 = a constant

$x_{1i}, x_{2i}, ... , x_{ki}$ = values of the independent variables $x_1, x_2, ... , x_k$

$\beta_1, \beta_2, ... , \beta_k$ = partial regression coefficients for independent variables $x_1, x_2, ... , x_k$

ε_i = random error, or residual

16.7 d/p/m When there are two independent variables, the regression equation can be thought of in terms of a geometric plane. When there are three or more independent variables, the regression equation becomes a mathematical entity called a hyperplane; it is impossible to visually summarize a regression with three or more independent variables because it will be in four or more dimensions.

16.9 p/a/e

a. The y-intercept is 300, the partial regression coefficients are 7 for x_1 and 13 for x_2.

b. If 3 people live in a 6-room home, the estimated bill is $\hat{y} = 300 + 7(3) + 13(6) = 399$.

16.11 p/c/m The Minitab printout is shown below.

```
Regression Analysis: Visitors versus AdSize, Discount

The regression equation is Visitors = 10.7 + 2.16 AdSize + 0.0416 Discount

Predictor        Coef      SE Coef          T         P
Constant       10.687        3.875       2.76     0.040
AdSize         2.1569       0.6281       3.43     0.019
Discount      0.04157      0.04380       0.95     0.386

S = 3.375      R-Sq = 71.6%     R-Sq(adj) = 60.3%

Analysis of Variance
Source           DF           SS         MS         F         P
Regression        2       143.92      71.96      6.32     0.043
Residual Error    5        56.95      11.39
Total             7       200.87

Predicted Values for New Observations
New Obs     Fit      SE Fit         95.0% CI              95.0% PI
1         24.59       1.74   (  20.12,   29.06)  (  14.83,   34.35)

Values of Predictors for New Observations
New Obs    AdSize  Discount
1            5.00      75.0
```

a. The regression equation is Visitors = 10.687 + 2.1569*AdSize + 0.04157*Discount.

b. The y-intercept indicates that about 10 or 11 visitors (10.687) would come to the clubs if there were neither ads nor discounts. The partial regression coefficient for the ad data indicates that, holding the level of the discount constant, increasing the ad size by one column inch will bring in about 2 new visitors (2.1569). Finally, the partial regression coefficient for the discount data indicates that, holding the size of the ad constant, an additional $1 discount will add 0.04157 to the number of visitors.

c. If the size of the ad is 5 column-inches and a $75 discount is offered, the estimated number of new visitors to the club is 24.59. See the "Fit" column in the printout.

The corresponding Excel multiple regression printout is shown below.

	A	B	C	D	E	F	G
14	SUMMARY OUTPUT			Visitors	Col-Inches	Discount	
15	*Regression Statistics*			23	4	100	
16	Multiple R	0.8465		30	7	20	
17	R Square	0.7165		20	3	40	
18	Adjusted R Square	0.6031		26	6	25	
19	Standard Error	3.3749		20	2	50	
20	Observations	8		18	5	30	
21				17	4	25	
22	ANOVA			31	8	80	
23		*df*	*SS*	*MS*	*F*	*Significance F*	
24	Regression	2	143.924	71.962	6.318	0.043	
25	Residual	5	56.951	11.390			
26	Total	7	200.875				
27							
28		*Coefficients*	*Standard Error*	*t Stat*	*P-value*	*Lower 95%*	*Upper 95%*
29	Intercept	10.687	3.875	2.758	0.040	0.726	20.648
30	Col-Inches	2.157	0.628	3.434	0.019	0.542	3.771
31	Discount	0.042	0.044	0.949	0.386	-0.071	0.154

16.13 p/c/m The Minitab printout is shown below.

```
Regression Analysis: Crispness versus OvenTime, Temp

The regression equation is Crispness = - 127 + 7.61 OvenTime + 0.357 Temp

Predictor        Coef      SE Coef         T        P
Constant      -127.19        61.33     -2.07    0.072
OvenTime        7.611         3.873      1.97    0.085
Temp           0.3567        0.1177      3.03    0.016

S = 15.44      R-Sq = 58.6%      R-Sq(adj) = 48.2%

Analysis of Variance
Source           DF          SS          MS        F        P
Regression        2      2696.4      1348.2     5.65    0.029
Residual Error    8      1907.3       238.4
Total            10      4603.6

Predicted Values for New Observations
New Obs     Fit      SE Fit          95.0% CI              95.0% PI
1        -17.79       29.01   ( -84.69,    49.11) ( -93.57,   57.99) XX
X  denotes a row with X values away from the center
XX denotes a row with very extreme X values

Values of Predictors for New Observations
New Obs  OvenTime      Temp
1           5.00        200
```

a. The regression equation is Crispness = -127.19 + 7.611*OvenTime + 0.3567*Temp.

b. The y-intercept indicates that a crust that is not cooked will receive a crispness rating of -127.19. (Caution should be used in interpreting this value since there were no such extreme values in the data used to estimate the regression.) The partial regression coefficient for OvenTime indicates that, for a given temperature, an additional minute in the oven will add 7.611 points to the crispness rating. Likewise, the partial regression coefficient for Temp indicates that, for a given cooking time, a one-degree increase in the oven temperature will result in a 0.3567 increase in the crispness rating.

c. The estimated crispness rating for a pie that is cooked 5 minutes at 200 degrees is -17.79. See the "Fit" column of the Minitab printout or substitute OvenTime = 5 and Temp = 200 into the regression equation. This estimate should be viewed cautiously since the oven temperature is well beyond the limits of the data used to estimate the regression.

The corresponding Excel multiple regression printout is shown below.

	A	B	C	D	E	F	G
12				Crispness	Time	Temp.	
13				68	6.0	460	
14				76	8.9	430	
15	SUMMARY OUTPUT			49	8.8	360	
16	*Regression Statistics*			99	7.8	460	
17	Multiple R	0.7653		90	7.3	390	
18	R Square	0.5857		32	5.3	360	
19	Adjusted R Square	0.4821		96	8.8	420	
20	Standard Error	15.4405		77	9.0	350	
21	Observations	11		94	8.0	450	
22				82	8.2	400	
23	ANOVA			97	6.4	450	
24		*df*	*SS*	*MS*	*F*	*Significance F*	
25	Regression	2	2696.3635	1348.1817	5.6549	0.0295	
26	Residual	8	1907.2729	238.4091			
27	Total	10	4603.6364				
28							
29		*Coefficients*	*Standard Error*	*t Stat*	*P-value*	*Lower 95%*	*Upper 95%*
30	Intercept	-127.1896	61.3267	-2.0740	0.0718	-268.6093	14.2300
31	Time	7.6111	3.8732	1.9651	0.0850	-1.3205	16.5428
32	Temp.	0.3567	0.1177	3.0315	0.0163	0.0854	0.6281

16.15 p/a/m

a. The multiple regression equation is

TEST04 = 11.98 + 0.2745*TEST01 + 0.37619*TEST02 + 0.32648*TEST03

The y-intercept indicates that an individual unit scoring 0 on the first three tests can expect to score 11.98 on the fourth test. (However, this is meaningless, since the test scores range from 200 to 800.) The partial regression coefficient for TEST01 indicates that, for a given set of scores on TEST02 and TEST03, a unit will gain 0.2745 points on TEST04 for an additional point on TEST01. Similarly, the partial regression coefficient for TEST02 implies, for a given set of scores on TEST01 and TEST03, a unit will gain 0.37619 points on TEST04 for an additional point on TEST02. Likewise, the partial regression coefficient for TEST03 indicates an improvement of 0.32648 points on TEST04 for each additional point on TEST03, given a set of scores for TEST01 and TEST02.

b. If an individual unit has scored 350, 400, and 600 on the first three tests, its estimated score on the fourth test is: TEST04 = 11.98 + 0.2745(350) + 0.37619(400) + 0.32648(600) = 454.419.

16.17 c/a/m

a. The mean of y is $\hat{y} = 5.0 + 1.0(25) + 2.5(40) = 130.0$.

b. The multiple standard error of the estimate is $s_e = \sqrt{\dfrac{173.5}{20 - 2 - 1}} = 3.195$.

c. The approximate 95% confidence interval for the mean of y whenever $x_1 = 20$ and $x_2 = 30$ can be found in several steps. First, we must find the midpoint of the approximate confidence interval. This will be $\hat{y} = 5.0 + 1.0(20) + 2.5(30) = 100$.

The degrees of freedom are 20 - 2 - 1 = 17. The appropriate t-value is 2.110. The approximate confidence interval for the mean of y is:

$$\hat{y} \pm t\frac{s_e}{\sqrt{n}} = 100 \pm 2.110\frac{3.195}{\sqrt{20}} = 100 \pm 1.507 = (98.493, \ 101.507)$$

d. The approximate 95% prediction interval for an individual y value when $x_1 = 20$ and $x_2 = 30$ is:

$$\hat{y} \pm ts_e = 100 \pm 2.110(3.195) = 100 \pm 6.741 = (93.259, 106.741)$$

16.19 p/a/m

a. The expected percentage in the retail sector is $\hat{y} = -42.12 + 2.065(13.0) + 4.486(11) = 34.071$.

b. With d.f. = 12 - 2 - 1 = 9, the t-value is 2.262 and the approximate 95% confidence interval is:

$$\hat{y} \pm t\frac{s_e}{\sqrt{n}} = 34.071 \pm 2.262\frac{1.506}{\sqrt{12}} = 34.071 \pm 0.983 = (33.088, \ 35.054)$$

c. The approximate 95% prediction interval is:

$$\hat{y} \pm ts_e = 34.071 \pm 2.262(1.506) = 34.071 \pm 3.407 = (30.664, \ 37.478)$$

16.21 p/c/m The solution can be obtained with formulas and calculator, but we will use Minitab and the printout below:

```
Regression Analysis: CalcFin versus MathPro, SATQ

The regression equation is CalcFin = - 26.6 + 0.776 MathPro + 0.0820 SATQ

Predictor        Coef     SE Coef         T         P
Constant       -26.62       17.18     -1.55     0.172
MathPro        0.7763      0.1465      5.30     0.002
SATQ          0.08202     0.02699      3.04     0.023

S = 4.027      R-Sq = 88.5%     R-Sq(adj) = 84.7%

Analysis of Variance
Source          DF          SS        MS         F         P
Regression       2      751.57    375.78     23.17     0.002
Residual Error   6       97.32     16.22
Total            8      848.89

Predicted Values for New Observations
New Obs      Fit      SE Fit        90.0% CI            90.0% PI
1          68.74        2.43   (  64.01,   73.46)  (  59.59,   77.88)

Values of Predictors for New Observations
New Obs   MathPro       SATQ
1            70.0        500
```

a. The regression equation is CalcFin = -26.6 + 0.776*MathPro + 0.0820*SATQ. For the population of entering freshmen who scored 70 on the math proficiency test and 500 on the quantitative portion of the SAT exam, we are 90% confident that their mean calculus final exam score will be within the interval from 64.01 to 73.46.

b. For an individual entering freshman who scored 70 on the math proficiency test and 500 on the quantitative portion of the SAT exam, we are 90% confident that his or her calculus final exam score will be within the interval from 59.59 to 77.88.

16.23 p/a/m

a. Using only the regression equation and summary information obtained in exercise 16.12, we can determine the approximate 95% confidence interval for the mean overall rating of cars that receive ratings of 8 on ride, 7 on handling, and 9 on driver comfort. First, the midpoint is:

$$\hat{y} = 35.63 + 3.675(8) + 2.892(7) - 0.110(9) = 84.284.$$

There were nine observations used to estimate the regression, so d.f. = 9 - 3 - 1 = 5. The appropriate t-value is 2.571, the multiple standard error of the estimate is 2.858, and the approximate 95% confidence interval is:

$$\hat{y} \pm t\frac{s_e}{\sqrt{n}} = 84.284 \pm 2.571\frac{2.858}{\sqrt{9}} = 84.284 \pm 2.449 = (81.835, \ 86.733)$$

b. The corresponding approximate 95% prediction interval is:

$$\hat{y} \pm ts_e = 84.284 \pm 2.571(2.858) = 84.284 \pm 7.348 = (76.936, 91.632)$$

The preceding are the approximate intervals that could be calculated based only on the information shown in the printouts for exercise 16.12. As discussed in the text, the exact intervals will tend to be wider than the approximate intervals. This is because the exact intervals take into account that the

specified values for x_1, x_2, and x_3 may differ from their respective means. The exact Minitab intervals corresponding to parts a and b of this exercise are:
95% confidence interval, (79.587, 88.980); 95% prediction interval, (75.562, 93.005).

16.25 d/p/e The coefficient of multiple determination (R^2) is analogous to the coefficient of determination in simple linear regression. It is the proportion of variation in y that is explained by the multiple regression equation.

16.27 d/p/m The adjusted R^2 has been adjusted for the degrees of freedom. If the number of variables is large relative to the number of observations, R^2 will exaggerate the strength of the linear relationship. The adjusted R^2 will get closer to R^2 as sample sizes become larger relative to the number of variables.

16.29 d/p/e The coefficient of multiple determination for exercise 16.19 is 0.878. This means that 87.8% of the variation in the percentage of employees who work at retail stores is explained by variation in the percentage of employees working in the wholesale products and transportation sectors.

16.31 p/c/m The coefficient of multiple determination for the regression equation obtained in exercise 16.11 is 0.716. This indicates that 71.6% of the variation in the number of new visitors to the club is explained by the regression equation.

16.33 d/p/m An ANOVA test is used to evaluate the overall significance of a multiple regression equation. If the F calculated in the ANOVA exceeds the critical F, the null hypothesis of no significance is rejected. Individual t-tests are used to evaluate the significance of the individual partial regression coefficients.

16.35 p/c/m We will base much of our discussion on the Minitab printout for exercise 16.11. The results will be similar if you refer to the Excel printout.
a. The appropriate null and alternative hypotheses are:
    H_0: $\beta_1 = \beta_2 = 0$ and H_1: $\beta_j \neq 0$, for j = 1 or 2
    From the ANOVA portion of the Minitab printout, we have:

```
Analysis of Variance
Source              DF           SS          MS         F        P
Regression           2       143.92       71.96      6.32     0.043
Residual Error       5        56.95       11.39
Total                7       200.87
```

The p-value for the ANOVA test of the overall significance of the regression equation is 0.043.
Since p-value = 0.043 is < α = 0.05 level of significance for the test, we reject H_0. At this level, there is evidence to suggest that the regression equation is significant.

b. From the upper portion of the Minitab printout:

```
The regression equation is Visitors = 10.7 + 2.16 AdSize + 0.0416 Discount

Predictor        Coef     SE Coef          T         P
Constant       10.687       3.875       2.76     0.040
AdSize         2.1569      0.6281       3.43     0.019
Discount      0.04157     0.04380       0.95     0.386

S = 3.375      R-Sq = 71.6%     R-Sq(adj) = 60.3%
```

Here we are asked to conduct two hypothesis tests. We will not test the y-intercept since this test is generally not of practical importance. The appropriate null and alternative hypotheses are:
    Test for β_1: H_0: $\beta_1 = 0$ and H_1: $\beta_1 \neq 0$
    Test for β_2: H_0: $\beta_2 = 0$ and H_1: $\beta_2 \neq 0$

The p-value for the test of β_1 is 0.019. Since p-value = 0.019 is $<\alpha = 0.05$ level of significance for the test, we reject H_0. At this level, there is evidence to suggest that β_1 is nonzero.

The p-value for the test of β_2 is 0.386. Since p-value = 0.386 is not $<\alpha = 0.05$ level of significance for the test, we do not reject H_0. At this level, there is no evidence to suggest that β_1 is nonzero.

c. The ANOVA test for the overall regression indicates that the regression explains a significant proportion of the variation in the number of new visitors to the club. The tests for the individual partial regression coefficients indicate that the size of the ad contributes to the explanatory power of the model, while the discount offered does not.

d. With d.f. = 8 - 2 - 1 = 5, the appropriate t-value for the 95% confidence interval will be 2.571.
The 95% confidence interval for population partial regression coefficient β_1 is:
$$b_1 \pm t\, s_{b_1} = 2.1569 \pm 2.571(0.6281) = 2.1569 \pm 1.6148 = (0.5421, 3.7717)$$

The 95% confidence interval for population partial regression coefficient β_2 is:
$$b_2 \pm t\, s_{b_2} = 0.04157 \pm 2.571(0.04380) = 0.04157 \pm 0.1126 = (-0.0710, 0.1542)$$

With Excel, we can obtain confidence intervals for the population regression coefficients along with the standard regression output. Excel will provide 95% confidence intervals, but we can also specify the inclusion of 90% or any other confidence levels we wish to see. The Excel printout for exercise 16.11 included 95% confidence intervals for β_1 and β_2.

16.37 p/a/m To determine the 90% confidence interval for each partial regression coefficient in exercise 16.15, we must determine the appropriate t. There are (12 - 3 - 1) = 8 degrees of freedom, so the appropriate t is t = 1.860.

The 90% confidence interval for population partial regression coefficient β_1 is:
$$b_1 \pm t\, s_{b_1} = 0.2745 \pm 1.860(0.1111) = 0.2745 \pm 0.2066 = (0.0679, 0.4811)$$

This confidence interval does not contain zero, so it is likely the variation in the scores on test 1 do contribute significantly to the explanation of the variation of the scores on test 4.

The 90% confidence interval for population partial regression coefficient β_2 is:
$$b_2 \pm t\, s_{b_2} = 0.37619 \pm 1.860(0.09858) = 0.37619 \pm 0.18336 = (0.1928, 0.5596).$$

This confidence interval does not contain zero, so it is likely the variation in the scores on test 2 do contribute significantly to the explanation of the variation of the scores on test 4.

The 90% confidence interval for population partial regression coefficient β_3 is:
$$b_3 \pm t\, s_{b_3} = 0.32648 \pm 1.860(0.08084) = 0.32648 \pm 0.15036 = (0.1761, 0.4768)$$

This confidence interval does not contain zero, so it is likely the variation in the scores on test 3 do contribute significantly to the explanation of the variation of the scores on test 4.

16.39 p/a/m To verify the 95% confidence interval for each partial regression coefficient in exercise 16.30, we must determine the appropriate t. There are (8 - 2 - 1) = 5 degrees of freedom, so the appropriate t is t = 2.571.

The 95% confidence interval for population partial regression coefficient β_1 is:
$$b_1 \pm t\, s_{b_1} = -0.0911 \pm 2.571(0.0967) = -0.0911 \pm 0.2486 = (-0.3397, 0.1575)$$

Our calculated interval differs slightly at the fourth decimal place, this because of rounded calculations and the use of a table t-value that had been rounded to three decimal places. This confidence interval does contain zero, so it is likely the variation in the price of crude oil does not contribute significantly to the explanation of the variation in price for natural gas.

The 95% confidence interval for population partial regression coefficient β_2 is:
$$b_2 \pm t\, s_{b_2} = 0.6239 \pm 2.571(0.3474) = 0.6239 \pm 0.8932 = (-0.2693, 1.5171)$$

Our calculated interval differs slightly at the fourth decimal place, this because of rounded calculations and the use of a table t-value that had been rounded to three decimal places.

This confidence interval does contain zero, so it is likely the variation in the price for bituminous coal does not contribute significantly to the explanation of the variation in the price of natural gas.

16.41 p/c/m Referring to the Minitab printout in the solution to exercise 16.21:
a. In the ANOVA test for overall significance, p-value $= 0.002$ is < 0.05 level of significance, so we conclude that the overall regression is significant.
b. In testing the partial regression coefficients for math proficiency test score and SAT quantitative score, the p-values are 0.002 and 0.023, respectively. Each p-value is < 0.05 level of significance, and each of the partial regression coefficients differs significantly from zero.

16.43 p/c/m The Minitab printout is shown below.

```
The regression equation is
$GroupRevenue = -40855482 + 44282 RetailUnits + 152760 NumDealrs

Predictor        Coef     SE Coef         T        P
Constant    -40855482    20217627     -2.02    0.046
RetailUn        44282        1290     34.33    0.000
NumDealr       152760     1943687      0.08    0.938

S = 176197662   R-Sq = 99.3%     R-Sq(adj) = 99.3%

Analysis of Variance
Source           DF          SS          MS         F        P
Regression        2  4.46877E+20  2.23439E+20   7197.11    0.000
Residual Error   95  2.94933E+18  3.10456E+16
Total            97  4.49827E+20
```

a. The regression equation is
$GroupRevenue = -40,855,482 + 44,282*RetailUnits + 152,760*NumDealrs
The partial regression coefficient for RetailUnits is 44,282. On average, with the number of dealers fixed, an increase of 1 in retail units sold is accompanied by an increase of $44,282 in revenue for the dealer group.
The partial regression coefficient for NumDealrs is 152,760. On average, with the number of retail units fixed, an increase of 1 in the number of dealers will be accompanied by an increase of $152,760 in revenue for the dealer group.
b. The p-value in the ANOVA section of the printout is (to three decimal places) 0.000. This is less than the 0.02 level of significance. At this level, the overall regression equation is significant.
c. The p-values for the tests of the two partial regression coefficients are 0.000 and 0.938, respectively. Using the 0.02 level of significance, the partial regression coefficient for the first independent variable (retail units) is significantly different from zero, but the partial regression coefficient for the second independent variable (number of dealers) does not differ significantly from zero.
d. The 98% confidence interval for each partial regression coefficient could be calculated using formulas and pocket calculator, as was demonstrated in the solution to exercise 16.37. We will rely on the Excel printout, shown below. The 98% confidence interval for population partial regression coefficient β_1 is from 41,229.4 to 47,333.8. The 98% confidence interval for population partial regression coefficient β_2 is from -4,446,472 to 4,751,992.

	J	K	L	M	N	O	P
1	SUMMARY OUTPUT						
2							
3	*Regression Statistics*						
4	Multiple R	0.9967					
5	R Square	0.9934					
6	Adjusted R Square	0.9933					
7	Standard Error	176197662					
8	Observations	98					
9							
10	ANOVA						
11		*df*	*SS*	*MS*	*F*	*Significance F*	
12	Regression	2	4.469E+20	2.234E+20	7.197E+03	1.960E-104	
13	Residual	95	2.949E+18	3.105E+16			
14	Total	97	4.498E+20				
15							
16		*Coefficients*	*Standard Error*	*t Stat*	*P-value*	*Lower 98.0%*	*Upper 98.0%*
17	Intercept	-40855482.0	20217627.3	-2.021	0.046	-88695270.0	6984305.9
18	RetailUnits	44281.6	1289.89	34.330	0.000	41229.4	47333.8
19	NumDealrs	152760.2	1943686.79	0.079	0.938	-4446472	4751992

16.45 d/p/m Residual analysis can be used to examine the residuals with respect to the assumptions underlying multiple regression analysis. We can do many things with residual analysis, including: (1) constructing a histogram of the residuals as a rough check to see if they are approximately normally distributed, (2) constructing a normal probability plot or applying the Lilliefors test to examine whether the residuals could have come from a normally-distributed population, (3) plotting the residuals versus each of the independent variables to see if they exhibit some cycle or pattern with respect to that variable, and (4) plotting the residuals versus the order in which the observations were recorded to look for autocorrelation.

16.47 p/p/m Referring to the output for exercise 16.19, we can determine the following:
a. The partial regression coefficient for WHOLEPCT is 2.065. This implies that a one percent increase in employment in the wholesale products sector will result in a 2.065% increase in employment in the retail sector, employment in other sectors held constant. The partial regression coefficient for TRANSPCT is 4.486. This implies that a one percent increase in employment in the transportation sector will result in a 4.486% increase in the retail sector, employment in other sectors held constant.
b. 87.8% of the total variation in y is explained by the regression equation.
c. The overall regression equation is significant at the 0.000 (rounded) level.
d. The partial regression coefficient for WHOLEPCT is significant at the 0.000 (rounded) level, and the partial regression coefficient for TRANSPCT is significant at the 0.010 level. It appears that both variables are very useful in explaining the variation in employment in the retail sector.

16.49 p/c/m

a. The histogram does not reveal any radical departures from a symmetric distribution. Of course, it is difficult to determine this with only eight data points.

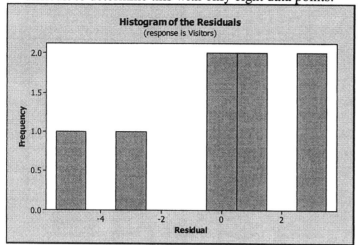

b. Normal probability plot. There are only eight points and they do not appear to deviate excessively from a straight line. The approximate p-value for the associated Lilliefors test (Minitab's Kolmogorov-Smirnov test for normality) is shown as > 0.15. At the 0.05 level of significance, we would conclude that the residuals could have come from a normally distributed population.

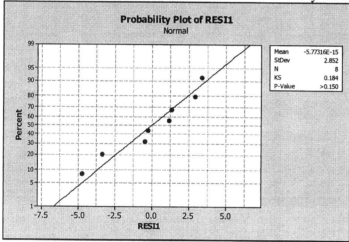

c. Plots of residuals versus the independent variables.
   Plot of residuals versus ad size.

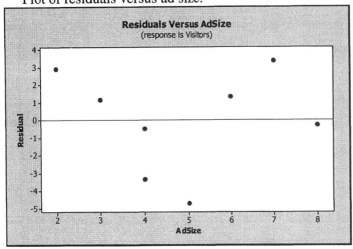

   Plot of residuals versus discount size.

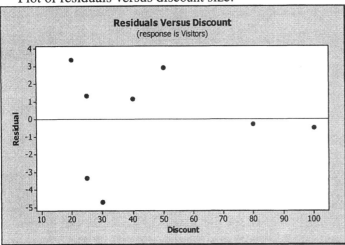

The plots above do not reveal any alarming problems. Overall, there is no strong evidence to indicate that any underlying assumptions of the multiple regression model have been violated.

16.51 p/c/m

a. This histogram does not appear to reveal any radical departures from a symmetric distribution, although there are relatively few data points.

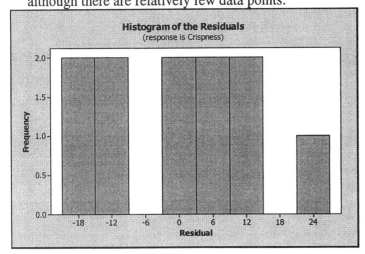

b. Normal probability plot. The points do not appear to deviate excessively from a straight line. The approximate p-value for the associated Lilliefors test (Minitab's Kolmogorov-Smirnov test for normality) is shown as > 0.15. At the 0.05 level of significance, we would conclude that the residuals could have come from a normally distributed population.

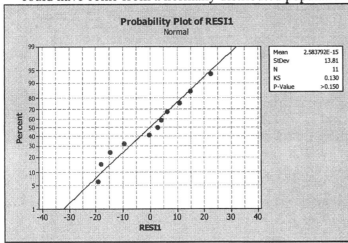

c. Plots of residuals versus the independent variables. The plots below do not reveal any alarming problems. Overall, there is no strong evidence to indicate that any underlying assumptions of the multiple regression model have been violated.
Plot of residuals versus time in oven.

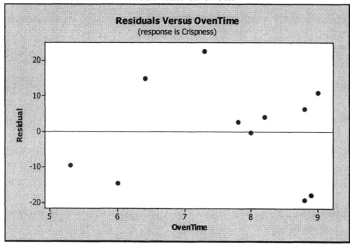

Plot of residuals versus oven temperature.

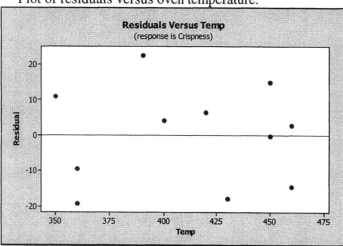

16.53 p/c/m

a. This histogram does not appear to reveal any radical departures from a symmetric distribution, although there are relatively few data points.

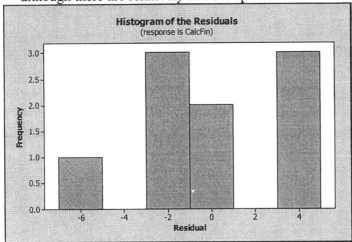

b. Normal probability plot. The points do not appear to deviate excessively from a straight line. The approximate p-value for the associated Lilliefors test (Minitab's Kolmogorov-Smirnov test for normality) is shown as > 0.15. At the 0.05 level of significance, we would conclude that the residuals could have come from a normally distributed population.

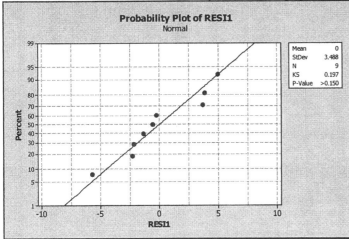

c. Plots of residuals versus the independent variables. The plots below do not reveal any alarming problems. Overall, there is no strong evidence to indicate that any underlying assumptions of the multiple regression model have been violated.

Plot of residuals versus math proficiency test.

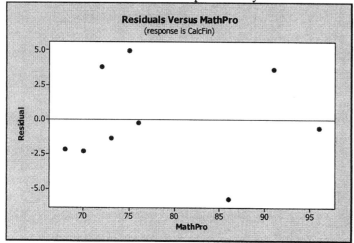

Plot of residuals versus SAT quantitative.

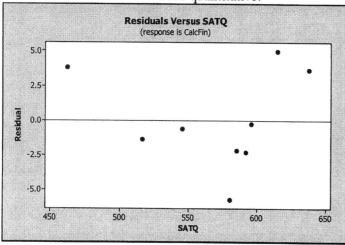

16.55 p/c/m The Minitab printout is shown below.

```
Regression Analysis: Distance versus Price, Sensitiv, Weight

The regression equation is
Distance = - 0.562 +0.000355 Price + 0.0112 Sensitiv - 0.0212 Weight

Predictor        Coef      SE Coef         T         P
Constant      -0.5617       0.8656     -0.65     0.545
Price       0.0003550    0.0005601      0.63     0.554
Sensitiv     0.011248     0.007605      1.48     0.199
Weight       -0.02116      0.02471     -0.86     0.431

S = 0.05167     R-Sq = 46.5%     R-Sq(adj) = 14.4%

Analysis of Variance
Source           DF           SS           MS         F        P
Regression        3     0.011590     0.003863      1.45    0.334
Residual Error    5     0.013349     0.002670
Total             8     0.024939
```

a. The estimated regression equation is:

Distance = -0.5617 + 0.0003550*Price + 0.011248*Sensitiv - 0.02116*Weight.

The partial regression coefficient for the price indicates that, holding the weight and sensitivity constant, a \$1 increase in price will result in a 0.0003550 mile increase in the warning distance. The partial regression coefficient for the sensitivity indicates that, holding the price and weight constant, a one unit increase in sensitivity will result in a 0.011248 mile increase in the warning distance. Finally, the partial regression coefficient for the weight indicates that, holding the price and sensitivity constant, a one ounce increase in weight will result in a 0.02116 mile decrease in the warning distance.

b. The appropriate degrees of freedom for this problem will be d.f. = 9 - 3 - 1 = 5. The t-value for a 95% confidence interval with 5 degrees of freedom is t = 2.571.

The 95% confidence interval for population partial regression coefficient β_1 is:

$b_1 \pm t\, s_{b_1} = 0.000355 \pm 2.571(0.0005601) = 0.000355 \pm 0.001440 = (-0.0011, 0.0018)$

The 95% confidence interval for population partial regression coefficient β_2 is:

$b_2 \pm t\, s_{b_2} = 0.011248 \pm 2.571(0.007605) = 0.011248 \pm 0.019552 = (-0.0083, 0.0308)$

The 95% confidence interval for population partial regression coefficient β_3 is:

$b_3 \pm t\, s_{b_3} = -0.02116 \pm 2.571(0.02471) = -0.02116 \pm 0.06353 = (-0.0847, 0.0424)$

c. The coefficient of multiple determination is 0.465. This indicates that 46.5% of the variation in the warning distance is explained by the regression equation. However, none of the partial regression coefficients is significant at the 0.10 level. (The coefficient for price is significant at the 0.554 level; for sensitivity, at the 0.199 level; and, for weight, at the 0.431 level.) The overall regression is only significant at the 0.334 level. The adjusted R-square is 0.144. Recall that this has been adjusted for the degrees of freedom. Thus, there are no significant relationships in this regression. Apparently the coefficient of multiple determination is as large as it is because of the limited size of the data set.

d. The residual analyses follow. First the histogram of the residuals is examined to see if it is symmetric about zero. Next the normal probability plot is graphed to examine whether the residuals could have come from a normally distributed population. Finally, the residuals are plotted against each of the independent variables to check for cyclical patterns.

In the following histogram of residuals, there seems to be a slight deviation from a symmetric distribution, but the number of data values is relatively small.

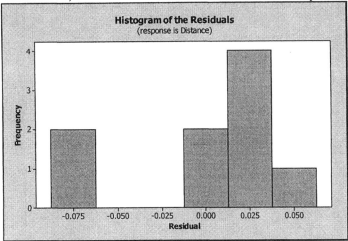

Normal probability plot. The points seem to deviate excessively from a straight line and the approximate p-value for the associated Lilliefors test (Minitab's Kolmogorov-Smirnov test for normality) is shown as 0.048. At the 0.05 level of significance, we would conclude that the residuals did not come from a normally distributed population.

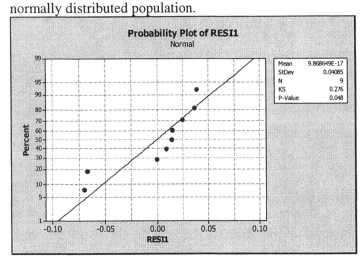

The plots of residuals versus the independent variables are shown below. Given the relatively small number of data points, none of the three plots shows any alarming patterns. In the third plot, most of the unusual pattern is due to the underlying data, with most of the weights clustered about the six ounce level, while one of the detectors weighs only 3.8 ounces.

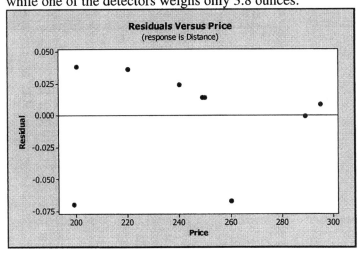

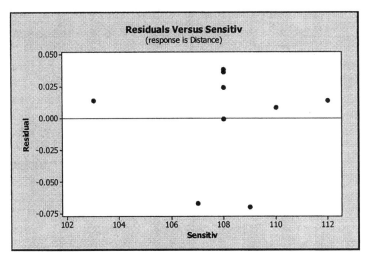

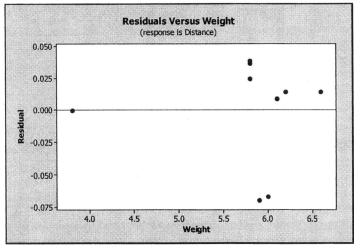

Overall, the residual analysis suggests that the residuals may not have come from a normally distributed population. If this is true, then one of the underlying assumptions has been violated and the multiple regression analysis may not be valid.

271

The Excel multiple regression solution for the data in this exercise is shown below. Note that Excel already provides 95% confidence intervals for the population regression coefficients.

	A	B	C	D	E	F	G
11			Distance	Price	Sensitivity	Weight	
12	SUMMARY OUTPUT		0.675	289	108	3.8	
13	*Regression Statistics*		0.660	295	110	6.1	
14	Multiple R	0.6817	0.640	240	108	5.8	
15	R Square	0.4647	0.560	249	103	6.6	
16	Adjusted R Square	0.1436	0.540	260	107	6.0	
17	Standard Error	0.0517	0.640	200	108	5.8	
18	Observations	9	0.540	199	109	5.9	
19			0.645	220	108	5.8	
20	ANOVA		0.670	250	112	6.2	
21		*df*	*SS*	*MS*	*F*	*Significance F*	
22	Regression	3	0.0116	0.0039	1.4471	0.3342	
23	Residual	5	0.0133	0.0027			
24	Total	8	0.0249				
25							
26		*Coefficients*	*Standard Error*	*t Stat*	*P-value*	*Lower 95%*	*Upper 95%*
27	Intercept	-0.56174	0.8656	-0.6490	0.5450	-2.7868	1.6633
28	Price	0.00035	0.0006	0.6338	0.5541	-0.0011	0.0018
29	Sensitivity	0.01125	0.0076	1.4790	0.1992	-0.0083	0.0308
30	Weight	-0.02116	0.0247	-0.8564	0.4309	-0.0847	0.0424

When generating the Excel printout, we can also specify a normal probability plot and plots of the residuals against the independent variables. Their appearance is essentially similar to those of Minitab, although the normal probability plot does not include a p-value to assist in interpretation.

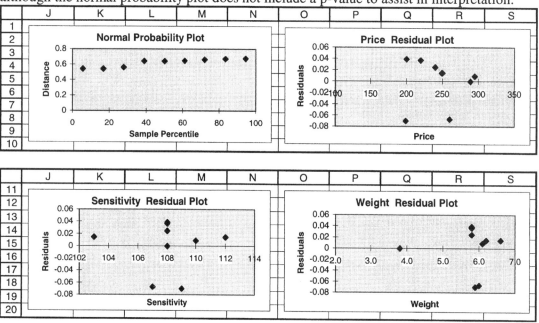

16.57 p/p/e The partial regression coefficient for x_1 implies that, holding the day of the week constant, a one degree Fahrenheit increase in the temperature will result in an increase of 8 in attendance. The partial regression coefficient for x_2 implies that the attendance increases by 150 people on Saturdays and Sundays (assuming a constant temperature).

16.59 d/p/m Multicollinearity is a situation in which two or more of the independent variables in a multiple regression are highly correlated with each other. When this happens, the two correlated x variables are really not saying different things about y. The standard errors for the partial regression coefficients become very large and the coefficients are statistically unreliable and difficult to interpret. Multicollinearity is a problem when we are trying to interpret the partial regression coefficients. There are several clues to the presence of multicollinearity: (1) an independent variable known to be an important predictor ends up having a partial regression coefficient that is not significant; (2) a partial regression coefficient exhibits the wrong sign; and/or, (3) when an independent variable is added or deleted, the partial regression coefficients for the other variables change dramatically.
A more practical way to identify multicollinearity is through the examination of a correlation matrix, which is a matrix that shows the correlation of each variable with each of the other variables.
A high correlation between two independent variables is an indication of multicollinearity.

16.61 p/c/m The Minitab printout is shown below.

```
Regression Analysis: Speed versus Occupnts, SeatBelt

The regression equation is Speed = 67.6 - 3.21 Occupnts - 6.63 SeatBelt

Predictor        Coef      SE Coef         T        P
Constant       67.629        5.017     13.48    0.000
Occupnts       -3.214        2.191     -1.47    0.170
SeatBelt       -6.629        3.200     -2.07    0.063

S = 5.465      R-Sq = 31.5%      R-Sq(adj) = 19.1%

Analysis of Variance
Source          DF          SS          MS        F        P
Regression       2      151.20       75.60     2.53    0.125
Residual Error  11      328.51       29.86
Total           13      479.71
```

The partial regression coefficient for Occupnts implies that, holding seat belt usage constant, the speed decreases by 3.214 miles per hour for each additional occupant in the car. The partial regression coefficient for SeatBelt implies that, for a given number of occupants, drivers who wear seat belts travel 6.629 miles per hour slower than those who do not. The p-value for Occupnts is 0.170; this implies that the partial regression coefficient for this variable is not significantly different from zero. The p-value for SeatBelt is 0.063; this implies that the partial regression coefficient for this variable is significantly different from zero at the 0.063 level. It appears that seat belt usage provides a much stronger explanation for the variation in speeds driven by various drivers than does the number of occupants in the car.
The Excel multiple regression solution for this exercise is shown below.

	D	E	F	G	H	I	J
2							
3	Regression Statistics						
4	Multiple R	0.5614					
5	R Square	0.3152					
6	Adjusted R Square	0.1907					
7	Standard Error	5.4649					
8	Observations	14					
9							
10	ANOVA						
11		df	SS	MS	F	Significance F	
12	Regression	2	151.2000	75.6000	2.5314	0.1246	
13	Residual	11	328.5143	29.8649			
14	Total	13	479.7143				
15							
16		Coefficients	Standard Error	t Stat	P-value	Lower 95%	Upper 95%
17	Intercept	67.6286	5.0172	13.4795	0.0000	56.5859	78.6713
18	Occupnts	-3.2143	2.1908	-1.4672	0.1703	-8.0363	1.6077
19	SeatBelt	-6.6286	3.1999	-2.0715	0.0626	-13.6715	0.4144

CHAPTER EXERCISES

16.63 p/c/m The Minitab printout is shown below.

```
Regression Analysis: AllFruit versus Apples, Grapes

The regression equation is AllFruit = 297 - 3.14 Apples - 15.9 Grapes

Predictor        Coef     SE Coef          T        P
Constant       296.51       88.01       3.37    0.043
Apples         -3.138        5.306      -0.59    0.596
Grapes        -15.92        14.72       -1.08    0.359

S = 9.335       R-Sq = 58.3%     R-Sq(adj) = 30.5%

Analysis of Variance
Source            DF           SS          MS        F        P
Regression         2       365.38      182.69     2.10    0.269
Residual Error     3       261.43       87.14
Total              5       626.81


Predicted Values for New Observations
New Obs      Fit     SE Fit          95.0% CI              95.0% PI
1         147.65      16.30   (  95.76,  199.54) (  87.86,  207.44) XX
X   denotes a row with X values away from the center
XX denotes a row with very extreme X values

Values of Predictors for New Observations
New Obs    Apples    Grapes
1            17.0      6.00
```

a. The regression equation is AllFruit = 296.51 - 3.138*Apples - 15.92*Grapes.
    The partial regression coefficient for apples implies that, holding the consumption of grapes constant, a one pound increase in the consumption of apples will result in a 3.138 pound decrease in the consumption of all fresh fruits. The partial regression coefficient for grapes implies that, holding apple consumption constant, a one pound increase in the consumption of grapes will result in a 15.92 pound decrease in the consumption of all fresh fruits.

b. The estimated per capita consumption of all fresh fruits during a year when 17 pounds of apples and 6 pounds of grapes are consumed is 147.65 pounds.

c. The 95% prediction interval for per capita consumption during a year like the one in part b is 87.86 to 207.44 pounds.

d. The 95% confidence interval for mean per capita consumption during all years like the one in part b is 95.76 to 199.54 pounds.

e. For this problem, the appropriate d.f. = 6 – 2 – 1 = 3. The t-value for a 95% confidence interval with 3 degrees of freedom is 3.182.

    The 95% confidence interval for population partial regression coefficient β_1 is:
    $$b_1 \pm t\,s_{b_1} = -3.14 \pm 3.182(5.306) = -3.14 \pm 16.88 = (-20.02, 13.74)$$

    The 95% confidence interval for population partial regression coefficient β_2 is:
    $$b_2 \pm t\,s_{b_2} = -15.92 \pm 3.182(14.72) = -15.92 \pm 46.84 = (-62.76, 30.92)$$

f. Neither the partial regression coefficient for apple consumption nor the coefficient for grape consumption is significant. (apples p-value, 0.596; grapes p-value, 0.359) Also, the overall regression is not significant; p-value = 0.269. The coefficient of multiple determination drops from 0.583 to 0.305 when adjusted for degrees of freedom. So, this regression does not appear to do a very good job of explaining the variation in fresh fruit consumption.

g. The residual analyses follow. First the histogram of the residuals is examined to see if it is symmetric about zero. Next the normal probability plot is graphed to examine whether the residuals could have come from a normally distributed population. Finally, the residuals are plotted against each of the independent variables to check for cyclical patterns.

The histogram of the residuals is shown below. This histogram offers no reason to believe that the residuals may not have come from a normally distributed population.

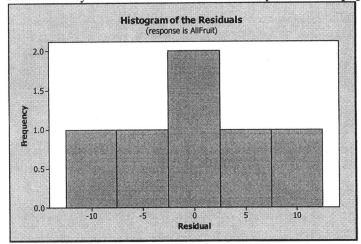

Normal probability plot. The points do not seem to deviate excessively from a straight line and the approximate p-value for the associated Lilliefors test (Minitab's Kolmogorov-Smirnov test for normality) is shown as > 0.15. At the 0.05 level of significance, we would conclude that the residuals could have come from a normally distributed population.

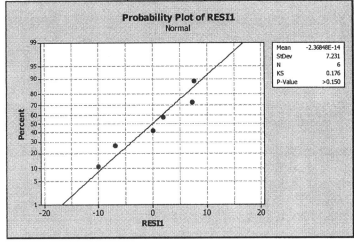

The plots for residuals versus the independent variables are shown below. For this small data set, no alarming patterns seem to be present. Overall, the residual analysis provides no evidence to suggest that the assumptions for the multiple regression model have not been satisfied.

Residuals versus per-capita apple consumption.

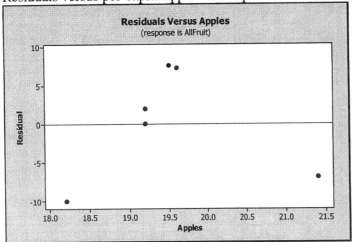

Residuals versus per-capita grape consumption.

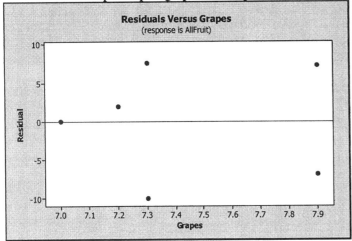

The Excel multiple regression solution for the data in this exercise is shown below. Note that Excel already provides 95% confidence intervals for the population regression coefficients.

	A	B	C	D	E	F	G
9	SUMMARY OUTPUT			All Fruits	Apples	Grapes	
10	Regression Statistics			96.7	21.4	7.9	
11	Multiple R	0.7635		116.5	19.6	7.9	
12	R Square	0.5829		113.2	18.2	7.3	
13	Adjusted R Square	0.3049		123.6	19.2	7.2	
14	Standard Error	9.3350		124.9	19.2	7.0	
15	Observations	6		126.7	19.5	7.3	
16							
17	ANOVA						
18		df	SS	MS	F	Significance F	
19	Regression	2	365.3849	182.6924	2.0965	0.2694	
20	Residual	3	261.4284	87.1428			
21	Total	5	626.8133				
22							
23		Coefficients	Standard Error	t Stat	P-value	Lower 95%	Upper 95%
24	Intercept	296.5136	88.0050	3.3693	0.0434	16.4420	576.5852
25	Apples	-3.1376	5.3064	-0.5913	0.5959	-20.0249	13.7497
26	Grapes	-15.9209	14.7188	-1.0817	0.3586	-62.7628	30.9209

Excel has generated the optional normal probability plot and plots of the residuals against the independent variables. Their appearance is essentially similar to those of Minitab, although the normal probability plot does not include a p-value to assist in interpretation.

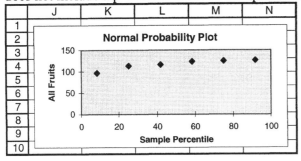

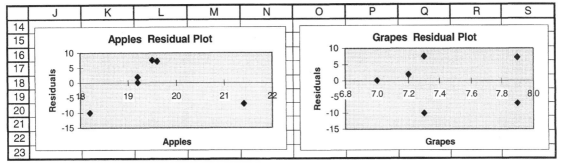

16.65 p/p/m

a. The partial regression coefficient for OVER65 is 0.7427 which indicates that, holding all other variables constant, an increase of one percentage point in this variable will result in a 0.7427 percentage point increase in the voting percentage. This coefficient is significant at the 0.100 level. The partial regression coefficient for CABLETV indicates that, holding all other variables constant, a one percentage point increase in the households with cable television will result in a 0.02458 percentage point increase in the percentage of the eligible voters voting. This coefficient is not significantly different from zero. The partial regression coefficient for UNEMPLOY indicates that, holding all other variables constant, a one percentage point increase in the unemployment rate will result in a 0.1471 percentage point decrease in the percentage of voters voting. This coefficient is not significantly different from zero. Finally, the partial regression coefficient for $/PUPIL indicates that, all other variables held constant, a $1000 increase in the expenditure per school pupil will result in a 2.005 percentage point increase in the percentage of voters voting. This coefficient is significantly different from zero at the 0.073 level.

b. The overall regression does not appear to be very significant in its ability to explain variation in the percentage of eligible voters who voted in the presidential election. The p-value is 0.265.

c. Persons over 65 would be retired and have more time to vote, so this coefficient could be positive. Persons with cable television would be more affluent and have more time to vote, so this coefficient could be positive. Unemployed persons could be disgruntled and apathetic, so this coefficient could be negative. Finally, states which spend more on education are likely to have more active electorates, so this coefficient could be positive.

d. Other variables that might help explain voting percentages would include statewide races of interest (a dummy variable that takes on the value of 1 when there is an important statewide race in conjunction with the presidential election); this coefficient will be expected to be positive because the statewide race will elevate interest. A weather variable (1 when the weather is good on election day) which should have a positive coefficient (since good weather is more conducive to leaving the house to vote) could help explain some of the variation. A time zone variable, which could be a dummy variable taking on the value of 1 for the Eastern and

Central time zones and 0 otherwise, might also explain some of the variation; this coefficient should be negative since the turnout will be reduced if people feel like the election is already decided.

16.67 p/c/m The estimated selling price of a house occupying a 0.1 acre lot with 100 square feet of living area and no central air conditioning is $38,699. This selling price does not seem reasonable. The problem with this estimate arises because the regression equation has been extrapolated far beyond the limits of the underlying data used to estimate it.

16.69 p/c/m The Minitab printout is shown below.

```
Regression Analysis: FrGPA versus SAT, HSRank

The regression equation is FrGPA = - 1.98 + 0.00372 SAT + 0.00658 HSRank

Predictor        Coef      SE Coef          T        P
Constant       -1.984        1.532      -1.30    0.218
SAT          0.003719     0.001562       2.38    0.033
HSRank       0.006585     0.008023       0.82    0.427

S = 0.4651     R-Sq = 45.2%     R-Sq(adj) = 36.8%

Analysis of Variance
Source            DF          SS          MS        F        P
Regression         2      2.3244      1.1622     5.37    0.020
Residual Error    13      2.8125      0.2163
Total             15      5.1370

Predicted Values for New Observations
New Obs      Fit     SE Fit        95.0% CI              95.0% PI
1          2.634      0.125    (  2.365,   2.904)    (  1.594,   3.674)

Values of Predictors for New Observations
New Obs      SAT    HSRank
1           1100      80.0
```

a. The regression equation is: FrGPA = -1.984 + 0.003719*SAT + 0.006585*HSRank.
   The partial regression coefficient for the SAT score indicates that, holding the rank constant, a 1 point increase in the SAT score will result in a 0.003719 point increase in the freshman GPA. The coefficient for the high school rank indicates that, holding the SAT score constant, a 1 point increase in the high school rank will result in a 0.006585 point increase in freshman GPA.
b. The estimated freshman GPA for a student who scored 1100 on the SAT and had a class rank of 80% is 2.634.
c. The 95% prediction interval for the GPA for a student like the one in part b is between 1.594 and 3.674.
d. The 95% confidence interval for the mean GPA for all students like the one in part b is 2.365 to 2.904.
e. For this problem, the appropriate d.f. = 16 - 2 - 1 = 13. The t-value for a 95% interval with 13 degrees of freedom is 2.160.
   The 95% confidence interval for population partial regression coefficient β_1 is:
   $$b_1 \pm t\,s_{b_1} = 0.003719 \pm 2.160(0.001562) = 0.003719 \pm 0.003374 = (0.000345, 0.007093)$$

   The 95% confidence interval for population partial regression coefficient β_2 is:
   $$b_2 \pm t\,s_{b_2} = 0.006585 \pm 2.160(0.008023) = 0.006585 \pm 0.017330 = (-0.010745, 0.023915)$$

f. The partial regression coefficient for the SAT score is significantly different from zero at the 0.033 level. The partial regression coefficient for the high school rank is not significantly different from zero (p-value = 0.427). The overall regression is significant at the 0.020 level.
g. The residual analyses follow. First the histogram of the residuals is examined to see if it is symmetric about zero. Next the normal probability plot is graphed to examine whether the residuals could have come from a normally distributed population. Finally, the residuals are plotted against each of the independent variables to check for cyclical patterns.

The histogram below seems to be fairly symmetric about zero.

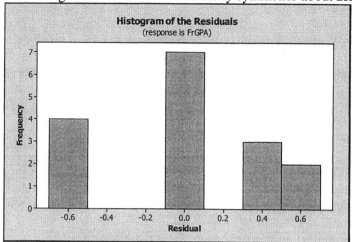

Normal probability plot. The points do not seem to deviate excessively from a straight line and the approximate p-value for the associated Lilliefors test (Minitab's Kolmogorov-Smirnov test for normality) is shown as > 0.15. At the 0.05 level of significance, we would conclude that the residuals could have come from a normally distributed population.

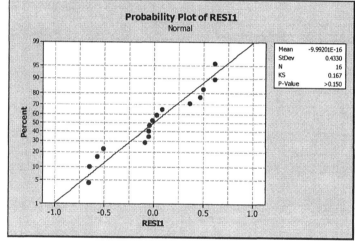

The plots for residuals versus the independent variables are shown below. Neither of the plots reveals any alarming patterns that suggest the underlying assumptions of the multiple regression analysis may have been violated. Overall, the residual analysis does not reveal anything to suggest that the assumptions underlying the multiple regression analysis have been violated.

Residuals versus SAT score.

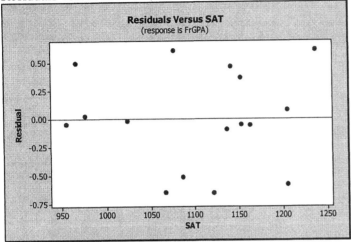

Residuals versus high school rank.

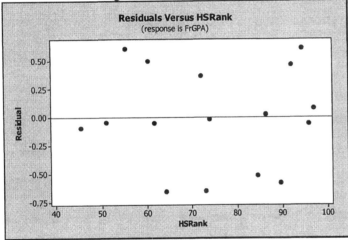

The Excel multiple regression solution for the data in this exercise is shown below. Note that Excel already provides 95% confidence intervals for the population regression coefficients.

	A	B	C	D	E	F	G
20	SUMMARY OUTPUT						
21	Regression Statistics						
22	Multiple R	0.6727					
23	R Square	0.4525					
24	Adjusted R Square	0.3683					
25	Standard Error	0.4651					
26	Observations	16					
27							
28	ANOVA						
29		df	SS	MS	F	Significance F	
30	Regression	2	2.3244	1.1622	5.3719	0.0199	
31	Residual	13	2.8125	0.2163			
32	Total	15	5.1370				
33							
34		Coefficients	Standard Error	t Stat	P-value	Lower 95%	Upper 95%
35	Intercept	-1.983878	1.53190	-1.29505	0.21783	-5.29334	1.32558
36	SAT	0.003719	0.00156	2.38050	0.03328	0.00034	0.00709
37	HS Rank	0.006585	0.00802	0.82074	0.42659	-0.01075	0.02392

Excel has generated the optional normal probability plot and plots of the residuals against the independent variables. Their appearance is essentially similar to those of Minitab, although the normal probability plot does not include a p-value to assist in interpretation.

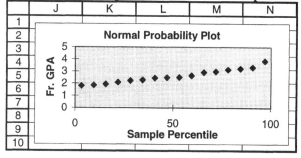

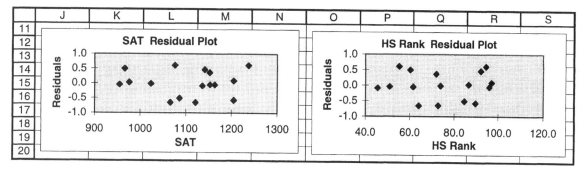

281

CHAPTER 17
MODEL BUILDING

SECTION EXERCISES

17.1 d/p/e The fifth-order polynomial model would be $E(y) = \beta_0 + \beta_1 x + \beta_2 x^2 + \beta_3 x^3 + \beta_4 x^4 + \beta_5 x^5$, a level of complexity that would not be very practical for use with most business data.

17.3 d/p/m The second-order polynomial model is $E(y) = \beta_0 + \beta_1 x + \beta_2 x^2$, and the form of the estimation equation fit to the data would be $\hat{y} = b_0 + b_1 x + b_2 x^2$. Assuming nonnegative values for both x and y, $b_1 < 0$ and $b_2 < 0$ would result in such a curvature.

17.5 d/p/m With an estimation equation of the form, $\hat{y} = b_0 + b_1 x + b_2 x^2 + b_3 x^3$, this curvature could be fit if $b_1 > 0$, $b_2 < 0$, and $b_3 > 0$. This is demonstrated in the Excel plot shown below. Note that Excel's default arrangement of the terms at the right side of the equation is the reverse of what we are used to seeing.

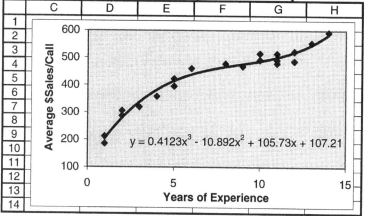

17.7 p/c/m The Minitab and Excel plots are shown below. The estimation equation can be expressed as $AvgRate = 286.094 - 8.239*\%Occup + 0.07709*\%Occup^2$, and it explains 49.1% of the variation in the average room rates. It is possible that hotels with a lower percentage occupancy might try to attract guests by offering low rates. Also, hotels enjoying a high percentage occupancy may not hesitate to charge the high rates that the market is apparently willing to pay for their rooms.

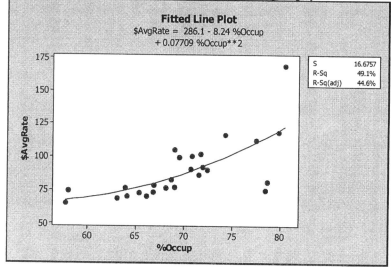

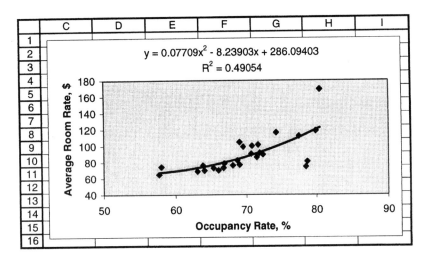

y = 0.07709x² - 8.23903x + 286.09403

R² = 0.49054

17.9 p/c/m The Minitab printout and the Minitab and Excel plots plot are shown below. The estimation equation can be expressed as 0-60 Time = 26.8119 - 0.153866*hp+0.0003083*hp². The equation explains 73.8% of the variation in 0-60 times. Substituting hp = 300, a car with 300 horsepower would be estimated as having a 0-60 time of 8.396 seconds. The p-value for the regression (to three decimal places) is 0.000, which is less than the 0.05 level of significance. At this level, the regression is significant.

```
Regression Analysis: 0-60 versus hp, hpSq

The regression equation is 0-60 = 26.8 - 0.154 hp +0.000308 hpSq

Predictor        Coef      SE Coef          T         P
Constant       26.812        2.241      11.97     0.000
hp           -0.15387      0.03183      -4.83     0.000
hpSq       0.00030827   0.00008718       3.54     0.002

S = 2.860      R-Sq = 73.8%     R-Sq(adj) = 70.9%

Analysis of Variance
Source            DF         SS          MS         F         P
Regression         2     415.54      207.77     25.39     0.000
Residual Error    18     147.27        8.18
Total             20     562.81

Predicted Values for New Observations
New Obs    Fit     SE Fit          95.0% CI            95.0% PI
1        8.396      1.418    ( 5.418,  11.375) ( 1.689,  15.104)

Values of Predictors for New Observations
New Obs       hp       hpSq
1            300      90000
```

284

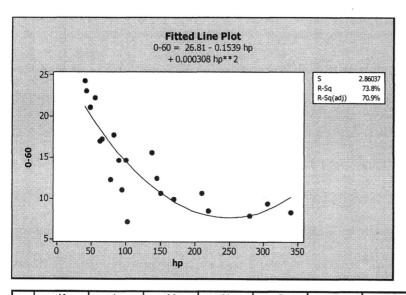

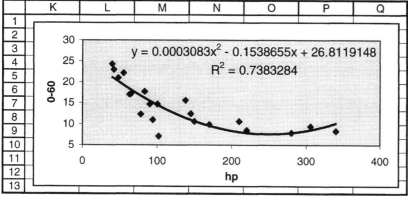

17.11 p/c/m Given only the information in the Excel printout, the estimation equation can be expressed as ForGross = 860.8444 - 4.1517*DomGross + 0.0077*DomGross^2.

For a film with domestic gross ticket sales of $400 million, the estimated foreign gross ticket sales will be ForGross = 860.8444 - 4.1517(400) + 0.0077(400)^2 = $432.16 million.

The regression is significant at the 0.05 level, since p-value = 0.00000471 is < 0.05 level of significance. Note that the Excel p-value is shown as 4.71E-06, scientific notation for 4.71(10)$^{-6}$.

A more accurate calculation can be made if the constant and coefficients in the estimation equation are not limited to four decimal places. The Minitab printout is shown below. As shown in the "Fit" column of the printout, a film with $400 million in domestic gross ticket sales would be predicted to generate $430.4 million in foreign gross ticket sales.

```
Regression Analysis: ForGross versus DomGross, DomGross_sq

The regression equation is ForGross = 861 - 4.15 DomGross + 0.00769 DomGross_sq

Predictor        Coef     SE Coef        T        P
Constant        860.8       245.0     3.51    0.003
DomGross       -4.152       1.439    -2.88    0.010
DomGross     0.007689    0.001920     4.00    0.001

S = 112.6      R-Sq = 76.4%     R-Sq(adj) = 73.6%

Analysis of Variance
Source           DF          SS          MS        F        P
Regression        2      697099      348550    27.49    0.000
Residual Error   17      215552       12680
Total            19      912651
```

```
New Obs     Fit     SE Fit      95.0% CI            95.0% PI
1          430.4      44.7  (  336.1,   524.7)  (  174.8,   686.0)

Values of Predictors for New Observations
New Obs  DomGross  DomGross
1            400    160000
```

17.13 d/p/m The most suitable model would be a second-order model with interaction, or $E(y) = \beta_0 + \beta_1 x_1 + \beta_2 x_2 + \beta_3 x_1^2 + \beta_4 x_2^2 + \beta_5 x_1 x_2$.

17.15 p/c/m The second-order model with interaction is $E(y) = \beta_0 + \beta_1 x_1 + \beta_2 x_2 + \beta_3 x_1^2 + \beta_4 x_2^2 + \beta_5 x_1 x_2$. Minitab has been used to obtain the estimation equation and printout shown below.

```
Regression Analysis: Avg$/call versus Yrs, Score, ...

The regression equation is
Avg$/call = 61 + 25.6 Yrs + 6.41 Score - 1.82 Yrs_sq - 0.058 Score_sq
          + 0.29 Yrs*Score

Predictor        Coef    SE Coef         T        P
Constant         61.2      132.2      0.46    0.651
Yrs             25.63      31.58      0.81    0.431
Score           6.409      8.174      0.78    0.446
Yrs_sq         -1.817      2.700     -0.67    0.512
Score_sq      -0.0582     0.1288     -0.45    0.658
Yrs*Scor        0.295      1.087      0.27    0.790

S = 29.85      R-Sq = 94.9%      R-Sq(adj) = 93.1%

Analysis of Variance
Source            DF        SS        MS        F        P
Regression         5    233075     46615    52.32    0.000
Residual Error    14     12473       891
Total             19    245548
```

The estimation equation is Avg\$/call = 61 + 25.6*Yrs + 6.41*Score - 1.82*Yrs^2 - 0.058*Score^2 + 0.29*Yrs*Score. As a decimal fraction, $R^2 = 0.949$, and 94.9% of the variation in average dollar sales per call is explained by the equation. Since p-value = 0.000 is < 0.05 level of significance, the model is significant at this level.

17.17 p/c/m

a. Minitab has been used to obtain the estimation equation and printout shown below.

```
Regression Analysis: OpRevenue versus Employees, Departures

The regression equation is OpRevenue = - 231 + 0.129 Employees + 0.00565 Departures

Predictor        Coef    SE Coef         T        P
Constant       -231.2      345.7     -0.67    0.511
Employee      0.12883    0.01277     10.09    0.000
Departur     0.005649   0.001243      4.55    0.000

S = 1220       R-Sq = 95.8%      R-Sq(adj) = 95.4%

Analysis of Variance
Source            DF          SS          MS        F        P
Regression         2   681281501   340640751   228.93    0.000
Residual Error    20    29759807     1487990
Total             22   711041308
```

As a decimal fraction, $R^2 = 0.958$, and 95.8% of the variation in operating revenue is explained by the equation. Since p-value = 0.000 is < 0.01 level of significance, the model is significant at this level.

b. Repeating part (a), but with an interaction term included, we obtain the printout below.

```
Regression Analysis: OpRevenue versus Employees, Departures, Emps*Deps

The regression equation is
OpRevenue = 399 + 0.0745 Employees + 0.00087 Departures +0.000000 Emps*Deps

Predictor         Coef      SE Coef        T        P
Constant         399.2        230.2     1.73    0.099
Employee       0.07453      0.01169     6.38    0.000
Departur      0.000874     0.001075     0.81    0.427
Emps*Dep    0.00000014   0.00000002     6.12    0.000

S = 726.3      R-Sq = 98.6%     R-Sq(adj) = 98.4%

Analysis of Variance
Source           DF          SS          MS         F        P
Regression        3   701019585   233673195    443.02    0.000
Residual Error   19    10021724      527459
Total            22   711041308
```

We find R^2 has increased from 0.958 to 0.986, and 98.6% of the variation in operating revenue is explained by the new equation. Since p-value = 0.000 is < the 0.01 level, the model is significant.

17.19 p/c/m Minitab has been used to obtain the estimation equation and printout shown below.

```
Regression Analysis: 0-60 versus hp, curbwt, hp*curbwt

The regression equation is
0-60 = 25.4 - 0.161 hp - 0.00030 curbwt +0.000028 hp*curbwt

Predictor         Coef      SE Coef        T        P
Constant        25.419        4.338     5.86    0.000
hp            -0.16109      0.03551    -4.54    0.000
curbwt       -0.000302     0.002233    -0.14    0.894
hp*curbw    0.00002818   0.00000946     2.98    0.008

S = 2.968      R-Sq = 73.4%     R-Sq(adj) = 68.7%

Analysis of Variance
Source           DF          SS          MS         F        P
Regression        3      413.09      137.70     15.64    0.000
Residual Error   17      149.71        8.81
Total            20      562.81
```

As a decimal fraction, $R^2 = 0.734$, and 73.4% of the variation in 0-60 mph time is explained by the equation. Since p-value = 0.000 is < 0.02 level of significance, the model is significant at this level.

17.21 p/p/m There will be two independent variables and the format for the linear regression equation for estimating the dollar amount of the order will be

Dollars = $b_0 + b_1$*Minutes + b_2*Visa

The coding of the qualitative variable: Visa = 1 if customer used a Visa card, otherwise = 0.

17.23 p/p/m There are three risk levels (low, medium, and high), so this information can be represented by just two qualitative variables. If the two variables are named Low and Medium:
For a person who specifies low risk, Low = 1 and Medium = 0.
For a person who specifies medium risk, Low = 0 and Medium = 1.
For a person who specifies high risk, Low = 0 and Medium = 0.

17.25 p/p/m Given that the estimated number of customers is $100 + 5x_1 + 50x_2$: For each additional one-degree increase in the high temperature, the estimated number of customers would increase by 5. If the day happens to fall on a weekend (Saturday or Sunday), the estimated number of customers would increase by 50. The estimated number of customers on an 80-degree Saturday would be $100 + 5(80) + 50(1) = 550$.

17.27 p/c/m Minitab has been used to obtain the estimation equation and printout shown below.

```
Regression Analysis: Price versus GB, 7200rpm?

The regression equation is Price = - 30.8 + 4.97 GB + 54.2 7200rpm?

Predictor        Coef     SE Coef        T        P
Constant       -30.77       54.89    -0.56    0.593
GB              4.975       1.140     4.36    0.003
7200rpm?        54.20       44.11     1.23    0.259

S = 66.39      R-Sq = 79.0%     R-Sq(adj) = 73.0%

Analysis of Variance
Source          DF          SS         MS        F        P
Regression       2      115899      57949    13.15    0.004
Residual Error   7       30851       4407
Total            9      146750
```

The regression equation is Price = - 30.77 + 4.975*GB + 54.20*7200rpm?. The partial regression coefficient for GB is 4.975 -- for a given rotational speed, an additional gigabyte of storage capacity would tend to raise the price by $4.975. The partial regression coefficient for the qualitative variable 7200rpm? is 54.20 -- for a given level of storage capacity, a drive with the higher rotational speed would tend to be priced $54.20 higher than one with the lower rotational speed.

The corresponding Excel printout is shown below.

	E	F	G	H	I	J	K
1	SUMMARY OUTPUT						
2							
3	*Regression Statistics*						
4	Multiple R	0.8887					
5	R Square	0.7898					
6	Adjusted R Square	0.7297					
7	Standard Error	66.3872					
8	Observations	10					
9							
10	ANOVA						
11		*df*	*SS*	*MS*	*F*	*Significance F*	
12	Regression	2	115898.8027	57949.4014	13.1486	0.0043	
13	Residual	7	30850.7973	4407.2568			
14	Total	9	146749.6000				
15							
16		*Coefficients*	*Standard Error*	*t Stat*	*P-value*	*Lower 95%*	*Upper 95%*
17	Intercept	-30.7680	54.8855	-0.5606	0.5926	-160.5516	99.0156
18	GB	4.9745	1.1404	4.3622	0.0033	2.2779	7.6711
19	7200rpm?	54.2024	44.1115	1.2288	0.2589	-50.1047	158.5094

17.29 p/c/d We must first devise a coding scheme for the assembly methods, then enter the data into the computer. Since there are three possible assembly methods, there will be two qualitative variables in the regression: MethA? and MethB?. If the unit was assembled by method A,
then MethA? = 1 and MethB? = 0. If the unit was assembled by method B, them MethA? = 0 and MethB? = 1. If the unit was assembled by method C (i.e., by neither method A nor method B), MethA? = 0 and MethB? = 0.

	A	B	C	D	E	F
1	productivity	YrsExper	MethA?	MethB?		Method
2	75	7	1	0		A
3	88	10	0	0		C
4	91	4	0	1		B
5	93	5	0	1		B
6	95	11	0	0		C
7	77	3	1	0		A
8	97	12	0	1		B
9	85	10	0	0		C
10	102	12	0	0		C
11	93	13	1	0		A
12	112	12	0	1		B
13	86	14	1	0		A

The Minitab printout is shown below.

```
Regression Analysis: productivity versus YrsExper, MethA?, MethB?

The regression equation is
productivity = 75.4 + 1.59 YrsExper - 7.36 MethA? + 9.73 MethB?

Predictor         Coef      SE Coef          T        P
Constant        75.369        6.307      11.95    0.000
YrsExper        1.5936       0.5139       3.10    0.015
MethA?          -7.360        4.372      -1.68    0.131
MethB?           9.734        4.491       2.17    0.062

S = 6.086      R-Sq = 74.1%     R-Sq(adj) = 64.4%

Analysis of Variance
Source           DF          SS          MS        F        P
Regression        3      847.33      282.44     7.63    0.010
Residual Error    8      296.33       37.04
Total            11     1143.67
```

The regression equation is
productivity = 75.369 + 1.5936*YrsExper - 7.360*MethA? + 9.734*MethB?.
The equation explains 74.1% of the variation in productivity. The partial regression coefficient for years of experience is 1.5936 -- for a given assembly method, an additional year of experience tends to increase productivity by 1.5936 units per hour. The interpretations of the other partial regression coefficients are as follows:
For a given number of years of experience, using method A tends to reduce productivity by 7.360 units per hour compared to method C.
For a given number of years of experience, using method B tends to increase productivity by 9.734 units per hour compared to method C.

The corresponding Excel printout is shown below.

	G	H	I	J	K	L	M
1	SUMMARY OUTPUT						
2							
3	*Regression Statistics*						
4	Multiple R	0.8608					
5	R Square	0.7409					
6	Adjusted R Square	0.6437					
7	Standard Error	6.0862					
8	Observations	12					
9							
10	ANOVA						
11		*df*	*SS*	*MS*	*F*	*Significance F*	
12	Regression	3	847.3324	282.4441	7.6250	0.0099	
13	Residual	8	296.3342	37.0418			
14	Total	11	1143.6667				
15							
16		*Coefficients*	*Standard Error*	*t Stat*	*P-value*	*Lower 95%*	*Upper 95%*
17	Intercept	75.3690	6.3073	11.9495	0.0000	60.8243	89.9136
18	YrsExper	1.5936	0.5139	3.1008	0.0146	0.4085	2.7787
19	MethA?	-7.3596	4.3721	-1.6833	0.1308	-17.4417	2.7224
20	MethB?	9.7340	4.4913	2.1673	0.0621	-0.6230	20.0909

17.31 c/a/m When we are converting the exponential estimation equation, $\hat{y} = b_0 b_1^{\,x}$, to linear form, the result can be expressed as $\log \hat{y} = \log b_0 + x \log b_1$.

In this problem, the conversion is reversed, and we can express the given equation as
$\log \hat{y} = -0.179 + x\, 0.140$.

Since $\log b_0 = -0.179$, then $b_0 = 10^{-0.179}$, or 0.66. Since $\log b_1 = 0.140$, then $b_1 = 10^{0.140}$, or 1.38. The exponential form of the equation is $\hat{y} = 0.66(1.38)^x$.

17.33 p/c/m We first use Excel to fit the equation $\hat{y} = 14.506581 e^{0.025683x}$ to the data.

Since $e = 2.71828$, $e^{0.025683} = 1.026016$, and we can express the equation as $\hat{y} = 14.5066(1.026016)^x$. To three decimal places, $R^2 = 0.509$. For a market in which the average occupancy rate is $x = 70$ percent, the estimated average room rate would be $\hat{y} = 14.5066(1.026016)^{70}$, or $87.57.

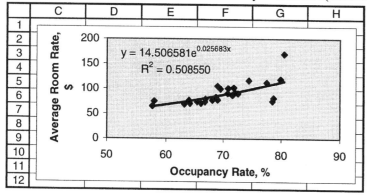

17.35 p/c/m Using Minitab, we transform the data so that we have the corresponding logarithm to the base 10, then perform a multiple regression using the transformed data. (As noted in exercise 17.17, operating revenue is expressed in millions of dollars.) The printout is shown below.

```
Regression Analysis: logRev versus logEmps, logDeps

The regression equation is logRev = - 0.128 + 1.00 logEmps - 0.112 logDeps

Predictor         Coef     SE Coef        T        P
Constant       -0.1285      0.2285    -0.56    0.580
logEmps         1.00396     0.09668    10.38    0.000
logDeps        -0.11212     0.07542    -1.49    0.153

S = 0.2112      R-Sq = 92.8%     R-Sq(adj) = 92.0%

Analysis of Variance
Source             DF        SS        MS        F        P
Regression          2    11.4312    5.7156   128.14    0.000
Residual Error     20     0.8921    0.0446
Total              22    12.3233
```

The equation and the result when it is converted back to multiplicative form are shown below:

Logarithmic form: log Revenue = -0.1285 + 1.0040*log Employees - 0.1121*log Departures

Multiplicative form: Revenue = $0.7439*\text{Employees}^{1.0040}*\text{Departures}^{-0.1121}$.

The estimated total operating revenue for an airline with 2000 employees and 10,000 departures would be $0.7439*(2000)^{1.0040}*(10,000)^{-0.1121} = \$546.2$ million.

17.37 d/p/m A correlation matrix is a matrix that shows the correlation of each variable in a multiple regression analysis with each of the other variables. A high correlation between two independent variables is an indication of multicollinearity.

17.39 d/p/m Multicollinearity is a problem when we are trying to interpret the partial regression coefficients. Multicollinearity is *not* a problem when we are simply substituting x values into the regression equation in order to estimate y.

17.41 d/p/m There is a very strong correlation (0.952) between predictor variables x_1 and x_3, suggesting the presence of multicollinearity.

17.43 p/c/m The Minitab and Excel correlation results are shown below. There are strong correlations between the pairs of independent variables, indicating that multicollinearity could be a problem.

```
Correlations: Aircraft, Employees, Departures

         Aircraft  Employee
Employee   0.948
           0.000

Departur   0.872     0.761
           0.000     0.000

Cell Contents: Pearson correlation
               P-Value
```

	G	H	I	J
1		*Aircraft*	*Employees*	*Departures*
2	Aircraft	1		
3	Employees	0.948	1	
4	Departures	0.872	0.761	1

17.45 p/c/m The Minitab and Excel correlation results are shown below. There are strong correlations between the pairs of independent variables, indicating that multicollinearity could be a problem.

```
Correlations: CrudeOil, NatGas, Coal, Nuclear

          CrudeOil   NatGas     Coal
NatGas     -0.971
            0.000

Coal       -0.883    0.897
            0.000    0.000

Nuclear    -0.971    0.954    0.907
            0.000    0.000    0.000

Cell Contents: Pearson correlation
               P-Value
```

	G	H	I	J	K
1		*CrudeOil*	*NatGas*	*Coal*	*Nuclear*
2	CrudeOil	1			
3	NatGas	-0.971	1		
4	Coal	-0.883	0.897	1	
5	Nuclear	-0.971	0.954	0.907	1

17.47 d/p/m Independent variables are selected for entry into the equation during stepwise regression based upon the amount of the remaining variation in y (the variation that has not already been explained by included variables) that a candidate variable can explain.

17.49 c/a/m
a. The first variable included in the equation is x_5. This is followed (in order) by x_2 and x_9.
b. The multiple regression equation after two independent variables have been introduced is:
$\hat{y} = 106.85 - 0.35x_5 - 0.33x_2$. This equation explains 19.66% of the total variation in y, and the multiple standard error of the estimate is 13.2.
c. The p-values for the tests of the partial regression coefficient x_5, x_2, and x_9 are 0.026, 0.010, and 0.044, respectively. At the 0.05 level, the partial regression coefficients for all three independent variables differ significantly from 0. At the 0.01 level, only the partial regression coefficient for x_2 differs significantly from 0.

17.51 p/c/m The correlation matrix for all of the variables is shown below, followed by the results of the conventional and stepwise regression analyses.

```
Correlations: TotalNR, CrudeOil, NatGas, Coal, Nuclear

          TotalNR  CrudeOil   NatGas     Coal
CrudeOil  -0.742
           0.009

NatGas     0.802    -0.971
           0.003     0.000

Coal       0.950    -0.883    0.897
           0.000     0.000    0.000

Nuclear    0.830    -0.971    0.954    0.907
           0.002     0.000    0.000    0.000

Cell Contents: Pearson correlation
               P-Value
```

Regression Analysis: TotalNR versus CrudeOil, NatGas, Coal, Nuclear

The regression equation is
TotalNR = - 0.94 + 2.52 CrudeOil + 0.803 NatGas + 0.0250 Coal + 0.0123 Nuclear

Predictor	Coef	SE Coef	T	P
Constant	-0.943	2.747	-0.34	0.743
CrudeOil	2.5158	0.1427	17.63	0.000
NatGas	0.8025	0.1098	7.31	0.000
Coal	0.0250308	0.0009866	25.37	0.000
Nuclear	0.012337	0.001045	11.80	0.000

S = 0.06806   R-Sq = 99.8%   R-Sq(adj) = 99.7%

Analysis of Variance

Source	DF	SS	MS	F	P
Regression	4	17.2613	4.3153	931.68	0.000
Residual Error	6	0.0278	0.0046		
Total	10	17.2891			

Stepwise Regression: TotalNR versus CrudeOil, NatGas, Coal, Nuclear

Alpha-to-Enter: 0.05 Alpha-to-Remove: 0.1

Response is TotalNR on 4 predictors, with N =   11

Step	1	2	3	4
Constant	41.0766	26.6354	15.8845	-0.9435
Coal	0.02291	0.03229	0.02722	0.02503
T-Value	9.10	7.54	9.94	25.37
P-Value	0.000	0.000	0.000	0.000
CrudeOil		0.70	1.85	2.52
T-Value		2.48	5.79	17.63
P-Value		0.038	0.001	0.000
Nuclear			0.0127	0.0123
T-Value			4.17	11.80
P-Value			0.004	0.000
NatGas				0.80
T-Value				7.31
P-Value				0.000
S	0.434	0.346	0.198	0.0681
R-Sq	90.21	94.46	98.41	99.84
R-Sq(adj)	89.12	93.07	97.73	99.73
C-p	358.6	201.9	56.4	5.0

All four of the independent variables were introduced during the stepwise regression, so the conventional multiple regression provides the same equation listed in step 4 of the stepwise regression. Coal was the independent variable with the strongest correlation with the dependent variable, and it was the first variable to enter the equation during the stepwise regression.

17.53 p/c/m The correlation matrix for the variables is shown below, followed by the results of the conventional and stepwise regression analyses.

```
Correlations: Avg$/call, Yrs, Score

        Avg$/cal      Yrs
Yrs       0.945
          0.000

Score     0.930     0.940
          0.000     0.000

Cell Contents: Pearson correlation
               P-Value
```

```
Regression Analysis: Avg$/call versus Yrs, Score

The regression equation is Avg$/call = 192 + 16.0 Yrs + 1.88 Score

Predictor       Coef      SE Coef         T        P
Constant      192.27        34.20      5.62    0.000
Yrs            15.991        5.690      2.81    0.012
Score           1.882        1.146      1.64    0.119

S = 36.58       R-Sq = 90.7%     R-Sq(adj) = 89.6%

Analysis of Variance
Source            DF          SS          MS        F        P
Regression         2      222805      111402    83.27    0.000
Residual Error    17       22743        1338
Total             19      245548
```

```
Stepwise Regression: Avg$/call versus Yrs, Score

  Alpha-to-Enter: 0.05  Alpha-to-Remove: 0.1

  Response is Avg$/cal on  2 predictors, with N =    20

      Step          1
Constant         241.3

Yrs               24.8
T-Value          12.24
P-Value          0.000

S                 38.3
R-Sq             89.27
R-Sq(adj)        88.67
C-p               3.7
```

In the conventional regression analysis, the two independent variables explain 90.7% of the variation in average dollar sales per call. The number of years of experience had the highest correlation with the dependent variable, and it was the first and only variable introduced in the stepwise regression.
The regression equation with number of years of experience as the only independent variable explains 89.27% of the variation in average dollar sales per call.

17.55 p/c/m

a. The correlation matrix shows that Memb7 and Memb8 are tied for being the most highly correlated with the overall rating. Note that Memb7 and Memb8 are also highly correlated with each other.

```
Correlations: Overall, Memb1, Memb2, Memb3, Memb4, Memb5, Memb6, Memb7, Memb8, M

          Overall    Memb1    Memb2    Memb3    Memb4    Memb5    Memb6    Memb7
Memb1      0.868
Memb2      0.843    0.709
Memb3      0.839    0.745    0.707
Memb4      0.895    0.704    0.719    0.725
Memb5      0.907    0.816    0.745    0.612    0.792
Memb6      0.889    0.810    0.649    0.710    0.795    0.788
Memb7      0.924    0.744    0.790    0.764    0.863    0.808    0.758
Memb8      0.924    0.714    0.725    0.778    0.852    0.838    0.846    0.830
Memb9      0.885    0.720    0.741    0.702    0.725    0.779    0.759    0.825

            Memb8
Memb9      0.772

Cell Contents: Pearson correlation
```

b. The stepwise regression is shown below. We would expect either Memb7 or Memb8 to be the first variable brought in.

```
Stepwise Regression: Overall versus Memb1, Memb2, ...

  Alpha-to-Enter: 0.05  Alpha-to-Remove: 0.1

Response is Overall  on  9 predictors, with N =    28

     Step        1        2        3        4        5        6        7
Constant    4.3600  -1.7404  -0.4649  -2.0891  -1.9432  -2.7894  -1.7702

Memb7        0.961    0.611    0.457    0.288    0.230    0.198    0.137
T-Value      12.36     7.72     6.20     5.49     5.00     5.21     4.03
P-Value      0.000    0.000    0.000    0.000    0.000    0.000    0.001

Memb6                 0.432    0.312    0.213    0.189    0.205    0.176
T-Value                5.82     4.72     4.92     5.23     6.95     7.23
P-Value                0.000    0.000    0.000    0.000    0.000    0.000

Memb5                          0.254    0.301    0.274    0.244    0.228
T-Value                         3.97     7.54     8.17     8.64    10.10
P-Value                        0.001    0.000    0.000    0.000    0.000

Memb3                                   0.231    0.216    0.191    0.182
T-Value                                  6.39     7.26     7.62     9.18
P-Value                                 0.000    0.000    0.000    0.000

Memb9                                            0.119    0.104    0.120
T-Value                                           3.51     3.77     5.42
P-Value                                          0.002    0.001    0.000

Memb2                                                     0.095    0.092
T-Value                                                    3.59     4.46
P-Value                                                   0.002    0.000

Memb4                                                              0.086
T-Value                                                            3.76
P-Value                                                           0.001

S             3.31     2.20     1.75     1.07    0.876    0.706    0.553
R-Sq         85.46    93.82    96.27    98.66    99.14    99.47    99.69
R-Sq(adj)    84.90    93.33    95.81    98.42    98.94    99.31    99.58
C-p         2693.1   1132.3    676.7    232.8    144.8     85.7     46.4

     Step        8        9
Constant    -1.432   -1.220

Memb7        0.135    0.132
T-Value       4.40     6.62
P-Value      0.000    0.000

Memb6        0.160    0.114
T-Value       7.00     6.56
P-Value      0.000    0.000
```

```
Memb5              0.209     0.152
T-Value             9.61      8.47
P-Value            0.000     0.000

Memb3              0.166     0.123
T-Value             8.75      8.25
P-Value            0.000     0.000

Memb9              0.119     0.126
T-Value             5.99      9.70
P-Value            0.000     0.000

Memb2              0.094     0.089
T-Value             5.05      7.27
P-Value            0.000     0.000

Memb4              0.073     0.081
T-Value             3.43      5.84
P-Value            0.003     0.000

Memb8              0.061     0.108
T-Value             2.39      5.73
P-Value            0.027     0.000

Memb1                        0.090
T-Value                       5.18
P-Value                      0.000

S                  0.498     0.324
R-Sq              99.76     99.90
R-Sq(adj)         99.66     99.86
C-p               34.9      10.0
```

c. From the results given in part b, it appears that the board could be reduced to four or five members without a tremendous loss in rating power. The coefficient of multiple determination for the regression using four board members is 0.9866; using five board members, it increases to 0.9914.

d. The following estimates the regression using only the data provided by board members 7 and 6. These were the first two board members introduced in the stepwise regression. Except for rounding, this equation is identical with the one in step two of the stepwise regression. The same holds for the t-ratios, p-values, and coefficients of multiple determination. Thus, the stepwise technique and the multiple regression technique yield the same results when using the same variables.

Regression Analysis: Overall versus Memb7, Memb6

The regression equation is Overall = - 1.74 + 0.611 Memb7 + 0.432 Memb6

```
Predictor        Coef     SE Coef         T        P
Constant       -1.740       4.032     -0.43    0.670
Memb7          0.61140     0.07923      7.72    0.000
Memb6          0.43151     0.07417      5.82    0.000

S = 2.202      R-Sq = 93.8%     R-Sq(adj) = 93.3%

Analysis of Variance

Source           DF          SS          MS        F        P
Regression        2     1840.60      920.30   189.88    0.000
Residual Error   25      121.17        4.85
Total            27     1961.77
```

17.57 p/c/d

a. Stepwise regression with the rating from board member 9 as the dependent variable and the ratings by the other 8 members as the set of independent variables from which to select.

```
Stepwise Regression: Memb9 versus Memb1, Memb2, ...

  Alpha-to-Enter: 0.05  Alpha-to-Remove: 0.1

  Response is  Memb9   on  8 predictors, with N =   28

      Step           1
Constant       3.551

Memb7           0.98
T-Value         7.45
P-Value        0.000

S               5.61
R-Sq           68.11
R-Sq(adj)      66.88
C-p             0.9
```

Only one variable is included in this regression, Memb7. It is not surprising that Memb7 enters in step one, because it had the highest correlation with Memb9. (See the correlation matrix in exercise 17.55.) However, no additional variables are included. The final regression equation is Memb9 = 3.551 + 0.98*Memb7.

The partial regression coefficient for board member 7 indicates that a one point increase in the rating from member 7 will tend to result in a 0.98 point increase in the rating from member 9.

The standard error of the estimate for this regression is 5.61, and the coefficient of determination is 0.6811.

b. Regression estimating Memb9 based on Memb7.

```
Regression Analysis: Memb9 versus Memb7

The regression equation is Memb9 = 3.55 + 0.980 Memb7

Predictor        Coef      SE Coef         T        P
Constant        3.551        9.914      0.36    0.723
Memb7          0.9803       0.1316      7.45    0.000

S = 5.606       R-Sq = 68.1%     R-Sq(adj) = 66.9%

Analysis of Variance

Source            DF          SS         MS        F        P
Regression         1      1744.9     1744.9    55.52    0.000
Residual Error    26       817.1       31.4
Total             27      2562.0
```

c. Except for slight differences due to rounding, the results of the regression in part b are the same as those provided by step two of the stepwise regression in part a.

17.59 p/c/m The second-order polynomial model shown below is a fairly good fit to the data. The equation is Pages = $-4.6 + 274.4x - 20.44x^2$. The model explains 89.2% of the variation in page count for Harry Potter books over the years. For 2008, the year code will be $x = 11$ and the predicted size of a Harry Potter book published that year will be 540.3 pages, as shown in the Minitab printout below.

```
Regression Analysis: Pages versus x, xSq

The regression equation is Pages = - 5 + 274 x - 20.4 xSq

Predictor        Coef      SE Coef         T        P
Constant         -4.6        249.2     -0.02    0.987
x               274.4        173.4      1.58    0.254
xSq            -20.44        23.20     -0.88    0.471

S = 120.5      R-Sq = 89.2%    R-Sq(adj) = 78.4%

Analysis of Variance
Source             DF          SS          MS        F        P
Regression          2      239285      119642     8.24    0.108
Residual Error      2       29029       14515
Total               4      268314

Unusual Observations
Obs        x       Pages        Fit      SE Fit     Residual     St Resid
  5     6.00       896.0      905.9       120.2         -9.9        -1.30 X
X denotes an observation whose X value gives it large influence.

Predicted Values for New Observations
New Obs     Fit      SE Fit          95.0% CI              95.0% PI
1         540.3      1187.3   ( -4568.5,   5649.0)   ( -4594.7,   5675.2) XX
X  denotes a row with X values away from the center
XX denotes a row with very extreme X values

Values of Predictors for New Observations
New Obs       x        xSq
1          11.0        121
```

17.61 p/c/m The second-order polynomial model shown below is a good fit to the data. The equation is Productivity = $10.705 + 0.974*Backlog - 0.015*Backlog^2$.

The size of the backlog explains 83.6% of the variation in productivity. If we were to fit a third-order model to the data, the percentage of variation explained would increase only slightly, to 84.0%.

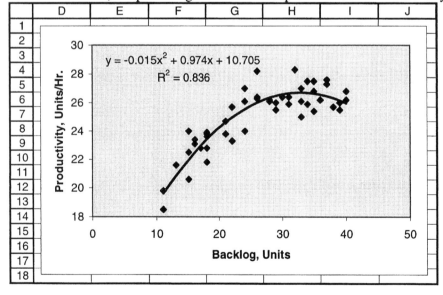

298

17.63 p/c/m The first-order polynomial estimation equation is shown below. The partial regression coefficient for room temperature is 1.06. For a given level of relative humidity, a one degree increase in the room temperature will tend to result in a 1.06 degree increase in the apparent temperature. The partial regression coefficient for relative humidity is 0.0925. For a given room temperature, a one percentage point increase in the relative humidity will tend to result in a 0.0925 degree increase in the apparent temperature. The equation explains 98.2% of the variation in apparent temperature.

```
Regression Analysis: ApparTemp versus RoomTemp, RelHumid

The regression equation is
ApparTemp = - 10.4 + 1.06 RoomTemp + 0.0925 RelHumid

Predictor        Coef     SE Coef          T        P
Constant     -10.4398      0.8599     -12.14    0.000
RoomTemp      1.05896      0.01264      83.80    0.000
RelHumid     0.092500     0.001842      50.21    0.000

S = 0.7728     R-Sq = 98.2%     R-Sq(adj) = 98.2%

Analysis of Variance
Source           DF          SS         MS        F        P
Regression        2      5699.9     2850.0  4771.91    0.000
Residual Error  173       103.3        0.6
Total           175      5803.2
```

The first-order polynomial estimation equation with interaction term is shown below. By including the interaction term, we have increased the explanatory power of the equation from $R^2 = 0.982$ to $R^2 = 0.994$, or from 98.2% to 99.4% of the variation in apparent temperature.

```
Regression Analysis: ApparTemp versus RoomTemp, RelHumid, RTemp*RelHumid

The regression equation is
ApparTemp = 3.90 + 0.847 RoomTemp - 0.194 RelHumid + 0.00425 RTemp*RelHumid

Predictor        Coef     SE Coef          T        P
Constant       3.8994      0.9444       4.13    0.000
RoomTemp      0.84652      0.01396      60.65    0.000
RelHumid     -0.19428      0.01596     -12.17    0.000
RTemp*Re    0.0042487   0.0002359      18.01    0.000

S = 0.4563     R-Sq = 99.4%     R-Sq(adj) = 99.4%

Analysis of Variance
Source           DF          SS         MS        F        P
Regression        3      5767.4     1922.5  9233.80    0.000
Residual Error  172        35.8        0.2
Total           175      5803.2
```

17.65 p/c/m

The first-order polynomial estimation equation is shown below. The partial regression coefficient for backlog size is 0.211. For a given gender, an increase of one unit in the backlog size will tend to result in an increase of 0.211 units/hour in productivity. The partial regression coefficient for being a female is 0.577. For a given backlog size, a female will tend to have a productivity that is 0.577 units/hour higher. The R^2 value is 0.676, and the equation explains 67.6% of the variation in productivity. To three decimal places, the p-value for the overall relationship is 0.000. This is less than the 0.02 level of significance, so at this level the model is significant.

```
Regression Analysis: productivity versus backlogsize, female

The regression equation is productivity = 19.1 + 0.211 backlogsize + 0.577 female

Predictor       Coef      SE Coef        T        P
Constant     19.0897       0.6153    31.02    0.000
backlogs     0.21108      0.02197     9.61    0.000
female        0.5774       0.3863     1.49    0.142

S = 1.360      R-Sq = 67.6%     R-Sq(adj) = 66.2%

Analysis of Variance
Source           DF           SS          MS        F        P
Regression        2      181.440      90.720    49.08    0.000
Residual Error   47       86.882       1.849
Total            49      268.322
```

17.67 p/c/m Using Minitab, we transform the data so that we have the corresponding logarithm to the base 10, then perform a multiple regression using the transformed data. The printout is shown below.

```
Regression Analysis: LogAppar versus LogTemp, LogRelHum

The regression equation is LogAppar = - 0.280 + 1.10 LogTemp + 0.0545 LogRelHum

160 cases used 16 cases contain missing values

Predictor       Coef      SE Coef        T        P
Constant    -0.28048      0.02739   -10.24    0.000
LogTemp      1.09898      0.01492    73.66    0.000
LogRelHu    0.054483     0.001470    37.06    0.000

S = 0.005616    R-Sq = 97.7%     R-Sq(adj) = 97.7%

Analysis of Variance
Source           DF           SS          MS          F        P
Regression        2      0.21442     0.10721    3399.67    0.000
Residual Error  157      0.00495     0.00003
Total           159      0.21937
```

The equation and the result when it is converted back to multiplicative form are shown below:

Logarithmic form: log Appar = -0.28048 + 1.09898*log Temp + 0.054483*log RelHum
Multiplicative form: Appar = $0.524228*Temp^{1.09898}*RelHum^{0.054483}$

The estimated apparent temperature for a room in which the actual temperature is 72 degrees Fahrenheit and the relative humidity is 50 percent would be $0.524228*70^{1.09898}*50^{0.054483} = 71.3$ degrees.

Note: For data cases where the relative humidity is 0, the logarithm is undefined, since 10 would have to be raised to the minus infinity power to produce a result of 0. Minitab expresses this undefined logarithm as "*" (missing data) and, when the regression is performed, it is automatically based on the other 160 data cases, as shown in the printout above. On the other hand, Excel's attempt to determine the logarithm of 0 leads to "#NUM!" appearing in each of the target cells. When using Excel to do this analysis, it is advisable to first delete the 16 cases where the relative humidity is 0.

17.69 p/c/m The Minitab correlation table and the stepwise regression printout are shown below. The correlations with operating cost per hour have also been included in the table. There are strong correlations between the pairs of independent variables, indicating the presence of multicollinearity.

Correlations: OpCost/Hr., Seats, Speed, Range, Gal./Hr.

```
          OpCost/H   Seats    Speed    Range
Seats      0.928
           0.000

Speed      0.810     0.834
           0.000     0.000

Range      0.806     0.801    0.857
           0.000     0.000    0.000

Gal./Hr.   0.936     0.961    0.832    0.838
           0.000     0.000    0.000    0.000

Cell Contents: Pearson correlation
               P-Value
```

Stepwise Regression: OpCost/Hr. versus Seats, Speed, Range, Gal./Hr.

```
  Alpha-to-Enter: 0.05  Alpha-to-Remove: 0.1

Response is OpCost/H on  4 predictors, with N =    34

     Step           1
Constant         697.8

Gal./Hr.          1.80
T-Value          15.10
P-Value          0.000

S                  665
R-Sq             87.69
R-Sq(adj)        87.31
C-p                2.2
```

Gal./Hr. is the only variable introduced. The estimation equation is simply OpCost/Hr. = 697.8 + 1.80*Gal./Hr., and it explains 87.69% of the variation in operating cost.

17.71 p/c/m

a. The Minitab correlation matrix is shown below. Multicollinearity might be a problem if all of the independent variables are used to estimate the final exam score. Test 1 is highly correlated to all of the other tests. In a stepwise regression, test 1 will be the first variable introduced since it has the highest correlation with the final exam score.

```
Correlations: FinalExam, Test1, Test2, Test3, Test4

        FinalExa    Test1    Test2    Test3
Test1     0.926
          0.000

Test2     0.884    0.875
          0.000    0.000

Test3     0.854    0.809    0.659
          0.000    0.000    0.006

Test4     0.851    0.915    0.832    0.695
          0.000    0.000    0.000    0.003

Cell Contents: Pearson correlation
               P-Value
```

b. The Minitab stepwise regression is shown below.

```
Stepwise Regression: FinalExam versus Test1, Test2, Test3, Test4

  Alpha-to-Enter: 0.05  Alpha-to-Remove: 0.1

  Response is FinalExa on  4 predictors, with N =   16

       Step        1
  Constant     14.79

  Test1        0.885
  T-Value       9.15
  P-Value      0.000

  S            4.13
  R-Sq        85.68
  R-Sq(adj)   84.65
  C-p          8.8
```

c. Test1 is the first variable brought into the regression, and no other variables are included. The regression equation is FinalExam = 14.79 + 0.885*Test1. This regression explains 85.68% of the variation in the final exam score.

d. In the conventional simple regression analysis using only Test1, the output includes the standard deviation for the regression coefficient and the ANOVA table. The printout is shown below.

```
Regression Analysis: FinalExam versus Test1

The regression equation is
FinalExam = 14.8 + 0.885 Test1

Predictor       Coef     SE Coef        T       P
Constant      14.792       7.437     1.99   0.067
Test1        0.88533     0.09674     9.15   0.000

S = 4.131      R-Sq = 85.7%    R-Sq(adj) = 84.7%

Analysis of Variance
Source           DF          SS        MS       F       P
Regression        1      1429.5    1429.5   83.75   0.000
Residual Error   14       239.0      17.1
Total            15      1668.4
```

CHAPTER 18
MODELS FOR TIME SERIES AND FORECASTING

SECTION EXERCISES

18.1 d/p/e The four components of a time series are: (1) trend, an overall upward or downward tendency, (2) cycle, a periodic fluctuation that repeats over time, (3) seasonality, a periodic fluctuation that repeats over a period of a year or less, and (4) irregularity, random fluctuations resulting from chance events.

18.3 p/a/e The multiplicative model takes the following form: y = T*C*S*I. Therefore, the quantity of water consumed during the month is y = 400,000*1.20*0.70*1.10 = 369,600 gallons.

18.5 p/a/m The Minitab printout and plot are shown below.

```
Regression Analysis: Earnings versus YearCode

The regression equation is Earnings = 12.2 + 0.514 YearCode

Predictor        Coef     SE Coef         T         P
Constant      12.2400      0.0435    281.55     0.000
YearCode      0.51400     0.01587     32.38     0.001

S = 0.03550     R-Sq = 99.8%     R-Sq(adj) = 99.7%

Analysis of Variance
Source           DF          SS          MS         F         P
Regression        1      1.3210      1.3210   1048.40     0.001
Residual Error    2      0.0025      0.0013
Total             3      1.3235

Predicted Values for New Observations
New Obs    Fit     SE Fit          95.0% CI                95.0% PI
1       17.3800    0.1204   ( 16.8621, 17.8979)   ( 16.8400, 17.9200) XX
X   denotes a row with X values away from the center
XX denotes a row with very extreme X values

Values of Predictors for New Observations
New Obs  YearCode
1            10.0
```

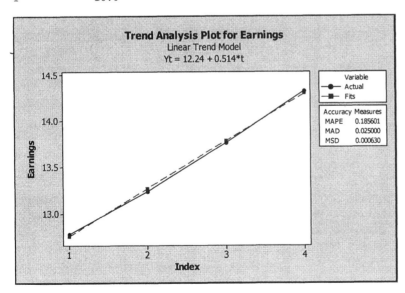

With x = the year code and x = 1 for 1998, the trend equation is Earnings = 12.24 + 0.514x. For 2007, x = 10 and the estimated wage for 2007 is 12.24 + 0.514(10) = $17.38. Also see the "Fit" portion of the printout. The Excel counterpart is shown below. Recall that Excel reverses the positions of the constant and the coefficient term compared to the more common format used in the textbook and by Minitab.

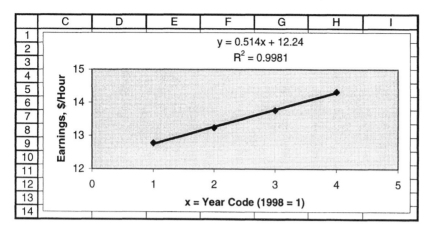

18.7 p/c/d

a. A portion of the Minitab printout is shown below, along with the plot. In this time series, 1990 is period 1 and 2007 is period 18. With x = the year code, the trend equation is Subs = -22.7782 + 11.0724x and the estimate for 2007 is -22.7782 + 11.0724(18) = 176.525 million.

Trend Analysis

Fitted Trend Equation

$Yt = -22.7782 + 11.0724*t$

Row	Period	Forecast
1	13	121.163
2	14	132.236
3	15	143.308
4	16	154.380
5	17	165.453
6	18	176.525

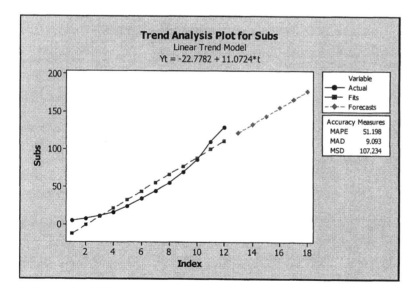

b. A portion of the Minitab printout is shown below, along with the plot. With x = the year code, the trend equation is Subs = 6.76295 - 1.58807x + 0.973884x^2. The estimate for 2007 can be calculated as 6.76295 - 1.58807(18) + 0.973884(18)^2 = 293.716 million subscribers.

Trend Analysis

Fitted Trend Equation

Yt = 6.76295 - 1.58807*t + 0.973884*t**2

Row	Period	Forecast
1	13	150.704
2	14	175.411
3	15	202.066
4	16	230.668
5	17	261.218
6	18	293.716

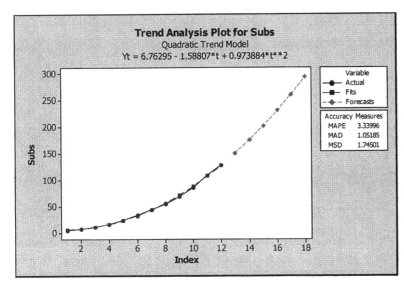

c. Based on the graphs shown in parts a and b, as well as the multiple plot shown below, the quadratic equation (part b) is by far the better fit to the data.

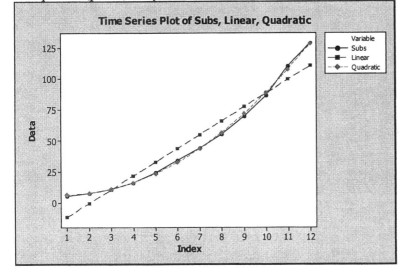

The Excel counterparts to the equations in parts a and b are shown below.

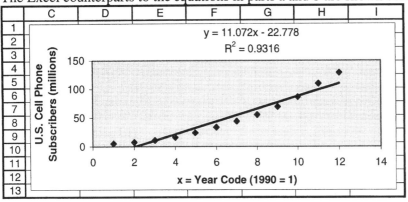

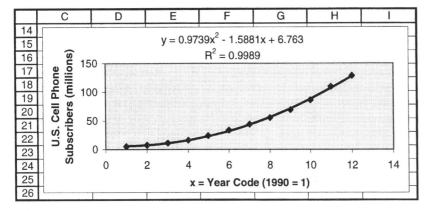

18.9 p/a/m

a. Using Excel to generate and graph the 3-year moving average.

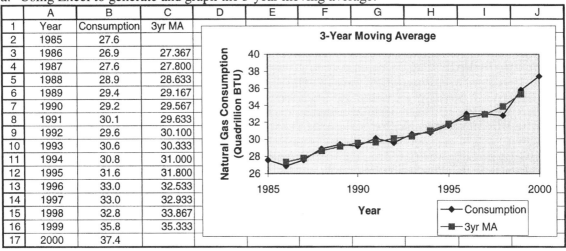

	A	B	C
1	Year	Consumption	3yr MA
2	1985	27.6	
3	1986	26.9	27.367
4	1987	27.6	27.800
5	1988	28.9	28.633
6	1989	29.4	29.167
7	1990	29.2	29.567
8	1991	30.1	29.633
9	1992	29.6	30.100
10	1993	30.6	30.333
11	1994	30.8	31.000
12	1995	31.6	31.800
13	1996	33.0	32.533
14	1997	33.0	32.933
15	1998	32.8	33.867
16	1999	35.8	35.333
17	2000	37.4	

b. Using Excel to generate and graph the 5-year moving average. The moving average is slightly "smoother" when N = 5 because a larger base provides greater damping.

	A	B	C	D	E	F	G	H	I	J
1	Year	Consumption	5yr MA							
2	1985	27.6								
3	1986	26.9								
4	1987	27.6	28.08							
5	1988	28.9	28.40							
6	1989	29.4	29.04							
7	1990	29.2	29.44							
8	1991	30.1	29.78							
9	1992	29.6	30.06							
10	1993	30.6	30.54							
11	1994	30.8	31.12							
12	1995	31.6	31.80							
13	1996	33.0	32.24							
14	1997	33.0	33.24							
15	1998	32.8	34.40							
16	1999	35.8								
17	2000	37.4								

5-Year Moving Average

Natural Gas Consumption (Quadrillion BTU) vs Year

— Consumption
— 5yr MA

18.11 p/a/m The underlying data, along with the 3-month and 5-month moving averages, are shown in the Excel printout below. The overall trend appears to be very slightly downward.

	A	B	C	D	E	F	G	H	I	J	K	L
1	Month	DegDays	3-mo.MA	5-mo.MA								
2	1	852										
3	2	644	699.667									
4	3	603	530.667	519.200								
5	4	345	366.667	358.000								
6	5	152	181.000	232.000								
7	6	46	70.667	116.200								
8	7	14	28.000	62.000								
9	8	24	37.333	91.800								
10	9	74	133.000	195.400								
11	10	301	313.000	357.000								
12	11	564	562.333	524.200								
13	12	822	748.667	674.800								
14	13	860	836.333	787.400								
15	14	827	850.333	748.200								
16	15	864	686.333	609.400								
17	16	368	453.333	445.000								
18	17	128	178.000	281.800								
19	18	38	59.000	110.800								
20	19	11	19.333	55.000								
21	20	9	36.333	89.800								
22	21	89	133.333	198.200								
23	22	302	323.667	360.800								
24	23	580	568.667	565.200								
25	24	824	811.667	710.000								
26	25	1031	889.333	768.400								
27	26	813	812.667	711.000								
28	27	594	566.667	581.000								
29	28	293	353.667	379.000								
30	29	174	162.667	217.600								
31	30	21	67.000	102.000								
32	31	6	14.333	56.400								
33	32	16	29.000	75.200								
34	33	65	116.333	166.800								
35	34	268	270.667	310.200								
36	35	479	490.000	472.200								
37	36	723	676.000	612.600								
38	37	826	772.000	667.600								
39	38	767	712.000	645.000								
40	39	543	558.667	534.000								
41	40	366	359.000	378.000								
42	41	168	193.333	227.200								
43	42	46	75.667	120.800								
44	43	13	23.333	67.000								
45	44	11	40.333	81.200								
46	45	97	115.667	191.600								
47	46	239	311.333	361.400								
48	47	598	566.333									
49	48	862										

3-Month & 5-Month Moving Averages

18.13 p/a/m

a. The overall trend is upward. b. The time series and the two centered moving averages are below.

	A	B	C	D	E	F	G	H	I	J	K	L
1	Month	Shipped	3-mo.MA	5-mo.MA								
2	1	4.5										
3	2	4.7	4.633									
4	3	4.7	4.800	4.780								
5	4	5.0	4.900	4.760								
6	5	5.0	4.800	4.760								
7	6	4.4	4.700	4.800								
8	7	4.7	4.667	4.720								
9	8	4.9	4.733	4.800								
10	9	4.6	4.967	4.880								
11	10	5.4	4.933	4.980								
12	11	4.8	5.100	5.080								
13	12	5.2	5.300	5.140								
14	13	5.4	5.167	5.120								
15	14	4.9	5.200	5.280								
16	15	5.3	5.267	5.320								
17	16	5.6	5.433	5.320								
18	17	5.4	5.467	5.440								
19	18	5.4	5.433	5.460								
20	19	5.5	5.433	5.480								
21	20	5.4	5.533	5.660								
22	21	5.7	5.800	5.640								
23	22	6.3	5.767	5.800								
24	23	5.3	5.967	5.880								
25	24	6.3	5.800	5.900								
26	25	5.8	5.967	5.960								
27	26	5.8	6.067	6.200								
28	27	6.6	6.300	6.180								
29	28	6.5	6.433	6.360								
30	29	6.2	6.467	6.540								
31	30	6.7	6.533	6.560								
32	31	6.7	6.700	6.640								
33	32	6.7	6.767	6.880								
34	33	6.9	7.000	6.860								
35	34	7.4	6.967	7.000								
36	35	6.6	7.133									
37	36	7.4										

3-Month & 5-Month Moving Averages

18.15 p/a/m

a. The overall trend is downward.

b. The time series and the two centered moving averages are shown below.

	A	B	C	D	E	F	G	H	I	J	K	L
1	Year	Aircraft	3-yr.MA	5-yr.MA								
2	1	282										
3	2	326	332.333									
4	3	389	324.667	279.600								
5	4	259	263.333	257.000								
6	5	142	190.000	220.800								
7	6	169	152.000	167.400								
8	7	145	145.333	140.000								
9	8	122	129.667	143.000								
10	9	122	133.667	140.600								
11	10	157	145.333	145.200								
12	11	157	160.667	158.000								
13	12	168	170.333	167.800								
14	13	186	175.000	176.000								
15	14	171	185.000	189.000								
16	15	198	197.000	204.600								
17	16	222	222.000	215.600								
18	17	246	236.333	251.000								
19	18	241	278.333	294.400								
20	19	348	334.667									
21	20	415										
22												
23												

3-Year & 5-Year Moving Averages

18.17 p/a/m Using Minitab and smoothing constant $\alpha = 0.4$. For the smoothed values, refer to the "Smooth" column of this partial printout.

Single Exponential Smoothing

Smoothing Constant Alpha: 0.4

Row	Time	Pieces	Smooth	Predict	Error
1	1	4.5	4.50000	4.50000	0.000000
2	2	4.7	4.58000	4.50000	0.200000
3	3	4.7	4.62800	4.58000	0.120000
4	4	5.0	4.77680	4.62800	0.372000
5	5	5.0	4.86608	4.77680	0.223200
6	6	4.4	4.67965	4.86608	-0.466080
7	7	4.7	4.68779	4.67965	0.020352
8	8	4.9	4.77267	4.68779	0.212211
9	9	4.6	4.70360	4.77267	-0.172673
10	10	5.4	4.98216	4.70360	0.696396
11	11	4.8	4.90930	4.98216	-0.182162
12	12	5.2	5.02558	4.90930	0.290703
13	13	5.4	5.17535	5.02558	0.374422
14	14	4.9	5.06521	5.17535	-0.275347
15	15	5.3	5.15912	5.06521	0.234792
16	16	5.6	5.33547	5.15912	0.440875
17	17	5.4	5.36128	5.33547	0.064525
18	18	5.4	5.37677	5.36128	0.038715
19	19	5.5	5.42606	5.37677	0.123229
20	20	5.4	5.41564	5.42606	-0.026063
21	21	5.7	5.52938	5.41564	0.284362
22	22	6.3	5.83763	5.52938	0.770617
23	23	5.3	5.62258	5.83763	-0.537630
24	24	6.3	5.89355	5.62258	0.677422
25	25	5.8	5.85613	5.89355	-0.093547
26	26	5.8	5.83368	5.85613	-0.056128
27	27	6.6	6.14021	5.83368	0.766323
28	28	6.5	6.28412	6.14021	0.359794
29	29	6.2	6.25047	6.28412	-0.084124
30	30	6.7	6.43028	6.25047	0.449526
31	31	6.7	6.53817	6.43028	0.269715
32	32	6.7	6.60290	6.53817	0.161829
33	33	6.9	6.72174	6.60290	0.297098
34	34	7.4	6.99304	6.72174	0.678259
35	35	6.6	6.83583	6.99304	-0.393045
36	36	7.4	7.06150	6.83583	0.564173

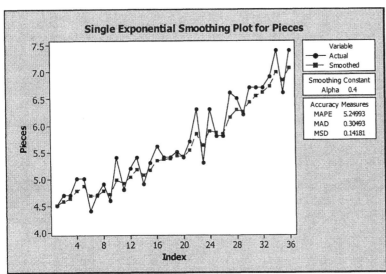

Using Minitab and smoothing constant α = 0.7. For the smoothed values, refer to the "Smooth" column of this partial printout.

Single Exponential Smoothing

Smoothing Constant Alpha: 0.7

Row	Time	Pieces	Smooth	Predict	Error
1	1	4.5	4.50000	4.50000	0.000000
2	2	4.7	4.64000	4.50000	0.200000
3	3	4.7	4.68200	4.64000	0.060000
4	4	5.0	4.90460	4.68200	0.318000
5	5	5.0	4.97138	4.90460	0.095400
6	6	4.4	4.57141	4.97138	-0.571380
7	7	4.7	4.66142	4.57141	0.128586
8	8	4.9	4.82843	4.66142	0.238576
9	9	4.6	4.66853	4.82843	-0.228427
10	10	5.4	5.18056	4.66853	0.731472
11	11	4.8	4.91417	5.18056	-0.380558
12	12	5.2	5.11425	4.91417	0.285832
13	13	5.4	5.31428	5.11425	0.285750
14	14	4.9	5.02428	5.31428	-0.414275
15	15	5.3	5.21728	5.02428	0.275717
16	16	5.6	5.48519	5.21728	0.382715
17	17	5.4	5.42556	5.48519	-0.085185
18	18	5.4	5.40767	5.42556	-0.025556
19	19	5.5	5.47230	5.40767	0.092333
20	20	5.4	5.42169	5.47230	-0.072300
21	21	5.7	5.61651	5.42169	0.278310
22	22	6.3	6.09495	5.61651	0.683493
23	23	5.3	5.53849	6.09495	-0.794952
24	24	6.3	6.07155	5.53849	0.761514
25	25	5.8	5.88146	6.07155	-0.271546
26	26	5.8	5.82444	5.88146	-0.081464
27	27	6.6	6.36733	5.82444	0.775561
28	28	6.5	6.46020	6.36733	0.132668
29	29	6.2	6.27806	6.46020	-0.260200
30	30	6.7	6.57342	6.27806	0.421940
31	31	6.7	6.66203	6.57342	0.126582
32	32	6.7	6.68861	6.66203	0.037975
33	33	6.9	6.83658	6.68861	0.211392
34	34	7.4	7.23097	6.83658	0.563418
35	35	6.6	6.78929	7.23097	-0.630975
36	36	7.4	7.21679	6.78929	0.610708

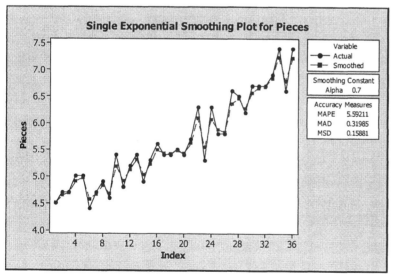

The smaller constant provides a smoother curve since it is weighted more heavily towards the previous smoothed value. The larger constant "catches up" with changes in direction in the original data series quicker because it gives more weight to the actual data series.

18.19 p/a/m

a. Partial printout and plot using Minitab with smoothing constant $\alpha = 0.6$.

```
Single Exponential Smoothing

Smoothing Constant Alpha: 0.6
```

Row	Time	Shipments	Smooth	Predict	Error
1	1	282	282.000	282.000	0.000
2	2	326	308.400	282.000	44.000
3	3	389	356.760	308.400	80.600
4	4	259	298.104	356.760	-97.760
5	5	142	204.442	298.104	-156.104
6	6	169	183.177	204.442	-35.442
7	7	145	160.271	183.177	-38.177
8	8	122	137.308	160.271	-38.271
9	9	122	128.123	137.308	-15.308
10	10	157	145.449	128.123	28.877
11	11	157	152.380	145.449	11.551
12	12	168	161.752	152.380	15.620
13	13	186	176.301	161.752	24.248
14	14	171	173.120	176.301	-5.301
15	15	198	188.048	173.120	24.880
16	16	222	208.419	188.048	33.952
17	17	246	230.968	208.419	37.581
18	18	241	236.987	230.968	10.032
19	19	348	303.595	236.987	111.013
20	20	415	370.438	303.595	111.405

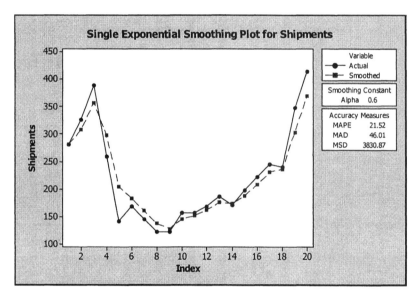

b. An exponentially-smoothed curve with $\alpha = 0.0$ simply results in a horizontal line that provides no useful information. An exponentially-smoothed curve with $\alpha = 1.0$ will exactly match the original time series; such a smoothing constant does not "smooth" the time series, it merely copies it.

18.21 p/a/m If the bookings were evenly distributed throughout the year, each quarter would have 25% of the bookings. However, the second quarter is 120% as strong as the "average" quarter, so it will have 1.20 * 25.0 = 30.0% of the year's bookings.

18.23 p/a/m For this seasonal index problem involving quarterly data, we will use Excel worksheet template tmquarts.xls, on the disk supplied with the text. The steps for its use are described within the worksheet template. Part a of the exercise is represented by columns D, E, and F; part b by columns G and H; and part c by column I. To get the chart shown here, it was necessary to format the axes for minimum and maximum values corresponding to the data.

	A	B	C	D	E	F	G	H	I
	Year	Qtr	Visitors	ctr mov tot	ctr mov av	pct of mv av	unadj index	adj index	Deseasoned
5									
6	1999	1	155				74.0789	74.7202	207.441
7		2	231				103.0857	103.9781	222.162
8		3	270	774.500	193.625	139.4448	122.6992	123.7615	218.162
9		4	105	800.000	200.000	52.5000	96.7030	97.5402	107.648
10	2000	5	182	834.500	208.625	87.2379	74.0789	74.7202	243.575
11		6	255	951.500	237.875	107.1992	103.0857	103.9781	245.244
12		7	315	1035.000	258.750	121.7391	122.6992	123.7615	254.522
13		8	294	1021.500	255.375	115.1248	96.7030	97.5402	301.414
14	2001	9	160	1001.500	250.375	63.9041	74.0789	74.7202	214.132
15		10	250	985.500	246.375	101.4713	103.0857	103.9781	240.435
16		11	280	1012.000	253.000	110.6719	122.6992	123.7615	226.242
17		12	297	1067.000	266.750	111.3402	96.7030	97.5402	304.490
18	2002	13	210	1139.500	284.875	73.7165	74.0789	74.7202	281.048
19		14	310	1201.000	300.250	103.2473	103.0857	103.9781	298.140
20		15	365	1227.500	306.875	118.9409	122.6992	123.7615	294.922
21		16	335	1242.500	310.625	107.8471	96.7030	97.5402	343.448
22	2003	17	225	1259.500	314.875	71.4569	74.0789	74.7202	301.123
23		18	325	1294.500	323.625	100.4249	103.0857	103.9781	312.566
24		19	384				122.6992	123.7615	310.274
25		20	386				96.7030	97.5402	395.734

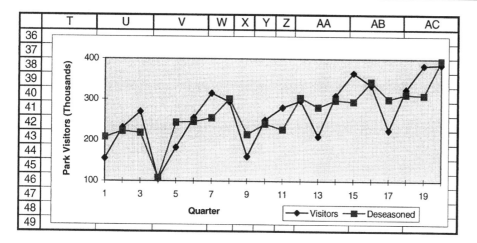

18.25 p/a/m For this seasonal index problem involving monthly data, we will use Excel worksheet template tmmonths.xls, on the disk supplied with the text. The steps for its use are described in the worksheet template. To get the chart shown here, it was necessary to format the axes for minimum and maximum values corresponding to the data.

	A	B	C	D	E	F	G	H	I
5		Month	Volume	ctr mov tot	ctr mov avg	% of mov av	unadj index	adj index	Deseasoned
6	J 1985	1	4.5				100.1260	100.4968	4.48
7	F	2	4.7				93.9862	94.3342	4.98
8	M	3	4.7				102.7778	103.1583	4.56
9	A	4	5.0				103.2140	103.5962	4.83
10	M	5	5.0				97.7004	98.0622	5.10
11	J	6	4.4				100.1757	100.5467	4.38
12	J	7	4.7	58.35	4.863	96.6581	97.8043	98.1665	4.79
13	A	8	4.9	58.90	4.908	99.8302	98.0220	98.3850	4.98
14	S	9	4.6	59.30	4.942	93.0860	96.5065	96.8638	4.75
15	O	10	5.4	59.90	4.992	108.1803	108.4395	108.8411	4.96
16	N	11	4.8	60.40	5.033	95.3642	92.8526	93.1964	5.15
17	D	12	5.2	61.10	5.092	102.1277	103.9680	104.3529	4.98
18	J 1986	13	5.4	62.00	5.167	104.5161	100.1260	100.4968	5.37
19	F	14	4.9	62.65	5.221	93.8547	93.9862	94.3342	5.19
20	M	15	5.3	63.45	5.288	100.2364	102.7778	103.1583	5.14
21	A	16	5.6	64.45	5.371	104.2669	103.2140	103.5962	5.41
22	M	17	5.4	65.15	5.429	99.4628	97.7004	98.0622	5.51
23	J	18	5.4	65.95	5.496	98.2563	100.1757	100.5467	5.37
24	J	19	5.5	66.70	5.558	98.9505	97.8043	98.1665	5.60
25	A	20	5.4	67.35	5.613	96.2138	98.0220	98.3850	5.49
26	S	21	5.7	68.45	5.704	99.9270	96.5065	96.8638	5.88
27	O	22	6.3	69.55	5.796	108.6988	108.4395	108.8411	5.79
28	N	23	5.3	70.40	5.867	90.3409	92.8526	93.1964	5.69
29	D	24	6.3	71.45	5.954	105.8083	103.9680	104.3529	6.04
30	J 1987	25	5.8	72.70	6.058	95.7359	100.1260	100.4968	5.77
31	F	26	5.8	73.95	6.163	94.1176	93.9862	94.3342	6.15
32	M	27	6.6	75.20	6.267	105.3191	102.7778	103.1583	6.40
33	A	28	6.5	76.35	6.363	102.1611	103.2140	103.5962	6.27
34	M	29	6.2	77.55	6.463	95.9381	97.7004	98.0622	6.32
35	J	30	6.7	78.75	6.563	102.0952	100.1757	100.5467	6.66
36	J	31	6.7				97.8043	98.1665	6.83
37	A	32	6.7				98.0220	98.3850	6.81
38	S	33	6.9				96.5065	96.8638	7.12
39	O	34	7.4				108.4395	108.8411	6.80
40	N	35	6.6				92.8526	93.1964	7.08
41	D	36	7.4				103.9680	104.3529	7.09

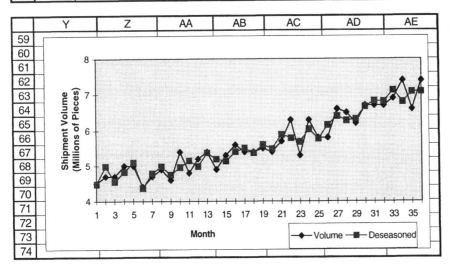

18.27 c/a/e For 2008, x = 15, so the forecasted sales for 2008 are 120 + 4.8(15) = $192.0 thousand. For 2010, x = 17, so the forecasted sales for 2010 are 120 + 4.8(17) = $201.6 thousand.

18.29 p/a/m For 2006, x = 7, so the forecasted number of rentals is $\hat{y} = 450 + 20(7) + 7.2(7)^2 = 942.8$. For 2008, x = 9, so the forecasted number of rentals is $\hat{y} = 450 + 20(9) + 7.2(9)^2 = 1213.2$.

18.31 p/a/m Using Minitab, the calculation summary is shown in the partial printout below. Refer to the "Predict" and "Forecast" columns. The years are coded so that 1994 is period 1.

```
Single Exponential Smoothing

Smoothing Constant Alpha: 0.7

Row   Time   Deficit    Smooth    Predict      Error
 1      1      14.0    14.0000    14.0000     0.0000
 2      2      18.1    16.8700    14.0000     4.1000
 3      3      21.7    20.2510    16.8700     4.8300
 4      4      15.5    16.9253    20.2510    -4.7510
 5      5      16.7    16.7676    16.9253    -0.2253
 6      6      32.1    27.5003    16.7676    15.3324
 7      7      51.9    44.5801    27.5003    24.3997
 8      8      52.8    50.3340    44.5801     8.2199

Row   Period  Forecast     Lower      Upper
 1      9      50.3340    31.3899    69.2781
```

Calculations are: (next period forecast) = 0.7(this period actual) + 0.3(this period forecast). Thus, the forecast for period 9 (2002) would be 0.7(actual in period 8) + 0.3(forecast for period 8), or 0.7(52.8) + 0.3(44.5801) = $50.334 billion.

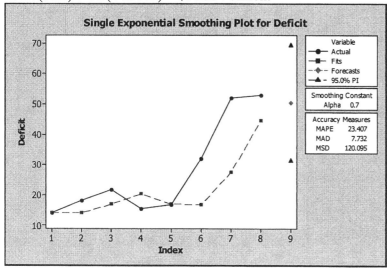

315

In using Excel, we would specify 0.3 as the "damping factor" (i.e., 1 minus our smoothing constant, $\alpha = 0.7$). Excel places the forecast for a period next to the preceding period, so we must shift the forecasts downward by one period. Also, we need to manually fill in the initial forecast for 1994 (14.000 in cell C2), click on the lower right corner of cell C9, then drag and autofill to cell C10 to obtain the forecast for 2002. The result, using ChartWizard, is shown below.

	A	B	C	D	E	F	G	H	I	J
1	Year	Deficit	Forecast							
2	1994	14.0	14.000							
3	1995	18.1	14.000							
4	1996	21.7	16.870							
5	1997	15.5	20.251							
6	1998	16.7	16.925							
7	1999	32.1	16.768							
8	2000	51.9	27.500							
9	2001	52.8	44.580							
10	2002		50.334							
11										
12										
13										
14										

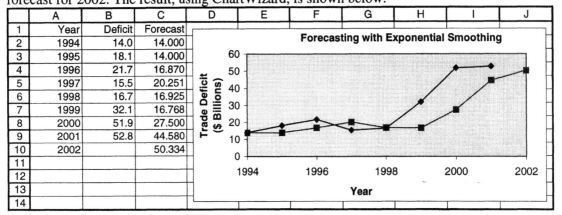

18.33 p/a/m In the partial Minitab printout below, refer to the "Predict" and "Forecast" columns.

Single Exponential Smoothing

Smoothing Constant Alpha: 0.6

```
Row    Time    NatGas    Smooth    Predict      Error
  1       1      27.6    27.6000    27.6000    0.00000
  2       2      26.9    27.1800    27.6000   -0.70000
  3       3      27.6    27.4320    27.1800    0.42000
  4       4      28.9    28.3128    27.4320    1.46800
  5       5      29.4    28.9651    28.3128    1.08720
  6       6      29.2    29.1060    28.9651    0.23488
  7       7      30.1    29.7024    29.1060    0.99395
  8       8      29.6    29.6410    29.7024   -0.10242
  9       9      30.6    30.2164    29.6410    0.95903
 10      10      30.8    30.5666    30.2164    0.58361
 11      11      31.6    31.1866    30.5666    1.03345
 12      12      33.0    32.2746    31.1866    1.81338
 13      13      33.0    32.7099    32.2746    0.72535
 14      14      32.8    32.7639    32.7099    0.09014
 15      15      35.8    34.5856    32.7639    3.03606
 16      16      37.4    36.2742    34.5856    2.81442

Row    Period    Forecast      Lower      Upper
  1        17      36.2742    33.8148    38.7337
```

The forecast for period 17 (2001) will be 0.6(actual in period 16) + 0.4(forecast for period 16), or 0.6(37.4) + 0.4(34.5856) = 36.2742 quadrillion Btu.

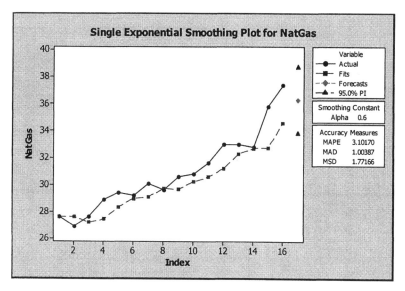

Single Exponential Smoothing Plot for NatGas

With Excel, we specify 0.4 as the "damping factor" (i.e., 1 - smoothing constant, α = 0.6). Excel places each forecast next to the preceding period, so we must shift the forecasts down by one period. Manually insert the initial forecast (25.600 in cell C2), click on the lower right corner of cell C17, then autofill to cell C18 to get the forecast for 2001. The result, using ChartWizard, is shown below.

	A	B	C	D	E	F	G	H	I
1	Year	Consumption	Forecast						
2	1985	27.6	27.600						
3	1986	26.9	27.600						
4	1987	27.6	27.180						
5	1988	28.9	27.432						
6	1989	29.4	28.313						
7	1990	29.2	28.965						
8	1991	30.1	29.106						
9	1992	29.6	29.702						
10	1993	30.6	29.641						
11	1994	30.8	30.216						
12	1995	31.6	30.567						
13	1996	33.0	31.187						
14	1997	33.0	32.275						
15	1998	32.8	32.710						
16	1999	35.8	32.764						
17	2000	37.4	34.586						
18	2001		36.274						

18.35 p/a/m The forecast for the second quarter is (560,000/4)(135/100) = 189,000 gallons.

18.37 d/p/e MAD considers the mean absolute error from the actual data; the direction of the error is not considered. In calculating MAD, the absolute values of the deviations are used. MSE considers the mean squared error from the actual data. In calculating MSE, the squared deviations from the actual data are used. The MSE approach penalizes large errors more since the errors are squared.

18.39 p/a/m Using Excel, the relevant calculation results are shown below.
Linear trend equation.

	A	B	C	D	E	F	G	H
1	year	x = yrcode	y = thefts	ycap = estimate	y - ycap	ly - ycapl		(y - ycap)sq
2	1992	1	6319	6263.50	55.50	55.50		3080.25
3	1993	2	7237	7201.50	35.50	35.50		1260.25
4	1994	3	7902	8139.50	-237.50	237.50		56406.25
5	1995	4	9224	9077.50	146.50	146.50		21462.25
6						475.00		82209.00
7					MAD =	118.75	MSE =	20552.25

Quadratic trend equation.

	A	B	C	D	E	F	G	H
1	year	x = yrcode	y = thefts	ycap = estimate	y - ycap	ly - ycapl		(y - ycap)sq
2	1992	1	6319	6364.50	-45.50	45.50		2070.25
3	1993	2	7237	7100.50	136.50	136.50		18632.25
4	1994	3	7902	8038.50	-136.50	136.50		18632.25
5	1995	4	9224	9178.50	45.50	45.50		2070.25
6						364.00		41405.00
7					MAD =	91.00	MSE =	10351.25

The quadratic equation is the better fit. This is true regardless of whether MAD or MSE is considered. In solving this exercise, we could also use Minitab's trend analysis capability -- the MAD and MSE values are automatically provided, and they will be slightly more accurate than the ones calculated here on the basis of the (rounded) equations given in the exercise.

18.41 p/c/m Referring to the Minitab plots shown in the solution to exercise 18.8, the MAD and MSE values are as shown below. Note that Minitab shows MSE as "MSD".
Linear trend equation: MAD = 5.2527   MSE = 41.3889
Quadratic trend equation: MAD = 5.0732   MSE = 40.1865
The MAD and MSE are also displayed in the standard Minitab Trend Analysis printout for a trend equation. These were not included in the partial printouts shown in the solution to exercise 18.8.
On the basis of either MAD or MSE, the quadratic trend equation is the better fit. These results could also be obtained by using the approach taken in exercises 18.38 and 18.39.

18.43 d/p/d The Durbin-Watson test is used to detect the presence of autocorrelation between adjacent residuals. If the test detects the presence of autocorrelation, there is the possibility in the given model that important variables may have been omitted. The inclusion of these variables can reduce the presence of autocorrelation and enhance the quality of confidence and prediction intervals.

18.45 p/a/m The null and alternative hypotheses are H_0: $\rho \leq 0$ and H_1: $\rho > 0$. Using Excel and the information provided, the relevant calculation results are shown below.

	A	B	C	D	E	F	G	H
1	t	(y)t	(ycap)t	(e)t	(e)t-sq	(e)t-1	(e)t - (e)t-1	[(e)t - (e)t-1]-sq
2	1	65.6	25.90	39.70	1576.09			
3	2	56.7	36.60	20.10	404.01	39.70	-19.60	384.16
4	3	79.5	47.30	32.20	1036.84	20.10	12.10	146.41
5	4	53.4	58.00	-4.60	21.16	32.20	-36.80	1354.24
6	5	59.7	68.70	-9.00	81.00	-4.60	-4.40	19.36
7	6	71.1	79.40	-8.30	68.89	-9.00	0.70	0.49
8	7	68.3	90.10	-21.80	475.24	-8.30	-13.50	182.25
9	8	79.1	100.80	-21.70	470.89	-21.80	0.10	0.01
10	9	104.2	111.50	-7.30	53.29	-21.70	14.40	207.36
11	10	104.2	122.20	-18.00	324.00	-7.30	-10.70	114.49
12	11	115.4	132.90	-17.50	306.25	-18.00	0.50	0.25
13	12	86.6	143.60	-57.00	3249.00	-17.50	-39.50	1560.25
14	13	141.3	154.30	-13.00	169.00	-57.00	44.00	1936.00
15	14	149.6	165.00	-15.40	237.16	-13.00	-2.40	5.76
16	15	177.1	175.70	1.40	1.96	-15.40	16.80	282.24
17	16	193.6	186.40	7.20	51.84	1.40	5.80	33.64
18	17	238.9	197.10	41.80	1747.24	7.20	34.60	1197.16
19	18	259.2	207.80	51.40	2641.96	41.80	9.60	92.16
20					12915.82			7516.23

The Durbin-Watson statistic is d = 7516.23/12,915.82 = 0.58.

In the Durbin-Watson table for $\alpha = 0.05$ for a directional test, the listed values for n = 18 observations and k = 1 independent variable are $d_L = 1.16$ and $d_U = 1.39$. We are testing for positive autocorrelation, so the listed value of interest is $d_L = 1.16$. Because d = 0.58 is less than $d_L = 1.16$, we conclude that positive autocorrelation exists and the linear model may not be appropriate for this series. (Refer to the "testing for positive autocorrelation" portion of Figure 18.7 in the text.) Minitab offers the Durbin-Watson statistic as an option in regression analysis, as shown below.

```
Regression Analysis: Yt versus t

The regression equation is Yt = 15.2 + 10.7 t

Predictor       Coef      SE Coef        T        P
Constant        15.20       13.97      1.09    0.293
t              10.701        1.291      8.29    0.000

S = 28.41       R-Sq = 81.1%      R-Sq(adj) = 79.9%

Analysis of Variance
Source          DF          SS          MS        F        P
Regression       1       55479       55479    68.73    0.000
Residual Error  16       12916         807
Total           17       68395

Durbin-Watson statistic = 0.58
```

We can also use Data Analysis Plus to examine the residuals and compute the Durbin-Watson statistic.

	A	B	C
1	Durbin-Watson Statistic		
2			
3	d = 0.5819		

18.47 p/a/m Given n = 16, k = 2, and d = 2.78:

a. The null and alternative hypotheses are H_0: $\rho = 0$ and H_1: $\rho \neq 0$.

From the table, with $\alpha = 0.05$, $d_L = 0.86$ and $d_U = 1.40$.

$4 - d_U = 4 - 1.40 = 2.6$ and $4 - d_L = 4 - 0.86 = 3.14$.

Thus, d = 2.78 is greater than $4 - d_U$, but d is less than $4 - d_L$.

Conclusion: The test is inconclusive.

b. The null and alternative hypotheses are H_0: $\rho \geq 0$ and H_1: $\rho < 0$.

From the table, with $\alpha = 0.01$, $d_L = 0.74$ and $d_U = 1.25$.

$4 - d_U = 4 - 1.25 = 2.75$ and $4 - d_L = 4 - 0.74 = 3.26$.

Thus, with d = 2.78, $(4 - d_U) < d < (4 - d_L)$.   Conclusion: the test is inconclusive.

18.49 p/c/m The Minitab printout is shown below.

```
Regression Analysis: Yt versus Yt-1

The regression equation is Yt = - 1.5 + 1.12 Yt-1

17 cases used 1 cases contain missing values

Predictor        Coef     SE Coef         T        P
Constant        -1.51       11.88     -0.13    0.901
Yt-1          1.11889     0.09859     11.35    0.000

S = 21.36      R-Sq = 89.6%      R-Sq(adj) = 88.9%

Analysis of Variance
Source           DF          SS        MS        F        P
Regression        1       58768     58768   128.79    0.000
Residual Error   15        6845       456
Total            16       65613

Predicted Values for New Observations
New Obs    Fit      SE Fit        95.0% CI              95.0% PI
1       288.51       15.74   ( 254.96,  322.05) ( 231.95,  345.06) X
X   denotes a row with X values away from the center

Values of Predictors for New Observations
New Obs      Yt-1
1             259
```

The first-order autoregressive forecasting equation is $\hat{y}_t = -1.51 + 1.119 y_{t-1}$.

The shipment value in period 18 was 259.2, and Minitab shows a forecast value of 288.51 for period 19. This can also be calculated as $-1.51 + 1.119(259.2) = 288.5$. (Note: Minitab carries more decimal places for the constant and coefficients than are shown in the printout, so Minitab calculations like these will tend to be slightly more accurate than the ones we are able to make.)

18.51 p/c/m The Minitab printout is shown below.

```
Regression Analysis: Yt versus Yt-1

The regression equation is Yt = 526 + 0.812 Yt-1

19 cases used 1 cases contain missing values

Predictor        Coef      SE Coef         T        P
Constant        525.6        171.8      3.06    0.007
Yt-1          0.81198      0.05127     15.84    0.000

S = 134.0      R-Sq = 93.7%     R-Sq(adj) = 93.3%

Analysis of Variance
Source          DF          SS        MS        F        P
Regression       1     4504124   4504124   250.78    0.000
Residual Error  17      305326     17960
Total           18     4809450

Predicted Values for New Observations
New Obs     Fit      SE Fit       95.0% CI            95.0% PI
1        2895.5        36.4   ( 2818.8,  2972.2) ( 2602.5,  3188.4)

Values of Predictors for New Observations
New Obs     Yt-1
1           2919
```

The first-order autoregressive forecasting equation is $\hat{y}_t = 525.6 + 0.812y_{t-1}$.

The carrying cost in period 20 was $2918.6 thousand and Minitab shows a forecast value of $2895.5 thousand for period 21. This can also be calculated as $525.6 + 0.812(2918.6) = $2895.5 thousand. In carrying out the regression, the residuals were stored into a separate column and the mean of their absolute values is MAD = $100.5 thousand.

18.53 d/p/e An index number is a percentage that expresses a measurement in a given period in terms of the corresponding measurement for a selected base period. The value of any index number for the base period is always 100.0.

18.55 p/a/d To determine which earner groups fared best and worst in terms of their percentage change in spending power during this period, we must compute ratios of both earnings and the CPI, then compare the ratios. Of course, one problem with these data is that we have only average hourly earnings; we do not know whether the number of hours worked per week increased, decreased, or remained fairly constant. By 2001, the CP1 was 2.149 times its size in 1980. In 2001, the average hourly wage in manufacturing was 2.041 times as large as its 1980 value. Corresponding multiples for mining, services, and retail trade were 1.925, 2.497, and 2.012, respectively. To examine changes in spending power in each sector, we can look at the ratio of the change in wages to the change in the CPI:

For manufacturing, the ratio is (2.041/2.149)*100 = 94.97. Therefore, workers in this industry experienced a 5.03% decline in spending power between 1980 and 2001.

For mining, the ratio is (1.925/2.149)*100 = 89.58. So, workers in this industry experienced a 10.42% decline in spending power between 1980 and 2001.

For services, the ratio is (2.497/2.149)*100 = 116.19. Therefore, workers in the services industry experienced a 16.19% increase in spending power between 1980 and 2001.

For retail trade, the ratio is (2.012/2.149)*100 = 93.62. Therefore, workers in this sector experienced a 6.38% decline in spending power between 1980 and 2001.

The wage-earners who fared best were in the services industry; those who faried worst were in the mining industry.

18.57 p/a/m To get the executive's 1990 salary into 2001 equivalent dollars, we must take the ratio of the 2001 CPI to the 1990 CPI and multiply this by her 1990 salary. Therefore, the 1990 salary in its 2001 equivalent dollars is: (177.1/130.7)*80,000 = $108,401.

CHAPTER EXERCISES

18.59 p/c/m With x = 1 for 1994, the linear trend equation is Military = 1.6086 - 0.0379x. For 2010, x = 17, and the forecast for 2010 can be calculated as 1.6086 - 0.0379(17) = 0.964 million.

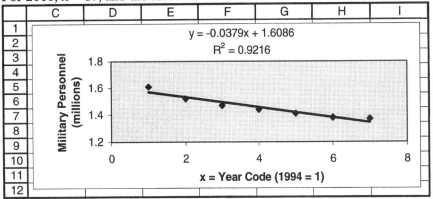

18.61 p/c/m The Minitab quadratic trend equation and plot are shown below. For 1985 coded as x = 1, the quadratic trend equation is Restaurants = -1408.05 + 1206.50x - 9.92905x^2. For 2010, x = 26 and the forecast for 2010 is -1408.05 + 1206.50(26) - 9.92905$(26)^2$ = 23,249 Subway restaurants.

```
Trend Analysis for Restaurants

Data      Restaurants
Length    18
NMissing  0

Fitted Trend Equation

Yt = -1408.05 + 1206.50*t - 9.92905*t**2

Accuracy Measures
MAPE        13
MAD        402
MSD     216409

Forecasts
Period  Forecast
19        17931.1
20        18750.4
21        19549.8
22        20329.4
23        21089.1
24        21828.9
25        22548.9
26        23249.0
```

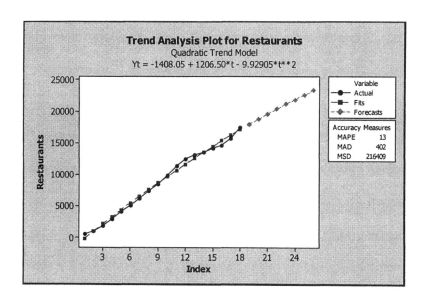

Trend Analysis Plot for Restaurants
Quadratic Trend Model
Yt = -1408.05 + 1206.50*t - 9.92905*t**2

18.63 p/c/m

a. Linear trend equation. With 1990 coded as x = 1, the equation is AvgBill = 16.8086 + 1.15786x. The forecast for 2010 (period 21) is AvgBill = 16.8086 + 1.15786(21) = $41.124. The Minitab trend analysis and plot are shown below.

Trend Analysis

```
Data        AvgBill
Length      7.00000
NMissing    0

Fitted Trend Equation
Yt = 16.8086 + 1.15786*t

Accuracy Measures
MAPE:        2.66790
MAD:         0.585714
MSD:         0.452353

Row   Period   Forecast
  1       8     26.0714
  2       9     27.2293
  3      10     28.3871
  4      11     29.5450
  5      12     30.7029
  6      13     31.8607
  7      14     33.0186
  8      15     34.1764
  9      16     35.3343
 10      17     36.4921
 11      18     37.6500
 12      19     38.8079
 13      20     39.9657
 14      21     41.1236
```

323

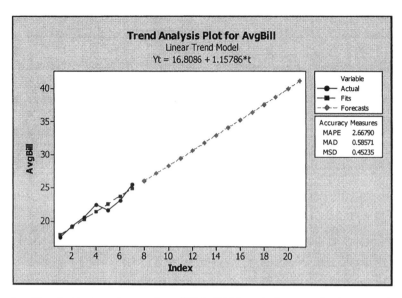

Trend Analysis Plot for AvgBill
Linear Trend Model
$Y_t = 16.8086 + 1.15786 * t$

Variable
— ● — Actual
— ■ — Fits
— ◆ — Forecasts

Accuracy Measures
MAPE 2.66790
MAD   0.58571
MSD   0.45235

b. Quadratic trend equation. With 1990 coded as x = 1, the equation is
AvgBill = $16.6486 + 1.26452x - 0.0133x^2$. As shown in the Minitab forecast below, the forecast for
2010 (period 21) is $16.6486 + 1.26452(21) - 0.0133(21)^2 = \$37.324$. The printout and plot are below.

Trend Analysis

```
Data        AvgBill
Length      7.00000
NMissing    0

Fitted Trend Equation
Yt = 16.6486 + 1.26452t  - 1.33E-02*t**2

Accuracy Measures
MAPE:          2.61600
MAD:           0.578095
MSD:           0.450220

Row    Period   Forecast
  1        8    25.9114
  2        9    26.9493
  3       10    27.9605
  4       11    28.9450
  5       12    29.9029
  6       13    30.8340
  7       14    31.7386
  8       15    32.6164
  9       16    33.4676
 10       17    34.2921
 11       18    35.0900
 12       19    35.8612
 13       20    36.6057
 14       21    37.3236
```

324

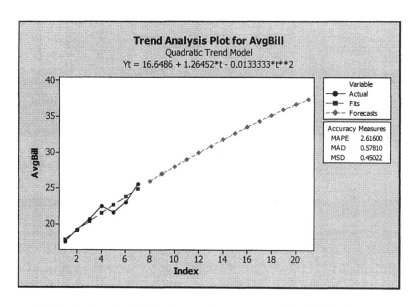

Trend Analysis Plot for AvgBill
Quadratic Trend Model
$Yt = 16.6486 + 1.26452 \cdot t - 0.0133333 \cdot t \cdot \cdot 2$

Variable
● Actual
■ Fits
◆ Forecasts

Accuracy Measures
MAPE 2.61600
MAD 0.57810
MSD 0.45022

c. Referring to the Minitab plots in parts a and b, the MAD and MSE values are as shown below. Note that Minitab shows MSE as "MSD".

Linear trend equation: MAD = 0.58571   MSE = 0.45235
Quadratic trend equation: MAD = 0.57810   MSE = 0.45022

On the basis of either MAD or MSE, the quadratic trend equation is the better fit to the data. The comparable Excel trend equations and their plots are shown below.

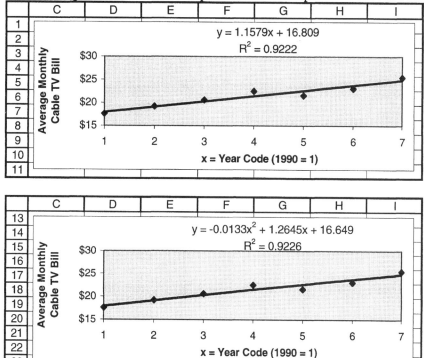

$y = 1.1579x + 16.809$
$R^2 = 0.9222$

$y = -0.0133x^2 + 1.2645x + 16.649$
$R^2 = 0.9226$

325

18.65 p/a/m

a. The graph is shown along with the 3-period centered moving average in part b. The overall trend appears to be downward.

b. Three-period centered moving average. Using Minitab, the underlying values and graph for the moving average with N = 3 are shown below. Refer to the "MA" column for the centered moving average values.

Moving average

```
Data        Sales
Length      18.0000
NMissing    0
```

Moving Average Length: 3

```
Accuracy Measures
MAPE:   11.133
MAD:     9.644
MSD:   134.711
```

Row	Period	Sales	MA	Predict	Error
1	1	105	*	*	*
2	2	126	110.667	*	*
3	3	101	111.333	*	*
4	4	107	103.667	110.667	-3.6667
5	5	103	96.333	111.333	-8.3333
6	6	79	87.333	103.667	-24.6667
7	7	80	81.333	96.333	-16.3333
8	8	85	85.000	87.333	-2.3333
9	9	90	85.000	81.333	8.6667
10	10	80	81.667	85.000	-5.0000
11	11	75	75.667	85.000	-10.0000
12	12	72	75.000	81.667	-9.6667
13	13	78	80.000	75.667	2.3333
14	14	90	89.333	75.000	15.0000
15	15	100	96.000	80.000	20.0000
16	16	98	97.000	89.333	8.6667
17	17	93	98.333	96.000	-3.0000
18	18	104	*	97.000	7.0000

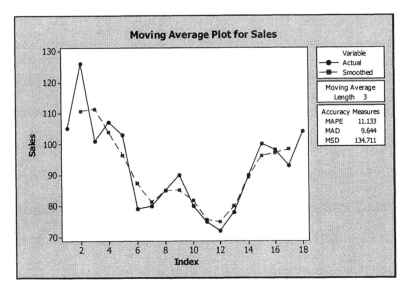

326

c. Exponentially smoothed curve, using Minitab and with smoothing constant α = 0.3. Refer to the "Smooth" column for the smoothed values.

Single Exponential Smoothing

```
Data          Sales
Length        18.0000
NMissing      0

Smoothing Constant Alpha: 0.3

Accuracy Measures
MAPE:    11.177
MAD:      9.992
MSD:    154.534
```

Row	Time	Sales	Smooth	Predict	Error
1	1	105	105.000	105.000	0.0000
2	2	126	111.300	105.000	21.0000
3	3	101	108.210	111.300	-10.3000
4	4	107	107.847	108.210	-1.2100
5	5	103	106.393	107.847	-4.8470
6	6	79	98.175	106.393	-27.3929
7	7	80	92.723	98.175	-18.1750
8	8	85	90.406	92.723	-7.7225
9	9	90	90.284	90.406	-0.4058
10	10	80	87.199	90.284	-10.2840
11	11	75	83.539	87.199	-12.1988
12	12	72	80.077	83.539	-11.5392
13	13	78	79.454	80.077	-2.0774
14	14	90	82.618	79.454	10.5458
15	15	100	87.833	82.618	17.3821
16	16	98	90.883	87.833	10.1674
17	17	93	91.518	90.883	2.1172
18	18	104	95.263	91.518	12.4820

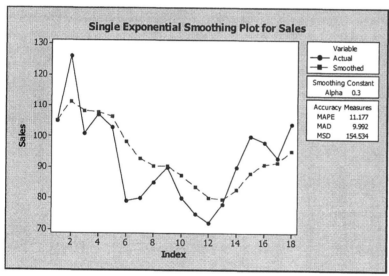

d. Fitting an exponentially smoothed curve with α = 0.7. As expected, the larger smoothing constant leads to a curve that follows the data more closely, but is not as smooth.

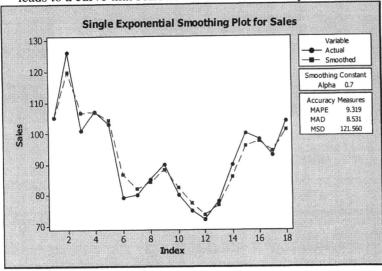

18.67 p/a/m For this seasonal index problem involving quarterly data, we will use Excel worksheet template tmquarts.xls, on the disk supplied with the text. The steps for its use are described in the worksheet template. The graphical summary of the solution is shown below the calculation results. The seasonal indexes are: I, 95.14; II, 89.69; III, 95.64; IV, 119.53

	A	B	C	D	E	F	G	H	I
5	Year	Qtr	Sales	ctr mov tot	ctr mov av	pct of mv av	unadj index	adj index	Deseasoned
6	2000	1	7.53				95.0039	95.1389	7.91
7		2	7.21				89.5614	89.6886	8.04
8		3	7.54	31.85	7.96	94.7087	95.5089	95.6445	7.88
9		4	9.57	31.84	7.96	120.2261	119.3585	119.5280	8.01
10	2001	5	7.52	31.94	7.98	94.1913	95.0039	95.1389	7.90
11		6	7.21	32.02	8.00	90.0828	89.5614	89.6886	8.04
12		7	7.73	32.11	8.03	96.3090	95.5089	95.6445	8.08
13		8	9.54	32.21	8.05	118.4909	119.3585	119.5280	7.98
14	2002	9	7.73	32.27	8.07	95.8165	95.0039	95.1389	8.12
15		10	7.20	32.35	8.09	89.0400	89.5614	89.6886	8.03
16		11	7.87				95.5089	95.6445	8.23
17		12	9.55				119.3585	119.5280	7.99

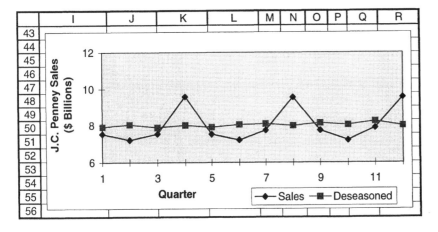

18.69 p/a/m For this seasonal index problem involving quarterly data, we will use Excel worksheet template tmquarts.xls, on the disk supplied with the text. The procedure is described in the worksheet template. The graphical summary of the solution is shown below the calculation results. The seasonal indexes are: I, 182.240; II, 68.707;   III, 34.210; IV, 114.843.

	A	B	C	D	E	F	G	H	I
5	Year	Qtr	Gas	ctr mov tot	ctr mov av	pct of mv av	unadj index	adj index	Deseasoned
6	1991	1	2048				182.5073	182.2403	1123.8
7		2	757				68.8075	68.7069	1101.8
8		3	398	4537.00	1134.25	35.0893	34.2598	34.2097	1163.4
9		4	1342	4559.50	1139.88	117.7322	115.0114	114.8432	1168.6
10	1992	5	2032	4595.50	1148.88	176.8687	182.5073	182.2403	1115.0
11		6	818	4647.50	1161.88	70.4034	68.8075	68.7069	1190.6
12		7	409	4816.00	1204.00	33.9701	34.2598	34.2097	1195.6
13		8	1435	4994.50	1248.63	114.9264	115.0114	114.8432	1249.5
14	1993	9	2276	5064.50	1266.13	179.7611	182.5073	182.2403	1248.9
15		10	931	5087.50	1271.88	73.1990	68.8075	68.7069	1355.0
16		11	436	5172.00	1293.00	33.7200	34.2598	34.2097	1274.5
17		12	1454	5175.50	1293.88	112.3756	115.0114	114.8432	1266.1
18	1994	13	2426	5083.50	1270.88	190.8921	182.5073	182.2403	1331.2
19		14	788	5017.50	1254.38	62.8201	68.8075	68.7069	1146.9
20		15	395				34.2598	34.2097	1154.6
21		16	1363				115.0114	114.8432	1186.8

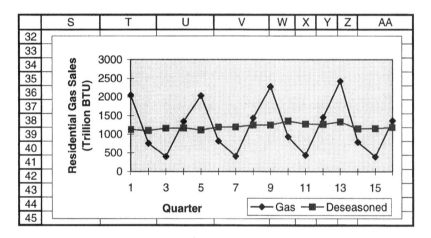

18.71 p/a/m For this seasonal index problem involving quarterly data, we will use Excel worksheet template tmquarts.xls, on the disk supplied with the text. The procedure is described in the worksheet template. The graphical summary of the solution is shown below the calculation results. The seasonal indexes are: I, 94.68; II, 93.54; III, 111.05; IV, 100.73.

	A	B	C	D	E	F	G	H	I
5	Year	Qtr	Rev.	ctr mov tot	ctr mov av	pct of mv av	unadj index	adj index	Deseasoned
6	1998	1	1.260				94.6555	94.6777	1.331
7		2	1.244				93.5193	93.5411	1.330
8		3	1.504	5.62	1.40	107.0653	111.0286	111.0546	1.354
9		4	1.524	5.79	1.45	105.3214	100.7030	100.7266	1.513
10	1999	5	1.434	5.97	1.49	96.0241	94.6555	94.6777	1.515
11		6	1.408	6.09	1.52	92.4567	93.5193	93.5411	1.505
12		7	1.711	6.18	1.54	110.7892	111.0286	111.0546	1.541
13		8	1.553	6.33	1.58	98.1281	100.7030	100.7266	1.542
14	2000	9	1.577	6.53	1.63	96.6521	94.6555	94.6777	1.666
15		10	1.571	6.72	1.68	93.4841	93.5193	93.5411	1.679
16		11	1.940	6.88	1.72	112.8071	111.0286	111.0546	1.747
17		12	1.715	6.99	1.75	98.0771	100.7030	100.7266	1.703
18	2001	13	1.729	7.04	1.76	98.2805	94.6555	94.6777	1.826
19		14	1.650	7.02	1.76	93.9569	93.5193	93.5411	1.764
20		15	1.946	6.86	1.72	113.4528	111.0286	111.0546	1.752
21		16	1.684	6.65	1.66	101.2856	100.7030	100.7266	1.672
22	2002	17	1.433	6.54	1.63	87.6654	94.6555	94.6777	1.514
23		18	1.525	6.48	1.62	94.1794	93.5193	93.5411	1.630
24		19	1.847				111.0286	111.0546	1.663
25		20	1.660				100.7030	100.7266	1.648

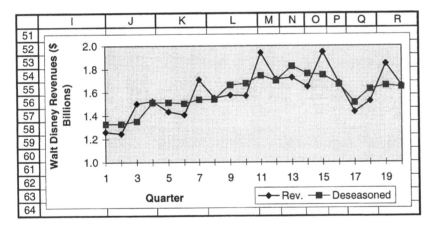

18.73 p/a/e The seasonal index = (actual data / deseasonalized data) * 100. The seasonal indexes are:

I: (1682/2320) * 100 = 72.50
II: (1963/2362) * 100 = 83.11
III: (2451/2205) * 100 = 111.16
IV: (3205/2414) * 100 = 132.77

18.75 p/a/m If x = 1 for 1996, then x = 15 for 2010. Therefore, the forecasted unit sales for the filtering system component for all of 2010 is $150 + 12(15) + 0.01(15)^2 = 332.25$ (or 332 since this is a discrete variable). Using the seasonal indexes provided yields the estimates for quarterly sales:

I: (332/4) * (85/100) =70.55, or 71
II: (332/4) * (110/100) = 91.30, or 91
III: (332/4) * (125/100) = 103.75, or 104
IV: (332/4) * (80/100) =66.40, or 66

18.77 p/c/m Using Minitab, the calculation summary is shown below. Refer to the "Predict" column of the printout. The years are coded so that 1975 is period 1.

Single Exponential Smoothing

```
Data        Spending
Length      23.0000
NMissing    0
```

Smoothing Constant Alpha: 0.5

```
Accuracy Measures
MAPE: 16.7594
MAD:   2.0625
MSD:   6.0614
```

Row	Time	Spending	Smooth	Predict	Error
1	1	4.80	4.8000	4.8000	0.00000
2	2	5.33	5.0650	4.8000	0.53000
3	3	5.92	5.4925	5.0650	0.85500
4	4	6.69	6.0913	5.4925	1.19750
5	5	7.50	6.7956	6.0913	1.40875
6	6	7.37	7.0828	6.7956	0.57437
7	7	8.25	7.6664	7.0828	1.16719
8	8	8.10	7.8832	7.6664	0.43359
9	9	9.38	8.6316	7.8832	1.49680
10	10	12.34	10.4858	8.6316	3.70840
11	11	13.28	11.8829	10.4858	2.79420
12	12	14.48	13.1815	11.8829	2.59710
13	13	16.50	14.8407	13.1815	3.31855
14	14	17.93	16.3854	14.8407	3.08927
15	15	17.14	16.7627	16.3854	0.75464
16	16	13.73	15.2463	16.7627	-3.03268
17	17	10.56	12.9032	15.2463	-4.68634
18	18	10.32	11.6116	12.9032	-2.58317
19	19	11.25	11.4308	11.6116	-0.36159
20	20	14.07	12.7504	11.4308	2.63921
21	21	17.23	14.9902	12.7504	4.47960
22	22	17.75	16.3701	14.9902	2.75980
23	23	19.34	17.8550	16.3701	2.96990

Row	Period	Forecast	Lower	Upper
1	24	17.8550	12.8019	22.9082

Calculations are: (next period forecast) = 0.5(this period actual) + 0.5(this period forecast). Thus, the forecast for period 24 (1998) would be 0.5(actual in period 23) + 0.5(forecast for period 23), or 0.5(19.34) + 0.5(16.3701) = \$17.855 billion. The plot is shown below.

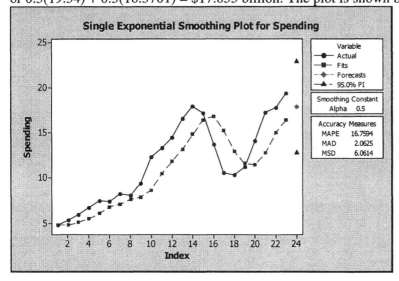

18.79 p/c/m

Quadratic trend equation.

With 1993 coded as x = 1, the equation is Cost = $38.3268 + 0.794341x + 0.0538258x^2$.

The Minitab trend analysis plot is shown below.

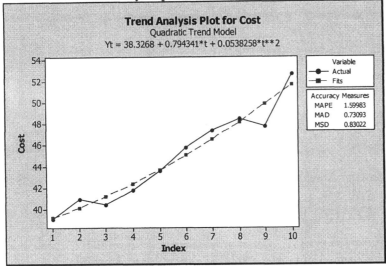

Linear trend equation.

With 1993 coded as x = 1, the equation is Cost = $37.1427 + 1.38642x$. The Minitab trend analysis plot is shown below.

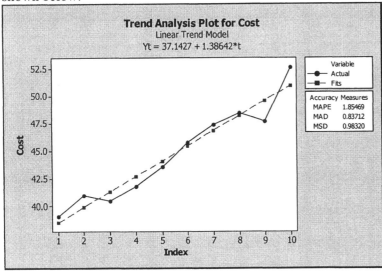

As shown in the information accompanying their respective plots, the quadratic trend equation has MSE = 0.83022 while the linear trend equation has MSE = 0.98320. (Recall that Minitab shows MSE as "MSD".) The quadratic trend is the better fit to the data. The comparable Excel trend equations and their graphs are shown below.

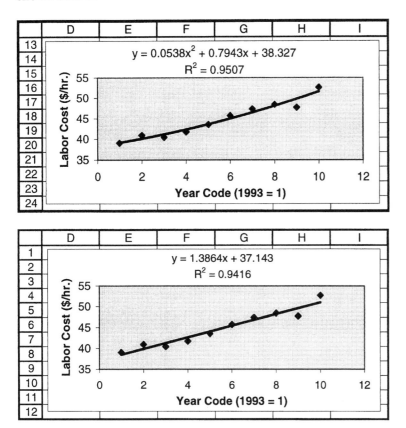

18.81 p/c/m The deseasonalized values are shown in column I of the Excel printout for part b of exercise 18.80. The chart of the original and deseasonalized quarterly values is shown below.

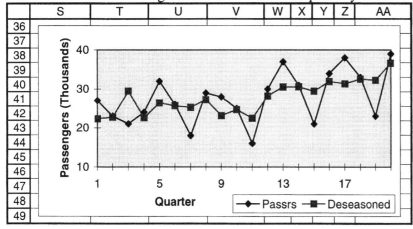

18.83 p/c/m The deseasonalized values are shown in column I of the Excel printout for part b of exercise 18.82. The chart of the original and deseasonalized monthly values is shown below.

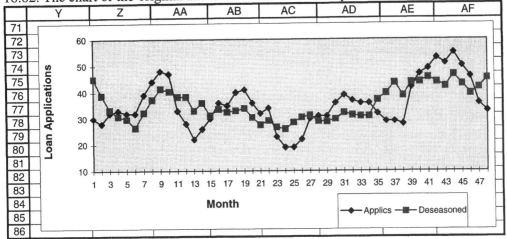

18.85 p/a/m The null and alternative hypotheses are $H_0: \rho \leq 0$ and $H_1: \rho > 0$.

As shown in the Minitab printout below, the Durbin-Watson statistic is d = 0.39. In the Durbin-Watson table for $\alpha = 0.05$ for a directional test, the listed values for n = 23 observations and k = 1 independent variable are $d_L = 1.26$ and $d_U = 1.44$. We are testing for positive autocorrelation, so the listed value of interest is $d_L = 1.26$. Because d = 0.39 is less than $d_L = 1.26$, there is evidence of positive autocorrelation at this level. (Refer to the "testing for positive autocorrelation" portion of Figure 18.7 in the text.)

```
Regression Analysis: Spending versus YrCode

The regression equation is Spending = 5.06 + 0.554 YrCode

Predictor        Coef      SE Coef         T         P
Constant        5.059        1.127      4.49     0.000
YrCode        0.55402      0.08217      6.74     0.000

S = 2.614     R-Sq = 68.4%     R-Sq(adj) = 66.9%

Analysis of Variance
Source            DF          SS          MS         F         P
Regression         1      310.62      310.62     45.46     0.000
Residual Error    21      143.48        6.83
Total             22      454.11

Durbin-Watson statistic = 0.39
```

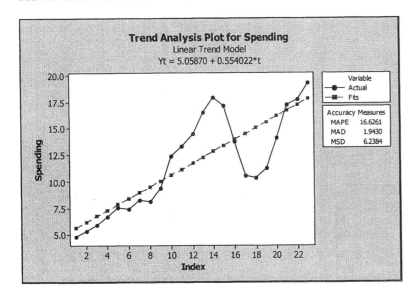

334

We can also use Data Analysis Plus to examine the residuals and compute the Durbin-Watson statistic.

	A	B	C
1	**Durbin-Watson Statistic**		
2			
3	d = 0.3913		

18.87 p/c/m The Minitab printout is shown below.

```
Regression Analysis: CPI versus CPIt-1

The regression equation is CPI = 1.90 + 1.02 CPIt-1

40 cases used 1 cases contain missing values

Predictor        Coef     SE Coef          T        P
Constant       1.9002      0.6617       2.87    0.007
CPIt-1        1.02015     0.00658     155.07    0.000

S = 2.004       R-Sq = 99.8%      R-Sq(adj) = 99.8%

Analysis of Variance
Source             DF          SS         MS        F        P
Regression          1       96577      96577  24048.12    0.000
Residual Error     38         153          4
Total              39       96730

Predicted Values for New Observations
New Obs     Fit     SE Fit        95.0% CI               95.0% PI
1       182.570      0.665   ( 181.224, 183.915)   ( 178.295, 186.844)

Values of Predictors for New Observations
New Obs    CPIt-1
1             177
```

The first-order autoregressive forecasting equation is $\hat{y}_t = 1.9002 + 1.02015y_{t-1}$. The Consumer Price Index in period 41 (2001) was 177.1 and Minitab shows a forecast value of 182.570 for period 42 (2002). This can also be calculated as $1.9002 + 1.02015(177.1) = 182.57$. In carrying out the regression, the residuals were stored into a separate column and the mean of their absolute values is MAD = 1.466.

18.89 p/a/m Between 1996 and 2001, the median weekly earnings of union workers grew to 1.1672 times its 1996 value. The CPI grew to 1.1287 times its 1996 value over this same period. Therefore, real earnings changed by $(1.1672/1.1287)*100 = 103.41$, which implies an increase of 3.41%.

CHAPTER 19
DECISION THEORY

SECTION EXERCISES

19.1 d/p/e Risk is the condition when all the possible states of nature and their associated probabilities are known. Uncertainty is the condition when the possible states of nature are known, but the exact probabilities for the states of nature are not known. Ignorance is the condition in which neither the possible states of nature nor any of the associated probabilities are known. Most business decisions are made under conditions of uncertainty.

19.3 d/p/e The decision alternatives for the airport manager are to extend the runway or not to extend the runway. The possible states of nature would include increased complaints from local residents, reduced complaints from local residents, or no change in complaints from area residents.

19.5 d/p/m The trainee should consider the possible levels of sales he might make as his states of nature. Of course, the various levels of sales would affect his income under the different types of income plans.

19.7 d/p/m The rows of a payoff table depict the alternatives the decision maker has available.
In this case, the orange grower may choose to install the heating system or not to install the system. The columns of a payoff table represent the possible states of nature that might occur; in this case, the winter will be cold enough to require the system to save the crop or it will be mild enough not to require the system.

19.9 p/a/m If Dave uses the maximin criterion, he will be assuming that the worst possible outcome will happen. Thus, he could make $0 profit if he orders no shirts, -$200 if he orders 1000 shirts, or -$400 if he orders 2000 shirts. To make the best of a bad situation, Dave should order no shirts and avoid any losses. If Dave uses the maximax criterion, he will be assuming that the best possible outcome will happen. Thus, he could make $0 if he orders no shirts, $800 if he orders 1000 shirts, or $1600 if he orders 2000 shirts. To make the best of this situation, Dave should order 2000 shirts.

For Dave to use the minimax. regret criterion, he must first construct a regret matrix:

	The team		
	wins	loses	ties
Order no shirts	$1600	$0	$300
Order 1000 shirts	$800	$200	$100
Order 2000 shirts	$0	$400	$0

If Dave orders no shirts, his maximum possible regret is $1600. If he orders 1000 shirts, his maximum possible regret is $800. Finally, if he orders 2000 shirts, his maximum possible regret is $400. Therefore, Dave should order 2000 shirts to minimize his maximum possible regret.

19.11 d/p/m
a. This is an example of maximax reasoning. Maximax is a very optimistic criterion, and this trucker is exhibiting optimistic behavior.
b. This is an example of maximax reasoning. The retailer is being very optimistic in assuming that the public relations firm can turn things around for his store.
c. This is an example of maximin reasoning. The owner of the roller skating rink is assuming that bad things might happen, and he is going to try to make the best of the bad situation.

19.13 p/a/m To help the collector decide whether or not she should purchase the insurance, one should first construct a payoff table. The collector faces two alternatives: purchase the insurance or do not purchase the insurance, and there are two possible outcomes: the shipment arrives with no problems or the shipment is damaged in transit. Her payoff table will look like:

	Shipment is not lost	Shipment is lost
Purchase insurance	$500	$500
Do not purchase insurance	$0	$50,000

Note: these numbers represent costs. (If the goods are damaged in transit, the collector will recover her $50,000, but she will still be short the $500 she paid for the insurance.)

Using the maximin criterion, the collector is adopting a pessimistic outlook. If she purchases the insurance, the most the shipment might cost her is $500. If she does not purchase the insurance, the shipment might cost her $50,000. Therefore, she will choose to purchase the insurance.
Using the maximax criterion, the collector is adopting an optimistic outlook. If she purchases the insurance, the least it will cost her is $500. If she does not purchase the insurance, the least it might cost her is $0. Therefore, being optimistic, the collector will choose not to purchase the insurance.

To use the minimax regret criterion, one must first construct a regret matrix, as shown below.

	Shipment is not lost	Shipment is lost
Purchase insurance	$500	$0
Do not purchase insurance	$0	$49,500

If the collector purchases the insurance, her maximum possible regret is $500. If she does not purchase the insurance, her maximum possible regret is $49,500. To minimize her maximum possible regret, the collector will choose to purchase the insurance.

19.15 d/p/e NonBayesian decision criteria ignore the probabilities for the respective states of nature, while Bayesian decision criteria take them into account.

19.17 p/a/m Recall the payoff table from exercise 19.4:

	Return audited (probability = 0.1)	Return not audited (probability = 0.9)
Makes claims	$18,000	$12,000
Does not make claims	$14,000	$14,000

If the probability that his return will be audited is 0.1, the probability that his return will not be audited must be 0.9 = (1 - 0.1). These probabilities have been included in the payoff table. Further recall that these are costs.
The expected cost if the consultant makes the travel claims is:
0.1($18,000) + 0.9($12,000) = $12,600
The expected cost if the consultant does not make the travel claims is:
0.1($14,000) + 0.9($14,000) = $14,000
To minimize his expected costs, the consultant will choose to make the travel claims.

The most the consultant should be willing to pay for more information regarding his likelihood of being audited is equal to the expected value of perfect information (EVPI). Since we are dealing with costs, the EVPI formula looks slightly different. (Note that we expect costs to decline as we acquire better information.) In this case, EVPI = expected cost with present information minus expected cost with perfect information. Expected cost with perfect information is 0.1($14,000) + 0.9 ($12,000) = $12,200. Therefore, EVPI = $12,600 - $12,200 = $400. The consultant should not pay more than $400 for information regarding his likelihood of being audited.

19.19 p/a/m Recall the payoff table from exercise 19.10. (The probabilities have been included.)

	Market Condition		
	I (p = 0.1)	II (p = 0.6)	III (p = 0.3)
Design A	$25	$14	$6
Design B	$15	$16	$18
Design C	$5	$10	$40

The expected profit for Design A is: 0.1($25) + 0.6($14) + 0.3($6) = $12.7 million
The expected profit for Design B is: 0.1($15) + 0.6($16) + 0.3($18) = $16.5 million
The expected profit for Design C is: 0.1($ 5) + 0.6($10) + 0.3($40) = $18.5 million
To maximize expected profit, the firm should choose Design C.

The most the firm should be willing to pay for a research study designed to reduce its uncertainty about market conditions is equal to the expected value of perfect information (EVPI). EVPI = the expected profit with perfect information minus the expected profit with present information. The expected profit with perfect information for this problem is:
0.1($25) + 0.6($16) + 0.3($40) = $24.1 million.

Therefore: EVPI = $24.1 million - $18.5 million = $5.6 million. The company should be willing to pay no more than $5.6 million for a research study designed to reduce its uncertainty about market conditions.

19.21 p/a/m Recall the payoff table from exercise 19.12. (The probabilities have been included.)

	Train at RR crossing (p = 0.1)	No train at RR crossing (p = 0.9)
Take direct route	35 min.	20 min.
Take longer route	30 min.	30 min.

Shorter travel times are desirable. The expected travel time if Fred takes the direct route is:
0.1(35 min.) + 0.9(20 min.) = 21.5 minutes.
The expected travel time for the longer route is: 0.1(30 min.) + 0.9(30 min.) = 30 minutes.
To minimize his expected travel time, Fred should take the direct route.

To determine how many minutes Fred would save if he could predict with certainty whether a train will be at the crossing, we simply need to calculate the expected value of perfect information (EVPI) for this problem. The expected travel time with perfect information is:
0.1(30 min.) + 0.9(20 min.) = 21 minutes.

Therefore: EVP1 = 21.5 min. - 21 min. = 0.5 minutes. Fred could save 0.5 minutes if he could predict with certainty whether a train will be at the crossing.

19.23 d/p/e Rarely will we ever obtain perfect information, yet we can usually purchase "better" information for a price. The expected value of perfect information (EVPI) tells us the upper limit for what "perfect" information would be worth. Certainly we would not want to pay more than the EVPI for "imperfect" information.

19.25 p/a/m To use the expected opportunity loss criterion, we must first construct a regret matrix.

	The state of the economy will be:		
	weak (p = 0.2)	moderate (p = 0.4)	strong (p = 0.4)
Invest in company A	$64	$0	$0
Invest in company B	$44	$10	$15
Invest in company C	$0	$2	$30

The EOL from investing in company A is: 0.2($64) + 0.4($0) + 0.4($0) = $12.8 thousand.
The EOL from investing in company B is: 0.2($44) +0.4($10) + 0.4($15) = $18.8 thousand.
The EOL from investing in company C is: 0.2($0) + 0.4($2) + 0.4($30) = $12.8 thousand.

Using the expected opportunity loss approach, the investor should select either company A or company C, and the minimum EOL will be $12.8 thousand.
The expected value of perfect information is equal to the minimum expected opportunity loss, so the EVPI = $12,800.

19.27 p/a/m Recall the regret matrix from exercise 19.12. (The probabilities have been included.)

	Train at RR crossing (p = 0.1)	No train at RR crossing (p = 0.9)
Take direct route	5 min.	0 min.
Take longer route	0 min.	10 min.

The EOL for the direct route is: 0.1 (5 min.) + 0.9(0 min.) = 0.5 minutes.
The EOL for the longer route is: 0.1 (0 min.) + 0.9(10 min.) = 9.0 minutes.

On average, if Fred travels the direct route it will cost him 0.5 minutes; if he selects the longer route, it will cost him 9 minutes -- with each of these expressed relative to his having perfect information.

19.29 p/a/m The manager has a marginal loss (ML) of $0.60 for each cod that is not sold and a marginal profit (MP) of $0.40, or ($1.00 - $0.60) for each cod that is sold. Therefore, he should stock cod so that the probability of selling the next one is:

$$p_c = \frac{0.60}{0.40 + 0.60} = 0.60$$

The z-score associated with this probability is 0.25. Using the equation $0.25 = \frac{x - 300}{50}$ and solving for x, we obtain x = 312.5 cod that the manager should stock. Since the manager cannot stock 0.5 cod, he should stock 313.

19.31 p/a/m The marginal loss (ML) for the athletic director is $0.40, or ($0.50 - $0.10), and the marginal profit (MP) is $2.50. Therefore, she should order enough programs so that the probability of selling the next program is $p_c = \frac{0.40}{2.50 + 0.40} = 0.1379$.

The z-score associated with this probability is -1.09. Solving the equation $-1.09 = \frac{x - 3000}{500}$

yields x = 2455 programs the athletic director should order.

CHAPTER EXERCISES

19.33 p/a/d The marginal profit (MP) is $5, and the marginal loss (ML) is $4. Therefore, the retailer

should stock so that the probability of selling the next unit is $p_c = \dfrac{4}{5+4} = 0.4444$.

We have to take the given distribution and determine where the probability of selling the next unit is 44.44%. We can use cumulative probabilities to do this. The probability of selling at least 0 units is 1.00, and the probability of selling at least 1 unit is 0.99. The probability of selling at least 7 units is 0.50, or $(0.18 + 0.14 + 0.09 + 0.05 + 0.02 + 0.01 + 0.01)$, and the probability of selling at least 8 units is 0.32. Since $0.50 > 0.4444 > 0.32$, the retailer should stock 7 units each day.

19.35 p/a/m With the pessimistic maximin criterion, the company will select the alternative with the best worst alternative. Choosing Dennis, the worst thing that could happen is a profit of $0.5 million, whereas the worst possibility in staying with the current spokesperson is a profit of $1.2 million.
Using the maximin criterion, the company will stay with the current spokesperson.

With the optimistic maximax criterion, the company will select the alternative with the best best alternative. Choosing Dennis, the best thing that can happen is a profit of $4.4 million, whereas the best thing that can happen if they stay with the current spokesperson is a profit of $1.2 million. Using the maximax criterion, the company will switch to Dennis.

With the minimax regret criterion, the company will choose the alternative with the minimum maximum regret. The regret matrix would be as shown below.

	Dennis:	
	stays clean	goes bad
Switch to Dennis in ads	$0	$0.7
Stay with current spokesperson	$3.2	$0

If the company switches to Dennis, the maximum possible regret will be $0.7 million.
If the company stays with the current spokesperson, the maximum possible regret will be $3.2 million.
The minimum of these maximum regrets is $0.7 million, so the company will switch to Dennis.

19.37 p/a/m With the pessimistic maximin criterion, the investor will select the alternative with the best worst alternative. Choosing DiskWorth, the worst thing that could happen is a profit of $20 thousand, whereas the worst possibility in selecting ComTranDat is a loss of $20 thousand. Using the maximin criterion, the investor will select DiskWorth.

With the optimistic maximax criterion, the investor will select the alternative with the best best alternative. Choosing DiskWorth, the best thing that can happen is a profit of $40 thousand, whereas the best thing that can happen if he selects ComTranDat is a profit of $90,000. Using the maximax criterion, the investor will select ComTranDat.

With the minimax regret criterion, the investor will choose the alternative with the minimum maximum regret. The regret matrix would be as shown below.

	Technology Sector:		
	weakens	stays same	strengthens
Select DiskWorth	$0	$0	$50
Select ComTranDat	$40	$20	$0

If the investor selects DiskWorth, the maximum possible regret will be $50 thousand.
If the investor selects ComTranDat, the maximum possible regret will be $40 thousand.
The minimum of these maximum regrets is $40 thousand, so the investor will select ComTranDat.

19.39 p/a/m Referring to exercise 19.38, the operator could use a NonBayesian decision criteion if he has no idea regarding probabilities for the various weather conditions for the upcoming winter. These include maximin, maximax, and minimax regret.

Using the maximin criterion, the operator finds that his lowest possible payoff if he does not lease the machine is $20,000, and his lowest possible payoff if he does lease the machine is $30,000.
Since $30,000 > $20,000, the operator will lease the snow-making machine.
Using the maximax criterion, the operator finds that his highest possible payoff if he does not lease the machine is $50,000 and his highest possible payoff if he does lease the machine is $40,000.
Since $50,000 > $40,000, the operator will not lease the machine.

To use the minimax regret criterion, the operator must first construct a regret matrix:

	The winter is:		
	mild	typical	severe
Does not lease snow machine	$10,000	$5,000	$0
Leases snow machine	$0	$0	$10,000

The maximum possible regret if he chooses not to lease the machine is $10,000, and the maximum possible regret if he chooses to lease the machine is $10,000. Therefore, the operator will be indifferent between the alternatives if he uses the minimax regret criterion.

19.40 p/p/e This driver is making a maximax choice. He is assuming an optimistic stance in thinking that only good things will happen to him.

CHAPTER 20
TOTAL QUALITY MANAGEMENT

SECTION EXERCISES

20.1 d/p/e Garvin's eight dimensions of quality applied to:

a. A pair of sunglasses:
1. Performance of functions such as blocking out sun rays and UVL.
2. Special features such as appearance, lightness, and color.
3. Reliability of the glasses and resistance to scratches, abuse and breakage.
4. Conformance to federal guidelines for ultraviolet light transmission.
5. Durability for different types of sports.
6. Serviceability of the frame, replacement of lens and nose pads.
7. Aesthetic appearance and comfort of the fit.
8. Perceived quality of the company that made them.

b. A pair of running shoes:
1. Performance of the running shoe for a particular sport.
2. Attractiveness of the shoe in terms of a function like a pump-up support system.
3. Reliability of the shoe for a particular sport.
4. Conformance to fire-resistance or other material standards.
5. Durability of the shoe over time.
6. Ease of replacing the laces.
7. Having a light-reflective heel.
8. Perceived quality of the company that made them.

20.3 d/p/m Insulation of a microwave oven to EPA standards of radiation emissions would be an important quality dimension.

20.5 d/p/m Random variation in a manufacturing process is the variation in the process that results from chance. The amount of cereal injected into a box may vary as a result of the temperature of the cereal, or the temperature of the room, or the different compositions of the grain.

20.7 d/p/m It is generally easier to find and eliminate the source for assignable variation than for random variation. Assignable variation can be traced to some identifiable cause, and the cause can be changed or eliminated. Random variation, on the other hand, occurs because of chance.

20.9 d/p/m SPC uses the sampled products to make inferences about the process from which they came, while acceptance sampling uses sampled products to make inferences about a larger population of the products themselves.

20.11 d/p/m Deming's PDCA(Plan-Do-Check-Act) cycle stresses the importance of continuous and systematic quality-improvement efforts at all levels of the organization.

20.13 d/p/m TQM stresses the prevention of defects through continual attention to, and improvements in, the processes through which the product is designed, produced, delivered, and supported.

20.15 d/p/m Kaizen focuses on small, ongoing improvements in quality rather than on large-scale innovations that come from advances in technology and equipment. Kaizen is the baseball equivalent of winning by hitting many singles.

20.17 d/p/m Quality audits involve evaluating a firm's process capabilities. The goal is to assess the firm's strategies, and success, in defect prevention. These audits can also be quantified to facilitate comparisons from one firm to the next.

20.19 d/p/m Competitive benchmarking involves studying and emulating the strategies and practices of organizations already known to generate products and services that are of world-class quality.

20.21 d/p/m JIT is not a gimmick for reducing inventory costs. It is a production philosophy that happens to have a favorable effect on inventory costs. With zero or minimal parts on hand to buffer irregularities in the production pace, the worker, machine, or supplier generating poor quality has nowhere to hide. This facilitates the timely identification and correction of weaknesses in the process that has allowed the defects or errors to occur.

20.23 d/p/m First, the individual worker must be responsible for the quality of output that he or she produces. Second, the worker must be provided with the authority to make changes deemed necessary for quality improvement.

20.25 d/p/m The cause and effect diagram provides a visual description of effects and their possible causes. It facilitates the identification and solution of problems in the production process.

20.27 d/p/m A check sheet is a mechanism for entering and retaining data that may be the subject of further analysis by another statistical tool.

20.29 d/p/m For a given individual and his/her tastes, the traditional approach might categorize the product as "good" as long as the mixture includes between 1.0 and 3.0 teaspoons of the chocolate mix. The Taguchi approach could recognize that it tastes best to that person when 2.0 teaspoons are used, and that quality decreases as the amount of chocolate gets further away from 2.0 teaspoons.

20.31 d/p/e A control chart is a chart that displays a sample statistic for each of a series of samples from a process. It has upper and lower control limits and is used in determining whether the process is in control.

20.33 d/p/e A control chart's upper and lower limits represent the boundaries for virtually all of the random variation. Values that fall outside these limits are considered as possibly having occurred as the result of assignable variation.

20.35 d/p/m A Type I error is rejecting a true null hypothesis; therefore, a Type I error in this situation will result in a belief that the process is not in control when it really is. A Type II error is failing to reject a false null hypothesis; therefore, a Type II error in this situation will result in a belief that the process is in control when it really is not.

20.37 c/a/e Given 3-sigma control limits for a control chart, there is a 0.0026 (or 0.26%) chance that a given sample will fall outside the control limits as the result of random variation alone. (Three-sigma limits include 99.74% of the area under the normal curve.)

20.39 d/p/m The range chart measures the variation in the measurements; it is similar to a standard error chart. The mean chart, on the other hand, measures the central tendency of the measurements. Both the range and the mean should be monitored to assure product quality. The mean is important for obvious reasons: we must be close to the specifications on average. The range is also important because we cannot tolerate too much overall deviation.

20.41 p/a/e Using the centerline and control limits developed in exercise 20.40, all of the samples given fall within the limits. Therefore, the process appears to be in control.

20.43 p/a/e Using the centerline and control limits developed in exercise 20.42, all of the samples except for number 5 fall within the limits. Therefore, the process may have been out of control when the items in sample 5 were produced.

20.45 p/a/m This exercise can be carried out with a pocket calculator, but we will use the computer. Using Minitab, we obtain the mean and range charts shown below.

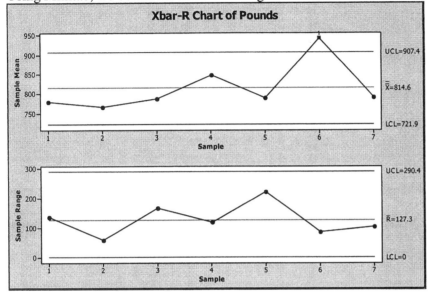

Minitab provides this warning:
```
Test Results for Xbar Chart of Pounds
TEST 1. One point more than 3.00 standard deviations from center line.
Test Failed at points:  6
```

For the seven samples, the mean of the sample means is 814.643 pounds, and this is the centerline for the mean chart. The mean of the sample ranges is 127.286 minutes. With n = 4 items in each sample, we use the control chart factors table in the appendix (also shown as Table 20.3 in the chapter) to find $A_2 = 0.729$. The 3-sigma control limits for the mean are calculated as:

814.643 ± 0.729(127.286), or 814.643 ± 92.791. The control limits are UCL = 907.43, LCL = 721.85.

Except for sample number 6 (sample mean = 943.25), each of the sample means falls within the control limits. On this basis, we suspect that the process may not be in control.

The centerline for the range chart is 127.286. With n = 4 items in each sample, the applicable control chart factors from the appendix table are $D_3 = 0$ and $D_4 = 2.282$. The 3-sigma control limits for the range are calculated as:

UCL = 2.282(127.286) = 290.47, and LCL = 0(127.286) = 0.

All of the sample ranges are within these limits and we conclude that the variability of the process is in control. However, because of the results for the mean chart, our overall conclusion is that the process is not in control.

We can also use the Data Analysis Plus add-in that accompanies the text. The mean and range charts shown below are comparable to those generated by Minitab. Note that the control limits we compute with the assistance of printed tables and a pocket calculator may sometimes differ very slightly from those generated with either Minitab or Data Analysis Plus.

	A	B	C	D	E	F	G	H
1	Statistical Process Control							
2								
3			*Pounds*					
4	Upper control limit		907.4341					
5	Centerline		814.6429					
6	Lower control limit		721.8516					
7	Pattern Test #1 Failed at Points: 6							
8								
9								
10								
11								
12								
13								
14								
15								
16								
17								
18								
19								
20								
21								
22								
23								

XBar-Chart (using R)

	A	B	C	D	E	F	G	H
1	Statistical Process Control							
2								
3			*Pounds*					
4	Upper control limit		290.466					
5	Centerline		127.2857					
6	Lower control limit		0					
7								
8								
9								
10								
11								
12								
13								
14								
15								
16								
17								
18								
19								
20								
21								
22								
23								
24								

R-Chart

20.47 c/a/e To calculate the upper and lower control limits, we use $p \pm z\sqrt{\dfrac{p(1-p)}{n}}$ and the control limits will be:

$$0.04 \pm 3\sqrt{\frac{(0.04)(0.96)}{150}} = 0.04 \pm 0.048 = (0,\ 0.088).$$ Since the calculated value of the lower limit is negative, LCL reverts to 0.

20.49 c/a/e The 3-sigma upper and lower control limits are:

$UCL = 7.5 + 3\sqrt{7.5} = 15.72$ and $LCL = 7.5 - 3\sqrt{7.5} = -0.72$. Since LCL is negative, it reverts to 0.

346

20.51 p/a/m This exercise can be carried out with a pocket calculator, but we will use the computer. Using Minitab, we obtain the p-chart shown below.

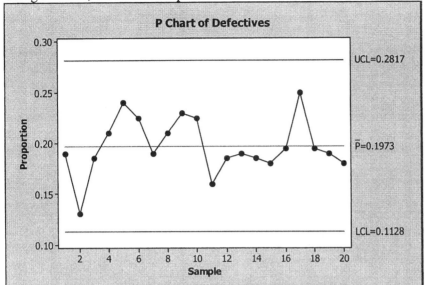

For each of the 20 samples, the sample proportion is the number of defectives divided by the sample size of 200. The sample sizes are equal, so the mean of the proportions is the sum divided by 20, or p-bar = 3.945/20 = 0.1973. The upper and lower control limits can be calculated as:

$$0.1973 \pm 3\sqrt{\frac{(0.1973)(1-0.1973)}{200}} = 0.1973 \pm 0.0844 \text{, or UCL} = 0.282 \text{ and LCL} = 0.113.$$

None of the sample proportions is outside the control limits; the process appears to be in control.

Excel and the Data Analysis Plus add-in can also be used to generate the p-chart for this exercise. The results are comparable to those generated by Minitab, and are shown below.

	A	B	C	D	E	F	G	H
1	Statistical Process Control							
2								
3			Defectives					
4	Upper control limit		0.2817					
5	Centerline		0.1973					
6	Lower control limit		0.1128					
7								
8			P-Chart					
9								
10								
11		0.30						
12		0.25						
13		0.20						
14		0.15						
15		0.10						
16								
17		0.05						
18								
19		0.00						
20		0 2 4 6 8 10 12 14 16 18 20 22						
21								
22								
23								

20.53 p/a/m This exercise could be solved with a calculator, but we will use Excel worksheet template tmspcch.xls, on the disk supplied with the text. With this worksheet template, we can easily view the sample data and calculation results, and automatically generate the statistical process control chart for the number of defects. Just follow the brief instructions within the worksheet template, entering or pasting the data. Shown below are the relevant portions of the printout.

	A	B	C	D	E
18	Sample #	Number of Defects, c	Centerline	UCL	LCL
19	1	1	4.450	10.779	0.000
20	2	2	4.450	10.779	0.000
21	3	4	4.450	10.779	0.000
22	4	6	4.450	10.779	0.000
23	5	1	4.450	10.779	0.000
24	6	5	4.450	10.779	0.000
25	7	1	4.450	10.779	0.000
26	8	3	4.450	10.779	0.000
27	9	11	4.450	10.779	0.000
28	10	12	4.450	10.779	0.000
29	11	8	4.450	10.779	0.000
30	12	5	4.450	10.779	0.000
31	13	5	4.450	10.779	0.000
32	14	4	4.450	10.779	0.000
33	15	2	4.450	10.779	0.000
34	16	6	4.450	10.779	0.000
35	17	2	4.450	10.779	0.000
36	18	3	4.450	10.779	0.000
37	19	4	4.450	10.779	0.000
38	20	4	4.450	10.779	0.000
39					
40		cbar:			
41		4.450			
42		(mean number of defects)			

For these 20 samples, the mean number of defects is c-bar = (1 + 2 + 4 + ... + 4)/20 = 89/20 = 4.450. The upper and lower 3-sigma control limits are:

$$4.450 \pm 3\sqrt{4.450}, = 4.450 \pm 6.329; \text{ and UCL} = 10.779, \text{LCL} = 0$$

Because the calculated value of the lower control limit is negative, LCL reverts to 0. Samples 9 and 10 (with 11 and 12 defects, respectively) exceed the upper control limit and we conclude that the process is not in control.

The c-chart portion of the Excel worksheet template is shown below, and we can readily observe that samples 9 and 10 exceed the upper control limit. Recall that LCL = 0.

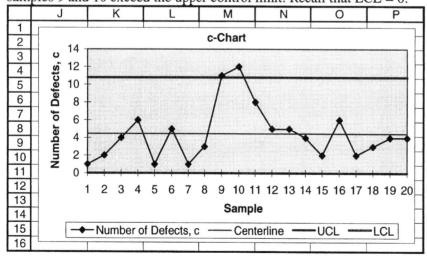

Minitab can also be used to generate the c-chart for this exercise. Numerical values are included for the centerline and control limits.

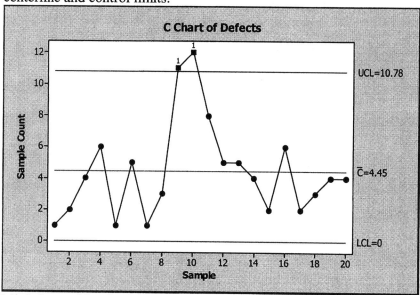

C Chart of Defects

Minitab provides the following warnings:

Test Results for C Chart of Defects
```
TEST 1. One point more than 3.00 standard deviations from center line.
Test Failed at points:  9, 10
```

20.55 p/a/m This Minitab c-chart shows all 30 samples, but with the control limits based only on samples 1 - 20. Note: In generating this chart, it may be useful to combine all 30 samples into a single file, then to specify that only samples 1 through 20 be used in estimating the centerline and control limits. Using the same centerline and control limits calculated in exercise 20.54, it appears the process is also in control during samples 21 - 30. No data points fall outside the limits.

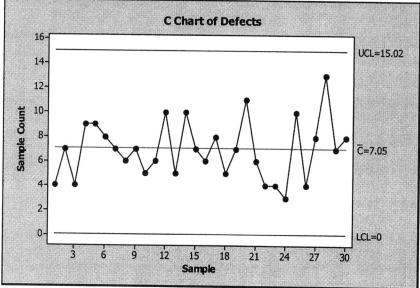

C Chart of Defects

20.57 p/c/m In using Minitab to again generate the mean and range control charts of exercise 20.45, it was specified that all eight tests be applied. Minitab reported that one of the six sample means was more than 3 sigmas from the centerline, a finding that we noted in the solution to exercise 20.45. None of the other seven tests found anything suspicious. The charts are repeated below.

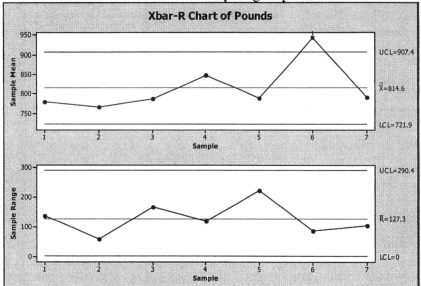

Minitab provides this warning:
```
Test Results for Xbar Chart of Pounds
TEST 1. One point more than 3.00 standard deviations from center line.
Test Failed at points:  6
```

20.59 p/c/m Minitab applied all eight tests to the mean and range control charts shown below. The mean chart failed test 1 (one point more than 3 sigmas from centerline) and test 5 (2 out of 3 points more than 2 sigmas from centerline, on the same side of the centerline).

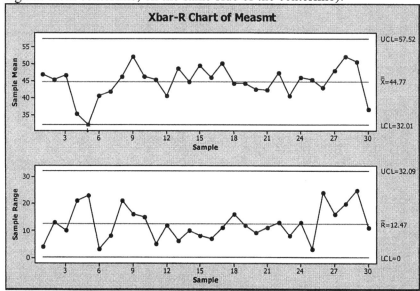

Minitab provided the following warnings:
```
Test Results for Xbar Chart of Measmt
TEST 1. One point more than 3.00 standard deviations from center line.
Test Failed at points:  5

TEST 5. 2 out of 3 points more than 2 standard deviations from center line (on
    one side of CL).
Test Failed at points:  5
```

CHAPTER EXERCISES

20.61 d/p/m Reliability of a unit refers to a unit having a low probability of breakdowns or requiring repair. The durability of a unit is the time before repair becomes necessary.

20.63 d/p/m For a given individual and his/her tastes: Traditional: a hamburger is "good" if it has neither < 1.0 or > 3.0 teaspoons of ketchup on it. Taguchi: it tastes best with 2.0 teaspoons of ketchup, and quality decreases as the amount of ketchup gets further away from 2.0 teaspoons.

20.65 d/p/m In Benchmarking, the new university should study and emulate strategies and procedures of those universities recognized as generating world-class students. Universities like MIT, Cal. Tech, and Carnegie Mellon could be useful as objects of the benchmarking effort.

20.67 d/p/m This firm should use a c-chart because it monitors the number of defects present in a product or a unit of production.

20.69 p/a/m This exercise can be carried out with a pocket calculator, but we will use the computer. Using Minitab, we obtain the p-chart shown below.

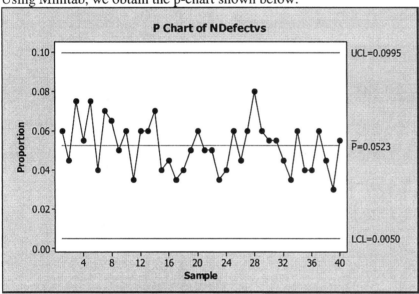

The mean of the sample proportions is p-bar = (0.06 + 0.045 + ... + 0.055)/40 = 2.090/40 = 0.05225. The upper and lower control limits are:

$$0.05225 \pm 3\sqrt{\frac{(0.05225)(1-0.05225)}{200}} = 0.05225 \pm 0.04721, \text{ or UCL} = 0.09946 \text{ and LCL} = 0.00504.$$

None of the sample proportions falls outside the control limits, and we have no reason to believe that the process is out of control. The corresponding Data Analysis Plus p-chart is shown below.

	A	B	C	D	E	F	G	H
1	Statistical Process Control							
2								
3			NDefectvs					
4	Upper control limit		0.0995					
5	Centerline		0.0523					
6	Lower control limit		0.005					
7								
8								

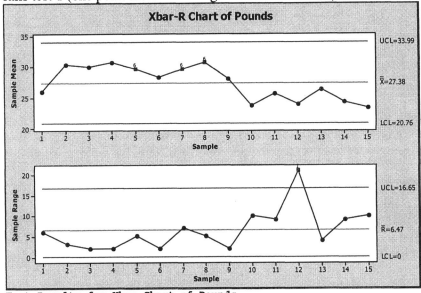

20.71 p/a/m Minitab is used to generate the mean and range charts shown below. The mean chart fails test 6 (4 out of 5 points more than 1 sigma from centerline, on one side of centerline) and the range chart fails test 1 (one point more than 3 sigmas from centerline). We conclude that the process is not in control.

Test Results for Xbar Chart of Pounds
TEST 6. 4 out of 5 points more than 1 standard deviation from center line (on one side of CL).
Test Failed at points: 5, 7, 8

Test Results for R Chart of Pounds
TEST 1. One point more than 3.00 standard deviations from center line.
Test Failed at points: 12

Using Data Analysis Plus, we obtain the comparable results shown below.

	A	B	C	D	E	F	G	H
1	**Statistical Process Control**							
2								
3			*Pounds*					
4	Upper control limit		33.9932					
5	Centerline		27.3778					
6	Lower control limit		20.7624					
7	Pattern Test #6 Failed at Points: 5, 6, 7, 8							
8								
9			XBar-Chart (using R)					
10								
11								
12								
13								
14								
15								
16								
17								
18								
19								
20								
21								
22								
23								
24								
25								

	A	B	C	D	E	F	G	H
1	**Statistical Process Control**							
2								
3			*Pounds*					
4	Upper control limit		16.6517					
5	Centerline		6.4667					
6	Lower control limit		0					
7	Pattern Test #1 Failed at Points: 12							
8								
9			R-Chart					
10								
11								
12								
13								
14								
15								
16								
17								
18								
19								
20								
21								
22								
23								
24								
25								

353

20.73 p/a/m Applying Minitab, we obtain the following p-chart for these samples. None of the sample proportions falls outside the control limits, Minitab finds no suspicious patterns, and we conclude that the process is in control.

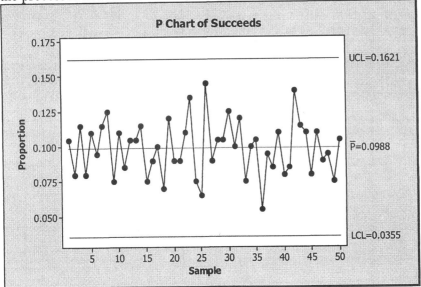

The mean of the sample proportions is p-bar = (0.105 + 0.08 + ... + 0.105)/50 = 4.94/50 = 0.09880. The upper and lower control limits are:

$$0.09880 \pm 3\sqrt{\frac{(0.09880)(1-0.09880)}{200}} = 0.09880 \pm 0.06330 \text{, or UCL} = 0.1621 \text{ and LCL} = 0.0355$$

Using Data Analysis Plus, we obtain the comparable results shown below.

	A	B	C	D	E	F	G	H
1	Statistical Process Control							
2								
3			Succeeds					
4	Upper control limit		0.1621					
5	Centerline		0.0988					
6	Lower control limit		0.0355					
7								
8								
9								
10								
11								
12								
13								
14								
15								
16								
17								
18								
19								
20								
21								
22								
23								